COASTAL PHYTOPLANKTON

PHOTO GUIDE FOR NORTHERN EUROPEAN SEAS

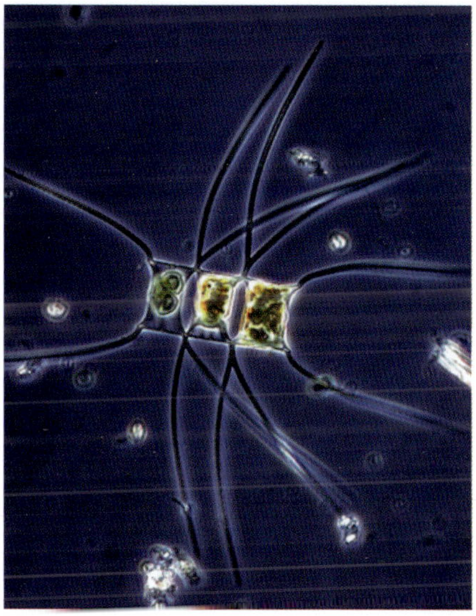

Alexandra Kraberg, Marcus Baumann
& Claus-Dieter Dürselen

COASTAL
PHYTOPLANKTON
PHOTO GUIDE FOR NORTHERN EUROPEAN SEAS

AWI

Alfred Wegener Institute for Polar and Marine Research
Handbooks on Marine Flora and Fauna
Edited by Karen H. Wiltshire and Maarten Boersma

Verlag Dr. Friedrich Pfeil · München 2010
ISBN 978-3-89937-113-0

Bibliographische Information der Deutschen Nationalbibliothek

Die Deutsche Nationalbibliothek verzeichnet diese Publikation in der deutschen Nationalbibliografie;
Detaillierte bibliografische Daten sind im Internet unter http://dnb.d-nb.de abrufbar.

Front cover:
The dinoflagellate *Protoperidinium depressum*:
A live cell collected from the Helgoland Roads long-term monitoring station.

Back cover:
Epifluorescence image of the diatom *Guinardia* flaccid.
The cell is seen under ultraviolet light, with the star-shaped chloroplasts fluorescing in red.

Page 1:
A chain of the centric diatom *Chaetoceros decipiens*.

Druckvorstufe: Verlag Dr. Friedrich Pfeil, München
Druck: Advantage Printpool, Gilching

Printed in the European Union

ISBN 978-3-89937-113-0

Verlag Dr. Friedrich Pfeil • Wolfratshauser Straße 27 • 81379 München, Germany
phone: +49 89 742827-0 • fax: +49 89 7242772
e-mail: info@pfeil-verlag.de • www.pfeil-verlag.de

Contents

Introduction

This is a book for the layperson and student as well as for environmental managers. It hopes to fulfil the needs of students of phytoplankton to successfully identify species where possible and to become aware of the difficulties and pitfalls involved in phytoplankton identification and enumeration.

Marine phytoplankton encompasses a vast array of different organisms, including diatoms, dinoflagellates and other flagellate and non-flagellate cells. In the face of climate change it becomes of paramount importance to assemble reliable and consistent information on the identity of and interactions between the component taxa of phytoplankton communities, so that changes in plankton biodiversity due to climate change or anthropogenic activities can be detected and marine ecosystems managed effectively.

Unfortunately, plankton taxonomy is not only a fascinating and intriguing area of research, it is also exceedingly difficult and the routine identification of plankton organisms is a daunting task, as both phytoplankton and zooplankton species diversity are very high and many very similar species co-occur. The presence of a large number of similar species makes the differentiation of individual species in a sample difficult with ordinary light microscopes, the standard tools of those involved in phytoplankton monitoring. Compounding this, phytoplankton identification is also often carried out in chemically preserved samples (e.g., Lugols iodine) which can reduce the number of visible morphological features usable for identification. However, with reliable literature and some experience a consistent identification is in many cases possible even for lay persons. In this book we pursue two objectives: We will provide a basic introduction to the taxonomic relationships between the different phytoplankton groups. We then take a practical approach and provide the reader with information, which might facilitate the identification of organisms even if the taxonomically important characteristics are not necessarily visible. We will earmark those species, which cannot be identified to species using the routine methods available to students or environmental managers.

It is impossible to produce a comprehensive treatment of all phytoplankton species. We, therefore, restrict ourselves to major phytoplankton species of the North Sea, describing many 'species lookalikes' that are easily confused. For a broader treatment of phytoplankton taxonomy and biogeography internet resources such as PLANKTON*NET (**http://planktonnet.awi.de**) or Algaebase (**www.algaebase.org**) are excellent sources of information. A detailed link list is provided in Appendix 1.

Acknowledgements

The authors gratefully acknowledge the funding provided by the Alfred Wegener Institute for Polar and Marine Research and specifically the support and patience of Karen Wiltshire and Maarten Boersma in the production of this book.

This book has been a team effort. We only had space to illustrate our texts with a small number of images. Further information and images can be found in PLANKTON*NET (**http://planktonnet.awi.de**), many of whose contributors also provided image material for the book and their contributions are gratefully acknowledged here and individually on the respective species pages:

Daniel Vaulot from the Station Biologique de Roscoff provided images of *Amphidinium carterae*.

Martin Löder, Silvia Peters and Nicole Aberle-Malzahn at the Alfred Wegener Institute for Polar and Marine Research, Biological Station Helgoland provided images of microzooplankton and Florian Hantzsche also of the AWI permitted the use of his FlowCam images. Further contributions from AWI colleagues came from Hanne Halliger of the Wadden Sea Station Sylt, who provided several images of live diatoms. Regina Hansen and Susanne Busch of the Institute of Baltic Sea Research and Gabriele Krauß of Landesamt für Umwelt, Naturschutz und Geologie provided images of different flagellates. In addition, we used Regina Hansen's images of *Prorocentrum minimum*. Last but not least Lars Edler provided images of live diatom species and Bo Sundstroem helped with the identification of some of the *Rhizosolenia*. Thanks to all of you for useful discussions and, simply, your generosity.

Finally thanks are due also to Hubert Hilpert at the Pfeil Verlag who turned the original manuscript into a book and dealt patiently with our many questions and revisions.

The diversity of phytoplankton and its classification

Estimates of plankton species diversity range from a few thousand (Sournia et al. 1991, Tett & Barton 1995) to hundreds of thousands, within the diatoms alone (Mann 1999). Since the work of Carl Linnaeus more than two hundred years ago, who was one of the first to attempt a systematic account of all animal and plant life, attempts have been made to describe and classify species diversity based on their morphological and biochemical properties. Linnaeus was also the first to introduce the system of naming a species with a genus name followed by the second part of the name to denote the species.

Based on the identified similarities and differences phytoplankton (as all other described living organisms) is hierarchically classified from Division and class with major differences down to genus and species with smaller morphological differences between individual species.

Division Bacillario**phyta** (Diatoms)
Class Coscinodisco**phyceae**
Order Coscinodisc**ales**
Family Coscinodisc**aceae**
Genus *Coscinodiscus*
Species *radiatus*

By convention each group level has a fixed word ending indicating the level, i. e. family names end in 'aceae', order names in 'ales' etc. This convention only applies to plants while different endings are used for the classification of animals. The level of 'division' is named 'Phylum' in the case of animals.

The classification of organisms is not fixed but the positioning of a species in this system (i. e., the genus, family etc. it belongs to) can change as new studies reveal novel diagnostic characters or cause a re-appraisal of the importance of existing ones. The increasing use of electron microscopes, for instance, has played an important role here as it has facilitated the visualization of minute structures and patterns that are not discernible when using light microscopy only, an example being species of the diatom genus *Thalassiosira*. The classification of organisms is, therefore, a very dynamic process and even at higher taxonomical levels there often is little agreement (see Williams 2007 for a summary of past and currently used systems for diatoms). As a result of this it is possible and indeed often the case that a given species name may change. While a species can only have one valid name, the previously used names are nevertheless important. They essentially provide the history of a species' taxonomy. This needs to be taken into account when using multiple literature sources as information on one species might be provided under different names in different books. All the names referring to the same species are called 'synonyms' of that species name.

While such re-evaluations have always been a basic feature of taxonomic research, the introduction of molecular techniques into the investigation of taxonomic relationships at all levels has greatly increased our ability to differentiate between species and to describe new ones (see Daugbjerg et al 2000 for an example involving the dinoflagellate genera *Gymnodinium* and *Gyrodinium*). They have also led to the distinction of different diversity types (intraspecific diversity = diversity within species) the ecological relevance of which are not yet well known. As a result the biological classification system as a whole is in constant flux and requires regular re-evaluations.

For this book we adopt, with few exceptions, the taxonomic classification from the online database AlgaeBase (**www.algaebase.org**, as per December 2008), which is currently the most extensive and up-to-date online collection of taxonomic information on microalgae. In Table 1 and 2 we briefly describe the basic features of the different groups of diatoms and dinoflagellates acknowledging that these groupings will be liable to change in future and that no universal agreement among taxonomists with respect to the classification of major phytoplankton species exists.

General taxonomic comments

We deal mainly with diatoms and dinoflagellates. However, we also provide a short section on ciliates and relevant flagellates that can be harmful or form blooms. We will try to keep specialized terminology to a minimum, although some basic concepts and terms for the different taxonomic groups, referred to throughout the book are introduced here. For further explanations of specific terms please refer to the extensive glossary. Traditionally the most basic way of grouping plankton is by size. The so-called picoplankton is by definition in the size range 0.2-2 µm. The nanoplankton is sized between 2-20 µm. Plankton larger than 20 µm but smaller than 200 µm is termed phytoplankton and even larger plankton sized between 200 µm and 2 mm is called mesoplankton. However, these groupings have little taxonomic relevance. Plankton groups such as diatoms and dinoflagellates will belong mostly to the microplankton, but have some members in at least three of these categories. Although we give size ranges for the different organisms described in this book, species identifications cannot be made on the basis of size alone.

1. Diatoms

Diatoms are one of the most diverse and beautiful groups of organisms in the plankton with several thousand species (Mann 1999) and sizes ranging from about 5-500 µm with a few exceptionally large species exceeding 1 mm in length. Diatoms are characterized by their silica cover called "frustule" or theca (Figure 1).

All diatoms have the same basic architecture consisting of two valves: the epivalve and the smaller hypovalve which fits into the larger one. Each valve consists of three basic components: the valve face (i. e. the top of the valve), valve mantle and the cingulum (also called girdle). The cingulum in turn consists of a series of bands, the number and morphology of which can be an important diagnostic feature but are often not easily seen using standard light microscopy.

Diatoms, have three major axes, as shown in Figure 2a-b. These are the apical, transapical and pervalvar axes. The transapical and apical axes describe the width and depth of the cell. In a centric diatom with a circular cross section, these would have the same width. Throughout the book the relevant axis is stated when giving length measurements or describing morphological characteristics as cells will often look fundamentally different depending on their orientation and it is important to be aware of this when enumerating plankton. In descriptions of a species the different morphological features are, therefore, usually described in relation to the organism's symmetry and orientation under the microscope, i. e. whether they are seen in girdle or valve view (Fig. 2c). Valve view is usually the most important for an examination of the morphological features needed for species identification. However, in genera such as *Rhizosolenia* and *Chaetoceros* it is the girdle view, which is more useful. In girdle view the girdle bands can be examined. In *Rhizosolenia* species these are visible in light microscopy in empty valves, but in most diatom species the characteristic girdle bands are very faint and cannot easily be seen without more elaborate microscopy tools, at least oil immersion. Establishing the orientation of the diatom on the slide should be the first step before proceeding to identifying the organism.

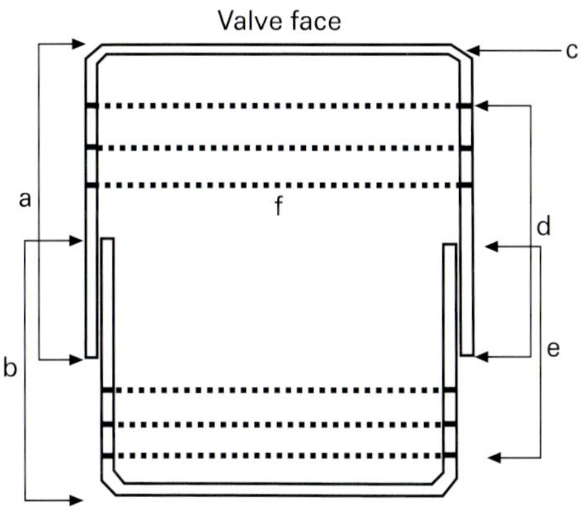

Fig. 1. Coss-section through a diatom frustule.
a, epitheca; b, hypotheca; c, valve mantle; d, epicingulum; e, hypocingulum; f, cingular bands.

Once the orientation of the cell has been established one can begin to examine the diatom cell itself. For the identification of diatoms several characteristics of the cell such as the size, morphology of the frustule, number and shape of the chloroplasts or the types of chain formation are used.

Chain/colony formation

The pattern of chain formation can be an important feature when identifying diatoms, although chain type alone is not usually enough to identify a cell to the level of species. Two points have to be taken into account: the mechanism of interlinkage of cells and the shape of the spaces or apertures formed between adjacent cells. Many diatom species are solitary, i. e., occur as single cells (e.g, the genus *Coscinodiscus*). In others individual cells are joined into colonies (e. g., *Chaetoceros socialis*) or chains (e. g., *Thalassiosira* species). Cells can be joined directly by abutting valve faces (e. g., *Achnanthes* or *Leptocylindrus*) or by interlinkage of their processes, e. g., the setae in *Chaetoceros* or strutted processes in *Thalassiosira*. In other species mucous secretions are involved. In *Odontella* and *Biddulphia* for instance the cells are linked via secretions from

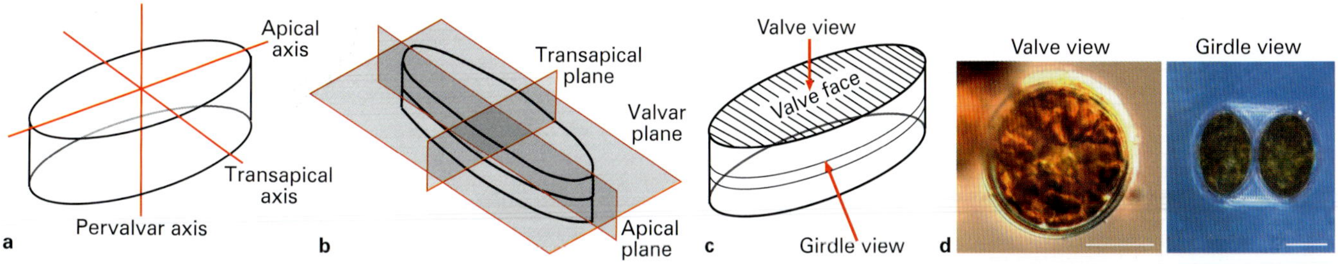

Fig. 2. Terminology used to describe the orientation/symmetry of a diatom cell.
a, Major axes and **b,** major planes respectively; **c,** Valve and girdle view; **d,** The diatom *Podosira stelligera* in valve and girdle view, illustrating the different appearance of the cell depending on orientation.

Table 1. Classification of the Bacillariophyta including diagnostic characteristics of the taxa described in the book.

Centric diatoms

Phylum **Bacillariophyta**

Class **Coscinodiscophyceae**

Order **Biddulphiales**

Family **Biddulphiaceae**

Chain forming diatoms interconnected by mucous secretions from valve elevations (at poles), presence of a pseudocellus.

Genus:

Biddulphia: Cells rectangular in girdle view with an elliptical or triangular cross section often forming zigzag chains. Valves with elevations at the valve poles, each pole with a pseudocellus. The valve face often shows ribs or ridges and sometimes spines in its central part.

Order **Chaetocerotales**

Family **Chaetocerotaceae**

Valves have long setae. Cells are solitary or connected into chains by their setae.

Genera:

Bacteriastrum: Cells with a circular cross-section. Several setae arranged around the margin. Setae of adjacent cells fused for a short distance then bifurcating. Terminal setae different from the intercalary setae, not branched, and often curved.

Chaetoceros: Cells with an elliptical to nearly circular-cross-section. Each valve (with very few exceptions, e.g., *C. throndsenii*) with two setae, one on each valve end along the apical axis. Setae of adjacent valves touching each other near their origin.

Order **Corethrales**

Family **Corethraceae**

Monogeneric family: see genus characteristics.

Genus:

Corethron: Cells with a circular cross-section, dome-shaped valves, and many girdle bands. Valve faces carrying a marginal ring of long and/or short hooked spines directed diagonally outwards.

Order **Coscinodiscales**

Family **Aulacodiscaceae**

Mono-generic family: see genus characteristics.

Genus:

Aulacodiscus: Cells solitary, with a circular cross-section. Valve face with three to six marginal labiate processes with distinct extermal extensions. In girdle view cells nearly rectangular with rounded or truncated corners, labiate processes clearly visible. Areolae on the valve face arranged in radial rows emanating from a hyaline area in the center.

Family **Coscinodiscaceae**

Cells solitary. Labiate processes, without external tubes, are located in one or more rings at the valve margin. Two larger labiate processes (macrorimoportulae) present.

Genus:

Coscinodiscus: Cells discoid with a circular cross-section. Valves flat or convex in girdle view. Areolation showing a radial pattern. One ring (for some species two) of small marginal labiate processes on the valve face. Two of the processes distinctly larger (termed macrorimoportulae).

Family **Heliopeltaceae**

Valve face divided into sectors, in some cases with alternating elevated and depressed sectors, one labiate process usually associated with each sector.

Genus:

Actinoptychus: Solitary, discoid cells with a circular cross-section. Valves divided into sectors with alternating elevated and depressed segments. Areolae arranged in radial rows. Labiate processes situated at the central valve margin of the elevated sectors.

Family **Hemidiscaceae**

Cells cylindrical with a short pervalvar axis. Areolation of valves radial and often fasciculate. One marginal ring of large labiate processes.

Genus:

Actinocyclus: Cells circular in cross-section. Radial areolation of valves usually fasciculate and forming sectors. Density of areolae higher near the valve margin (not following the division into sectors). Central annulus and a marginal pseudonodulus present and visible with light microscope.

Order **Cymatosirales**

Family **Cymatosiraceae**

Cells occurring singly, or in chains linked by spines. Valves with an elliptical to rhomboid outline in girdle view. They are slightly convex or undulating with two low elevations at the apical poles. Only one marginal process per cell (on one of the two valves) and only one chloroplast present.

Genera:

Brockmanniella: Cells broadly rectangular in girdle view and elliptical in valve view forming long, partly twisted ribbons. Valves slightly undulating. Adjacent cells in contact with each other with the central part of the valves and the two apical elevations.

Cymatosira: Cells rectangular or curved in girdle view and approximately rhomboid with rounded poles in valve view. Cells joined into slightly twisted ribbons. Minimum of one valve slightly convex in the centre. A ring of marginal spines linking the central portions of adjacent valves.

Table 1. (continued).

Lennoxia: Cells solitary and spindle-shaped with two long and fine hair-like extensions at the apical poles.

Plagiogrammopsis: Cells with a rhomboid outline in valve view. Cells roughly rectangular in girdle view but with central area of valve face convex and elevated valve poles with a constriction near the elevated margin resulting in large tear-drop shaped apertures between adjacent valves. The convex valves carrying a ring of spines linking adjacent valves. Valves of adjacent cells not directly touching. Long processes (termed pili) arising from the raised valve poles (Figure 6f).

Order **Hemiaulales**

Family **Bellerocheaceae**

Marine planktonic species with weakly silicified frustules, girdle bands delicate. Valves bi- to multipolar. Delicate costae radiating from a central annulus.

Genera:

Bellerochea: Cells bi-, tri-, or quadrangular in valve view and nearly rectangular in girdle view. Short elevations situated at the valve poles. Cells connected into ribbons. Valve centre and poles slightly raised, small apertures formed between adjacent cells. A single bilabiate process with an external tube situated on each valve. Cells containing numerous small, oval and slightly constricted chloroplasts.

Subsilicea: Cells with a small elliptical cross-section and a short pervalvar axis. Valves nearly rectangular with slightly rounded edges in girdle view. Cells connected into long tight ribbons. No processes or areolae visible on the valves. Each cell with one large chloroplast. Affiliation with the Bellerocheaceae still uncertain.

Family **Hemiaulaceae**

Cells with a bipolar symmetry with two marginal elevations on opposite valve poles. A single labiate process, numerous small disc-like chloroplasts.

Genera:

Cerataulina: Cells long cylindrical with a circular cross-section and torsion about the pervalvar axis forming straight chains with inconspicuous apertures between adjacent cells. Elevations small, low and with wing-like extensions.

Eucampia: Cells elliptical in cross-section and trapezoid in girdle view building helically coiled chains. Two pronounced elevations in valve corners causing large apertures between adjacent cells in a chain.

Order **Leptocylindrales**

Family **Leptocylindraceae**

Mono-generic family: see genus characteristics.

Genus:

Leptocylindrus: Cells long cylindrical with a circular cross-section, central valve face slightly convex or concave, forming tight chains. Girdle consisting of many intercalary bands. Ring of very small, inconspicuous spines located at the valve margin

Order **Lithodesmidales**

Family **Lithodesmiaceae**

Cells solitary or united to ribbons. Girdle consisting of several rows and bands. A single bilabiate process on each valve. Marginal ridges present.

Genera:

Ditylum: Solitary cells usually with triangular cross-section and quadrangular in girdle view, with the pervalvar axis much longer than the apical axis. Each valve with a conspicuous fringed (fimbrate) marginal ridge and a central bilabiate process with a long and strong external tube (Figure 6h).

Helicotheca: Cells narrow and elliptical in valve view, rectangular in girdle view with torsion along the pervalvar axis. Cells connected into ribbons without apertures between adjacent cells. Single bilabiate process of each valve positioned sub-centrally and with only a short external part. No marginal ridge is present.

Lithodesmium: Cells solitary or connected into ribbons with wide apertures between adjacent cells. Cells normally triangular in cross-section and rectangular to square in girdle view. Valves with conspicuous elevations at each valve pole and a central bilabiate process with a long external tube. Marginal ridge membraneous.

Order **Meloseirales**

Family **Hyalodiscaceae**

Cells lentil shaped, usually occurring in pairs united by girdle bands (copulae). A group containing benthic species often found attached to sand or algae.

Genus:

Podosira: Cells with convex valve face, either solitary or joined into pairs by a common girdle band. Girdle not conspicuous in single cells. Areolae on valves arranged in decussating radial sectors.

Family **Melosiraceae**

Pervalvar axis strongly developed, valve face with a ring of labiate processes, circular outline in valve view.

Genus:

Melosira: Cells with a circular cross-section and slightly convex valve face, and a high valve mantle in girdle view. Valve face with a fine structure of areolae and a marginal ring of labiate processes.

Family **Stephanopyxidaceae**

Mono-generic family: see genus characteristics

Genus:

Stephanopyxis: Cells higher than wide in girdle view and with a circular cross-section, convex valve face, and a high valve mantle. Adjacent cells connected by a concentric ring of labiate processes with long external tubes. Areolae on valve large and hexagonal.

Table 1. (continued).

Order **Paraliales**

Family **Paraliaceae**

Cells heterovalvate, broad cylindrical with a circular cross-section, forming straight and close-set chains, with adjacent cells connected by marginal, interlocking, ridges on one valve and grooves on the other valve.

Genus:

Paralia: Valves clearly differentiated into valve face and mantle. Valve face with a hyaline area in the centre and radial structures towards the margin. A ring of labiate processes bearing external spines located on the valve mantle. Terminal cells in a chain not with marginal spines and only reduced ridges.

Order **Rhizosoleniales**

Family **Rhizosoleniaceae**

Long cylindrical cells (extremely elongate pervalvar axis) usually building chains. A single internal labiate structure on the unipolar valves. Cells containing numerous small chloroplasts.

Genera:

Dactyliosolen: Cells circular in cross-section. Terminal parts of half of the girdle bands are wedge-shaped. Valve face with ribs radiating from the single process (normally not visible with the light microscope). The process located centrally or marginally, sometimes with an external tube.

Guinardia: Cells circular in cross-section. Valve face flat but with rounded edges. A small process with an external tube located near the valve margin. Numerous small open ligulate bands are forming the girdle.

Neocalyptrella: Cells with elliptic cross-section. Valves strongly conical ending with an external process. Cells sickle- or S-shaped. Two columns of girdle segments.

Proboscia: Cells circular in cross-section. Valves almost conical with the valve end tapering into a blunt tip (proboscis). External process absent.

Rhizosolenia: Cells circular or very slightly elliptical in cross-section. Valves conical with a straight labiate process at the apex. Girdle consisting of numerous bands arranged in two or several longitudinal columns.

Order **Thalassiosirales**

Family **Lauderiaceae**

Mono-generic family: see genus characteristics (formerly a member of the Thalassiosiraceae).

Genus:

Lauderia: Cells with a circular cross-section forming close-set straight chains. Valves with strutted processes both at the margin and scattered on the valve face, one large labiate process and several long occluded processes in a marginal position on the valve face.

Family **Skeletonemaceae**

Cells linked into chains by their marginal strutted processes (In *Thalassiosira* the cells are linked by their central strutted processes).

Genera:

Detonula: Cells cylindrical with a circular cross-section building tight chains. Valves have one central strutted process, one marginal ring of strutted processes and one marginal labiate process.

Skeletonema: Cells cylindrical with a circular cross-section. Adjacent cells connected by external tubes of a marginal ring of strutted processes. A single process on one valve connected to one or two processes on the adjacent valve. Valves with an additional central or subcentral labiate process.

Family **Thalassiosiraceae**

Cells with marginal rings of processes and one or more central strutted processes, the latter connecting cells into chains.

Genera:

Porosira: Cells cylindrical with a circular cross-section. Strutted processes not distributed in a definite pattern.

Thalassiosira: Cells cylindrical with a circular cross-section. Processes on the valve face showing distinct patterns. One or more marginal rings of strutted processes occurring in addition to central or subcentral strutted processes. Cells connected into chains by their central processes.

Order **Triceratiales**

Family **Triceratiaceae**

True ocelli present, i. e. areolated areas bordered by a structureless rim. Areolation of the ocellus different from that of the surrounding areolae.

Genera:

Cerataulus: Cells rectangular in girdle view with a circular or elliptical cross-section twisted around the pervalvar axis and two conspicuous large ocelli rising from the valve face on both valves. Valve surface covered with spines or granules. Areolae radiating from the valve center.

Odontella: Valves elliptical to lanceolate in valve view, with a conspicuous, stout horn on each of the two valve poles. Horns terminating in an ocellus. Adjacent cells in chains are joined by one or both of these elevations. Each valve with two or more labiate processes usually with long external tubes.

Triceratium: Cells usually triangular in valve view and wider than high in girdle view. Valves slightly convex and with an elevation at each corner terminating in an ocellus. Areolation coarse.

Table 1. (continued).

Pennate diatoms

Phylum **Bacillariophyta**

Class **Bacillariophyceae** (raphid forms)

Order **Bacillariales**

Family **Bacillariaceae**

Valves always elongated. In girdle view cell outline rectangular or spindle-shaped. Raphe system: canal raphe and more or less eccentric. Cells containing two plate-like chloroplasts, one toward each pole of the cell.

Genera:

Bacillaria: Cells elongate with pointed ends, joined into colonies by their flat valve faces. Cells able to slide alongside each other resulting in changing colony shapes. Raphe system with a small keel located in a central position.

Cylindrotheca: Cells solitary, spindle-shaped, often with needle-like elongated valve ends. Frustules somewhat twisted around the apical axis and only weakly silicified

Pseudo-nitzschia: Cells small elongate with pointed or rounded ends, connected into stepped chains with overlapping valve ends. The raphe system (a canal raphe) strongly eccentric running along one valve margin.

Order **Naviculales**

Family **Naviculaceae**

Cells elliptical to lanceolate in valve view and rectangular in girdle view. Cells isovalvar having a generally straight raphe system on both valves. One, two or four larger plate- or ribbon-like chloroplasts present.

Genera:

Meuniera: Cells small and elliptical in cross section with a moderately elongated pervalvar axis. Cells usually connected into chains by their flat or slightly concave valves. Four ribbon-like chloroplasts per cell lying in the girdle area, two on each side of the apical plane.

Navicula: Usually benthic, mostly solitary but sometimes forming chains or ribbons. Cells elliptical or, lanceolate in valve view with blunt, capitate or rostrate apices. One or two large chloroplasts lying along the girdle area.

Family **Pleurosigmataceae**

Cells solitary, elliptical in valve view (in some cases lanceolate to rhomboid and more or less sigmoid). Cells rectangular in girdle view with a short pervalvar axis. Raphe straight or sigmoid in the center of both valves (isovalvar).

Genera:

Gyrosigma: Cells sigmoid in valve view sometimes with rostrate apices. One transverse and one longitudinal striae system on the flat valve face. Two large plate-like chloroplasts lying in the girdle area, one on both sides of the apical plane.

Pleurosigma: One transverse and two oblique striae systems on the flat valve face. Two or four oblong chloroplasts present.

Class **Fragilariophyceae** (araphid forms)

Order **Fragilariales**

Family **Fragilariaceae**

Cells of the different genera with different geometric shapes. One labiate process on the valve face near to each apex. Each apex with a pore or slit field. Striae uniseriate with poroid areolae.

Genera:

Asterionellopsis: Heteropolar cells with an expanded basal part (foot pole) on one side and a needle-shaped part on the other. The basal part is rounded to elliptical in valve view and nearly triangular in girdle view. Cells connected into spiral chains by their valve faces on the foot pole. One or two chloroplasts located in the basal part.

Asteroplanus: Cells heteropolar with an anvil-shaped basal part on one side. The remainder is spindle-shaped. Cells joined into spiral chains by the valve faces of their basal part. Six to ten chloroplasts per cell.

Striatella: Cells elliptical in valve view and rectangular with a prolonged pervalvar axis in girdle view. Two sunken apical pore fields let the valve corners appear truncated. In girdle view bands clearly discernable. Several oblong chloroplasts arranged radially.

Order **Licmophorales**

Family **Licmophoraceae**

At the small base pole one or several frustules are attached to the substratum by mucilage stalks, sometimes arranged into large branched colonies (in Lugol-fixed samples they are often found singly).

Genus:

Licmophora: Cells heteropolar, wedge shaped in valve and in girdle view. Chloroplasts are discoid or plate-like.

Order **Rhaphoneidales**

Family **Rhaphoneidaceae**

Cells rectangular in girdle view and elliptical in valve view sometimes showing produced apices or central inflations. One or two pores as well as a labiate process located at each valve apex. Large poroid areolae forming uniseriate parallel or radial striae.

Genera:

Delphineis: Cells shallow with a linear or elliptical to nearly circular cross-section, solitary or connected into chains. Valve face flat with a distinct, wide sternum. Striae parallel to slightly radiate, two or three rows of areolae surrounding the apices. Several chloroplasts per cell.

Rhaphoneis: Shallow cells with a broadly elliptical cross-section and produced apices, usually solitary or connected into short chains. Often attached to sand grains or empty diatom frustules. Valve face flat with a narrow sternum. Striae parallel or radiate. Numerous small chloroplasts per cell.

Table 1. (continued).

Order **Thalassionemales**
Family **Thalassionemataceae**
Cells lanceolate to needle shaped (sometimes curved), solitary or forming colonies. One labiate process at each valve end and apical spines present at one or both ends. Marginal spines sometimes present. Numerous small chloroplasts per cell.
Genus:
Thalassionema: Small cells rectangular in girdle view and linear to spindle-shaped in valve view. Cells isopolar or heteropolar. Flat valve face with a wide sternum surrounded by one row of marginal circular areolae.

the ocelli/pseudocelli. Care has to be taken when examining chemically preserved samples as these, e.g., Lugol solution and formalin can cause the breakage of chains and colonies.

Valve structure

One of the most important diatom characteristics is the morphology of the frustule. The most basic and obvious feature of the frustule is its shape. A confusing array of terms is used to describe this. Some of the terms most commonly used are illustrated in Figure 3 and these will be used throughout the book. Again, the shape of the cell will depend on its orientation on the slide (see also Fig. 4).

Knowing the shape and symmetry of a diatom cell is rarely enough to identify the organism to species level. The next step is, therefore, usually a closer look at the morphology of the valve surface, particularly the patterns of valve ornamentation. The points about orientation and symmetry discussed above apply to the study of all diatoms. However, there are two large groups of diatoms each with their own unique valve features which is why we introduce them here. The first group is the pennate diatoms, which are bilaterally symmetrical. Features such as surface ornamentation are arranged in relation to a 'line'. This line can be a simple hyaline area, i.e., an area of silica without ornamentation (also known as a sternum) or a more complex structure called a raphe. A raphe is a slit through the valve, along the apical axis, which can be seen in valve view. The length and shape of this raphe is also a diagnostic feature. Particularly the raphe ends are important (Fig. 4). In the genus *Pseudo-nitzschia* the raphe structure is different. It is here called a canal raphe. This is not located in the central part of the cell but on one side running in an apical direction. Additional silicious structures called fibulae build a series of bridges between the areas on either side of the raphe canal.

In contrast, in centric diatoms surface ornamentation is arranged relative to a central point, areola or hyaline area. However, this does not mean that all centric diatoms have a circular outline in valve view. They can also have several poles, most commonly two or three. A tripolar cell essentially has a triangular outline in valve view (Fig. 5). The poles are usually areas with morphological features such as processes or ocelli. These features are introduced below. Not all of the relevant characteristics can be seen adequately when using light microscopy and we will point this out where necessary.

In centric diatoms, valve view reveals structures such as areolation patterns (i.e., pores in the valve surface) and the number and arrangement of valve processes. Valve processes

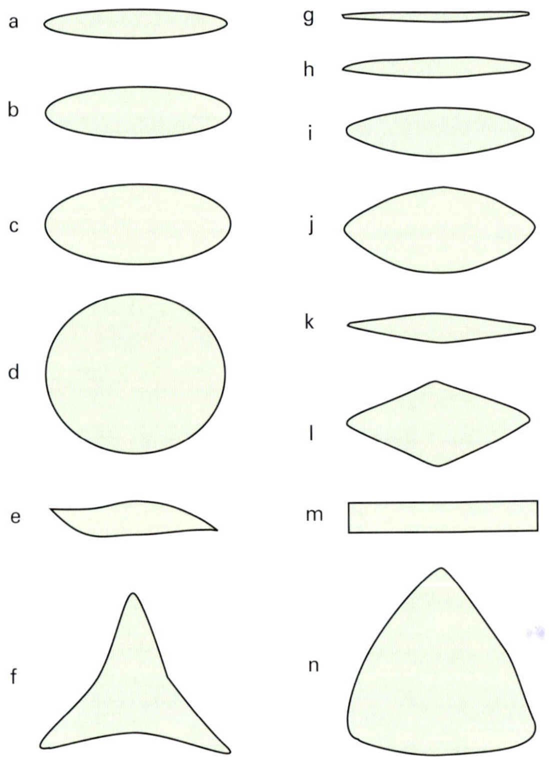

Fig. 3. Morphology of the frustule. **a-c,** Elliptic; **d,** Subcircular; **e,** Sigmoid lanceolate; **f,** Triangular with concave sides; **g,** Acicular (spindle-shaped); **h,** Fusiform; **i-j,** Lanceolate; **k-l,** Rhombic; **m,** Rectangular; **n,** Triangular with convex sides.

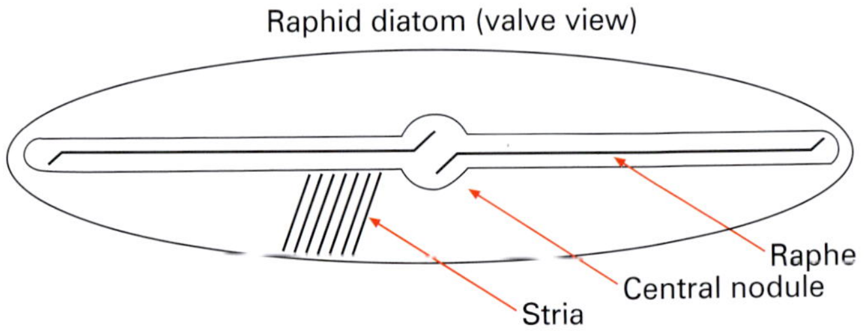

Fig. 4. Basic features on the valve face of a pennate diatom.

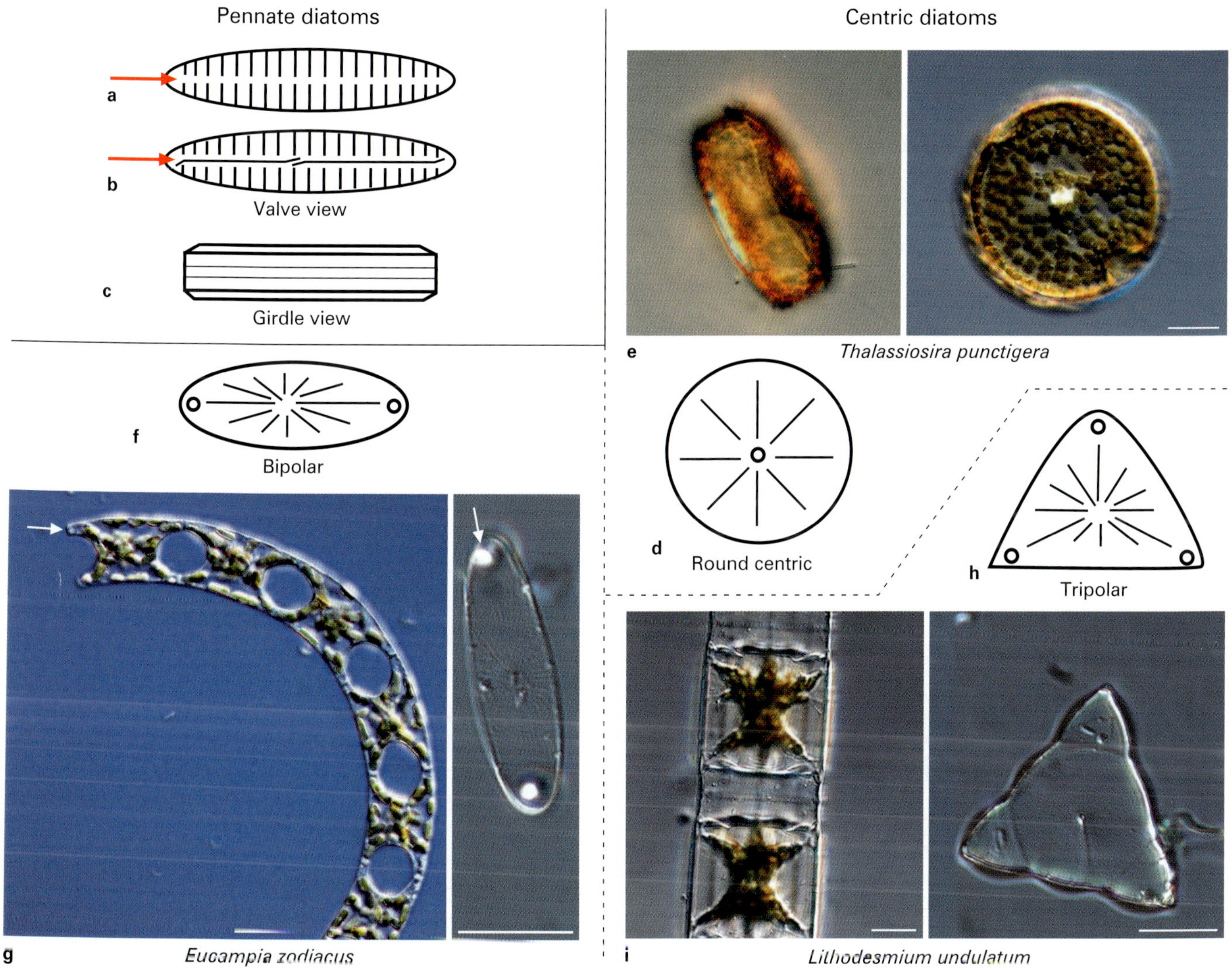

Fig. 5. **a-c,** Schematics of pennate diatoms in valve view (**a,b**) and girdle view (**c**). **a,** Pennate diatom with a sternum; **c,** diatom with a raphe. **d-i,** Centric diatoms: **d,** round centric; **f,** bipolar; **h,** tripolar. **e,g,i,** Examples of round, bipolar and tripolar centric diatoms: **e,** *Thalassiosira punctigera*; **g,** *Eucampia zodiacus* in valve and girdle view. The arrows are indicating the position of the ocellus in valve and girdle view; **i.** *Lithodesmium undulatum*. – d,f,h, valve view; e,g,i, light micrographs with girdle view on the left and valve view on the right. Scale bars = 20 µm.

include solid spines and tubular processes that penetrate the valve surface and whose external and internal extensions are important features for species identification (Fig. 6). Some of these processes are specific for certain diatom groups such as the strutted processes (Fig. 6a) in the Thalassiosirales (*Thalassiosira*, *Detonula*, *Skeletonema*, *Lauderia)*. The latter are processes with pores and struts around the tube opening on the inside of the valve. On the outside of the valve organic threads often arise from the process. Another process that occurs in the genus *Thalassiosira* is the occluded process. This process can have a long external tube (e.g., in *Thalassiosira punctigera*, but the process does not have an opening to the interior of the valve.

Other processes occur across a range of groups, including both centric and pennate species. An example is the labiate process (also called rimoportula) which is found across a host of diatom families (Fig. 6b). The labiate process is a tube through the valve with an opening to the outside and inside of the valve. On the interior part of the valve the opening is bordered by 'lips' (hence, the name labiate process). Often one or two labiate processes are present on the valve face or mantle. However,

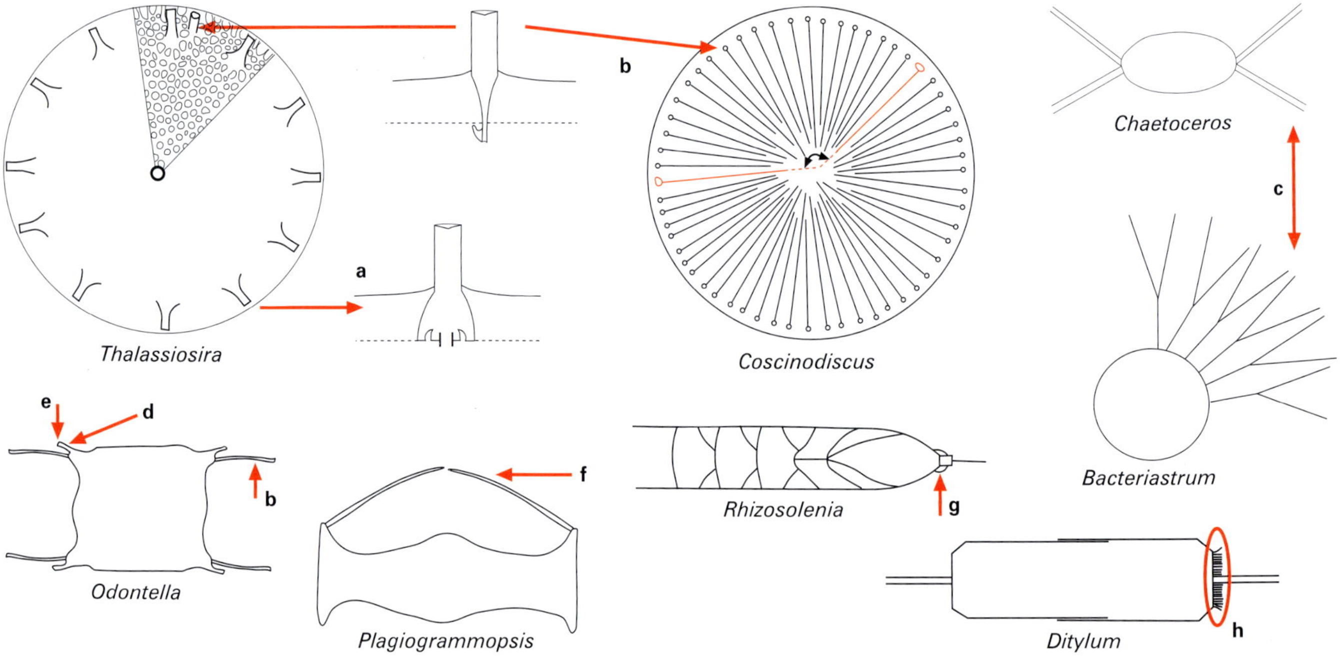

Fig. 6. Types of processes found in diatoms. **a,** strutted process; **b,** labiate process, red processes on *Coscinodiscus* graphic = macrorimoportulae; **c,** setae; **d,** hollow horns; **e,** Position of the ocellus in *Odontella*; **f,** pili; **g,** spines bordered by otaria; **h.** marginal ridges.

in the genera *Coscinodiscus* and *Stephanopyxis* many labiate processes are present, arranged in one or more rings around the valve margin. In addition, in the genus *Coscinodiscus*, two larger labiate processes (called macrorimoportulae) are also found within the marginal ring of processes. Although these structures are usually best seen in valve view, where the processes are located on the valve mantle they might only be visible in girdle view, e.g., in *Coscinodiscus wailesii*. Labiate and strutted processes are difficult to see in standard light microscopy, particularly in live cells, where the surface valve structures are usually obscured by the cell contents such as chloroplasts. On the exterior of the valve the visible parts of the different types of processes can look quite similar and it is, therefore, necessary to examine the inside of the valve for species identification. It is possible to clean valves using different acids, but to see minute details it is often necessary to view acid cleaned valves with more advanced techniques such as electron microscopy.

In addition to these processes, hollow horns also occur in several genera. These horns are essentially an extension of the valve proper. Prominent horns are found in the genus *Odontella* for instance (Fig. 6e). In this species the horns are topped by ocelli. These are small fields of pores (areolae) which have a pattern that is different from the valve area surrounding the ocellus. Further types of processes are shown in Figures 6f-h. In these cases, e.g., the pili (singular: pilus) the process types are not common to a large number of different diatom groups but specific for a diatom family or genus. The marginal ridge for instance is characteristic of the Lithodesmiaceae *(Lithodesmium* and *Ditylum)*.

Diatom reproduction

(see also Round, Crawford & Mann 1990)

Diatoms can reproduce both sexually and asexually and it is important to have at least a basic awareness of the different life cycle stages of diatoms and dinoflagellates (see below) so that they can be identified correctly when occurring in a sample. Asexual reproduction is usually most common. When diatom cells divide daughter cells receive half of the parent valve and a new hypovalve is produced to fit inside the parental valve. As a result, the average size of cells in a diatom population will decrease, i.e., individual cells become smaller and smaller. This decrease in size is also often accompanied by a change in cell proportions for instance an increase in cell height relative to the width.

Sexual reproduction is one way of restoring their original size. Normal vegetative diatom cells are diploid and, therefore, have to undergo a meiotic division to produce male and female gametes, which fuse to form a zygote. A common mode of reproduction, particularly in centric diatoms is oogamy, where large non-motile eggs and small, motile sperm are produced. The latter are initially formed inside a valve but are then liberated into the surrounding medium, where during routine identification of Lugol-preserved samples these gametes might be indistinguishable from other flagellates in the sample. The sperm fertilizes the egg within the parent theca. The zygote then exits the valve and expands into a large sphere bounded by an organic membrane the so-called auxospore. Within the auxospore a new, large, cell is formed. This is termed the initial cell. While oogamy is a common process in centric diatoms, in pennate diatoms, isogamy (i. e., indistinguishable male and female gametes) is more widespread. Here sexual reproduction is usually initiated by the pairing of two cells, which then undergo meiosis to produce gametes inside the original vegetative cell. Often, both gametes in a cell are motile and join (conjugate) to form auxospores. In some cases however only one of the gametes is motile.

An additional feature of the diatom life cycle is the production of resting stages, which are found in many centric and a few pennate species (McQuoid & Hobson 1995, 1996). In contrast to auxospore formation these are usually produced asexually. Both resting spores and resting cells have been described. The latter are morphologically very similar to the 'normal' vegetative cells, but are physiologically very different having a more condensed cytoplasm and greater lipid content (Edlund & Stoermer 1997, Sicko-Goad et al. 1989).

Resting spores on the other hand, have a different morphology from the vegetative cells and might, therefore, accidentally be recorded as separate species (Round et al. 1990). Differences in the morphology of resting spores are sometimes the only way to distinguish between otherwise very similar species. This is the case in the genus *Chaetoceros*, between for instance *C. teres* and *C. lauderi*. While the former is considered a cold water form the latter is typical of warmer waters and occurs later in the year in temperate regions. These two species can only be distinguished by their resting spores, which in *C. teres* are evenly vaulted while the primary valves are unevenly vaulted and spiny in *C. lauderi* (Rines & Hargraves 1988). Another example is *Chaetoceros lorenzianus* and *Chaetoceros decipiens*. While the vegetative stages closely resemble each other, *C. lorenzianus* produces resting spores with two pronounced, much branched protrusions. No resting spores are produced by *C. decipiens*.

2. Dinoflagellates

The dinoflagellates comprise a very diverse group of microorganisms that occur in marine as well as freshwater habitats. They can be found in all geographic regions from the tropics to the poles and, like the diatoms, they exhibit a huge variety of forms. About 2000 species of dinoflagellates are known and although they are usually regarded as phytoplankton, i. e., unicellular plants, 50 % of known species are considered heterotrophic, i. e., depend on organic food (Lessard 1991, Jeong 1999) and an increasing number of species are being discovered that are capable of both photosynthesis and the uptake of organic material (Stoecker 2007; Jeong et al. 2004, 2005; Carvalho 2008).

Dinoflagellate morphology

Dinoflagellates are divided into two large groups: Thecate dinoflagellates, which have an outer covering (theca) consisting of a series of cellulose plates and athecate (or naked dinoflagellates), which do not have a theca and, therefore, they often have a more variable shape. For both groups, as for diatoms, their orientation under the microscope is important when examining their morphology. The two most obvious characteristics in both naked and thecate dinoflagellates are the cingulum (transverse groove) and sulcus (longitudinal groove). These are two grooves that house the dinoflagellate's flagella. These flagella are characteristic of the group: One flagellum forms a very tightly undulating ribbon and is located in the transverse groove. The other flagellum is straight or ribbon-like but this is much less pronounced than in the transverse flagellum. Note: Although the flagella can be important features when identifying dinoflagellates, they are often difficult to see or become detached completely in the chemically preserved samples a plankton analyst is likely to encounter. This is particularly the case if the samples have been stored over extended periods of time.

By convention the side of the dinoflagellate on which the sulcus is located, is the ventral side of the cell (i. e., the ventral view). The side opposite this is the dorsal side. The part of the cell,

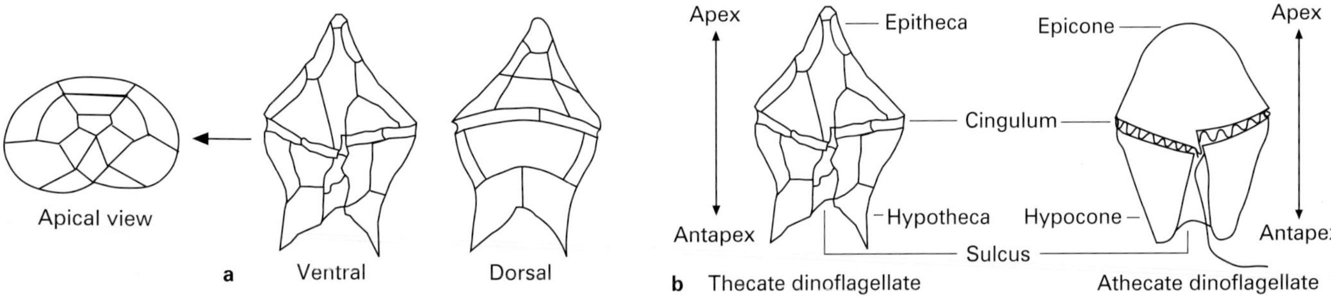

Fig. 7. a, Schematic drawing showing different cell orientations of a generalized thecate dinoflagellate. **b,** Ventral view of a thecate and athecate dinoflagellate. The two dissimilar flagella running through the cingulum and sulcus of both thecate and athecate dinoflagellates are shown only for the athecate species on the right.

Table 2. Classification of the Dinoflagellata, including family and genus characteristics (plate formulae following Steidinger & Tangen 1997).

Phylum **Myzozoa**

Class **Dinophyceae**

Order **Dinophysiales**

Family **Dinophysiaceae**

Description for temperate species as for genus.

Genera:

Dinophysis: Cells thecate and strongly laterally flattened (usually seen in lateral view). Cingulum located near the anterior end of the cell. Pronounced cingular lists obscuring the epitheca. The genus includes toxic species.

Phalacroma: As with *Dinophysis*, but cingular list not large enough to obscure the epitheca.

Order **Prorocentrales**

Family **Prorocentraceae**

Armoured dinoflagellates without a cingulum or sulcus. Flagella emerging from a small depression at the cell apex.

Genera:

Prorocentrum: Cells strongly laterally flattened. Thecal plates smooth or covered by spines and/ or pores.Two main thecal plates called valves.

Mesoporos: A central pore located in each of the main thecal plates.

Order **Gymnodiniales**

Family **Gymnodiniaceae**

Dinoflagellates without a theca but with clearly defined cingulum and sulcus. Cells possessing a single nucleus. Ocelli absent.

Genera:

Akashiwo: Cingulum lying in a median position. Apical groove running around the cell apex in a clockwise direction.

Amphidinium: Epicone much smaller than the hypocone, as cingulum located near the cell apex.

Gymnodinium: Cingulum lying in a median position. Cingulum circular or slightly displaced. Sulcus sometimes invading epicone.

Gyrodinium: Heterotrophic species with longitudinal surface striations and an elliptical apical groove bisected by a ridge.

Karenia: Genus resembling *Gymnodinium* and *Gyrodinium*. Cingulum in a median position with a small degree of displacement. Apical groove straight running from above the cingulum to the cell apex. The genus includes toxic species.

Katodinium: Cells spindle-shaped with a very short hypocone. Cingulum displaced by 3-4 times its width ventrally. Epicone striated longitudinally.

Lepidodinium: Cingulum weakly displaced. Cells pigmented. Main pigment chlorophyll b giving the cells a characteristic green colour.

Torodinium: Cells athecate with a very short hypocone as in *Katodinium*. Sulcus running along almost the length of the cell.

Family **Polykrikaceae**

Cells athecate, pseudocolonial, consisting of a number of integrated zooids each with a cingulum, but sharing a sulcus.

Genus:

Polykrikos: 2, 4, 8, or 16 zooids integrated into one colony. Shared sulcus straight or almost so. Species heterotrophic.

Family **Warnowiaceae**

Athecate cells with ocellus.

Genus:

Nematodinium: Cells with eyespots and nematocysts. Cingulum strongly displaced.

Order **Noctilucales**

Family **Noctilucaceae**

Description for temperate species as for genus.

Genus:

Noctiluca: Cells large (visible to the naked eye) spherical and without a theca. One flagellum and tentacle present but no sulcus. *Noctiluca* in temperate waters heterotrophic and food inclusions often visible within the cells. In tropical regions *Noctiluca* cells often appearing green due to the presence of endosymbionts.

Order **Peridinales**

Family **Ceratiaceae**

Descriptions for North Sea species as for genus.

Genera:

Ceratium: Cells with long apical and straight or curved antapical horns and apical horns of varying length. Cells are slightly dorso-ventrally compressed.

Plate formula: Po, 4', 6'', 5-6c, ?s, 6''', 1p, 1'''' (Fensome et al. 1993), Po, 4', 5'', 4-5c, ?s, 5''', 2'''' (Sournia 1986)

Family **Cladopyxiaceae**

Peridiniella: A small species with a delicate but distinctly ornamented theca, often with rows of pores, e.g., on the precingular plates.

Plate formula: Po, X, 4', 3-4a, 7'', 6c, 6-7s, 6''', 0p 2''''

Family **Gonyaulacaceae**

Thecate species with 6 precingular plates, antapical outline symmetrical, no dorso-ventral compression, girdle displacement.

Genera:

Alexandrium: Cells medium sized. Cingulum in a median position and only weakly displaced. By some authors this genus is assigned to the family Goniodomataceae.

Plate formula: Po, 4', 6'', 6c, 9(10)s, 5''', 2''''

Amylax: Cells with a strongly pointed epitheca and long apical horn. Antapical end of hypotheca blunt and with several spines. Cingulum only weakly displaced.

Plate formula: Po, 3', 3a, 6'', 6c, 7-8s, 6''', 0p, 2''''

Table 2. (continued).

Gonyaulax: Thecae with pronounced ornamentation, more or less well defined apical horns and a strongly displaced cingulum sometimes with an overhang, i.e. cingulum making more than a full turn around the cell. Antapical spines often present.
Plate formula: Po, 3', 2a, 6'', 6c, 7s, 6''', 0p, 2'''' (Steidinger & Tangen 1997), Po, 3', 2a, 6c, 4-8s, 5''', 1p, 1'''' (Dodge 1989).

Protoceratium: small oval to broadly biconical cells with heavy reticulations and areolation on the thecal plates.
Plate formula: Po, 3', 1a, 6'', 6c, 6s, 1p, 1'''' (Dodge 1989)

Family **Peridiniaceae** (subfamily: **Calciodinelloideae**)

Peridinian cells in which the cingulum has 4-6 plates plus a transitional plate. In addition, the Calciodinelloideae have a hexa second apical intercalary plate and calcareous cysts.

Genera:

Heterocapsa: Cells small and thecate. Cingulum circular. Hypotheca smaller than the epitheca.
Plate formula variable and still under discussion: Po, 4'-6', 2a-3a, 6-8'', 6c, 0-1p, 2'''', 7?s (Morrill & Loeblich (1981)

Scrippsiella: Theca delicate, hypotheca rounded and epitheca pointed.
Plate formula: Po, X, 4', 3a, 7'', 6c, 4-5s, 5''', 0p, 2''''

Family **Protoperidiniaceae**

Peridinian cells with 3 cingular plates plus a transitional plate.

Genera:

Diplopsalis: Heterotrophic thecate species with a lenticular outline in ventral view. Cells often pinkish in colour. Cingulum and sulcus with pronounced lists. Left sulcal list large and protruding beyond the cell antapex, 1 antapical plate.
Plate formula: Po, X, 3', 1a, 6'', 4c(3+t), 5s, 5''', 1''''

Oblea: Theca delicate. Cell outline almost round in ventral view. Just the antapex somewhat flattened. Cingulum in a slightly postmedian position and showing no displacement.
Plate formula: Po, X, 3', 1a, 6'', 4c(3+t), 6s, 5''', 2''''

Preperidinium: Thecate species resembling other members of the *Diplopsalis* group.
Plate formula: Po, X, 3', 2a, 7'', 4c(3+t), 5s, 5''', 1''''

Protoperidinium: Thecate species, usually with median cingulum that with or without displacement. Apical and antapical horns or spines usually present. Thecae often reticulate.
Plate formula: Po, X, 4', 2-3a, 7'', 4c, 6s, 5''', 2''''

Family **Pyrophacaceae**

Thecate species, either rounded or lenticular in ventral view. Thecae delicate and easily detaching from the cell body.

Genera:

Pyrophacus: Cells strongly flattened in an apical-antapical direction. Lens-shaped in outline when seen in ventral view. Cells strongly pigmented. Plate formula variable.

Fragilidium: Cells small, thecate, and rounded in ventral outline, as in *Pyrophacus* tabulation variable.
Plate formula: Po, 4-5', 7-9'', 9-11c, 6-8s, 7-8''', 1''''

which lies anterior when the cell is swimming is termed the apex and the side opposite the apex is the antapex (Fig. 7a). When looking at the ventral side the part of the cell to the right of the sulcus is the left side, the area to the left of the sulcus is the right side. A look at the cell from the top, the so called apical view is also instructive, as it can show whether the cell has a circular outline or is compressed in some way. Cells can be compressed either in a dorso-ventral direction or laterally. An extreme example of the latter is the genus *Dinophysis*. As its dorsal and ventral side is so narrow it almost always comes to lie in a lateral position on a microscope slide.

The cingulum itself is a transverse groove that divides the cell into two parts. These two parts are called epitheca (the anterior part of the cell) and hypotheca (posterior part of the cell) in thecate dinoflagellates and epicone/ hypocone in naked dinoflagellates (Lebour 1925) (Fig. 7b).

Two aspects of the morphology of the cingulum are important for species identification. Firstly, the cingulum can be displaced. This means that the two ends of the cingulum do not meet on the ventral side, but one side is at a greater distance from the cell apex than the other. If, looking at the ventral side of the cell, the portion of the cingulum on the left hand side of the cell lies below that on the right side the cingulum is descending. In the opposite case it is ascending (Fig. 8a). An example is the genus *Gyrodinium*. Secondly the cingulum's position: If the cingulum is situated half way between the apex and antapex, it is said to be in an equatorial or median position. This is the case in many *Protoperidinium* species. In other groups such as *Dinophysis* the cingulum has moved close to the anterior end of the cell (Fig. 8b) or the posterior, e.g., in the naked dinoflagellate *Katodinium glaucum*. Special cases are the species of the dinoflagellate family Prorocentraceae (Fig. 8c). In these, there are no separate grooves for the two flagella. Instead both arise from flagellar pores at the anterior part of the cell (the periflagellar area). The area containing the flagella is the periflagellar area.

The morphology of the sulcus or longitudinal groove, also differs between different groups of dinoflagellates. The sulcus is commonly straight, and runs from the cell's antapex to the the cingulum, e.g., in the athecate species *Akashiwo sanguinea*. However, it can also extend beyond the cingulum and extend into the epitheca or epicone. An example is the athecate dinoflagellate *Gyrodinium undulans*. The form and position of the cingulum and sulcus can often be discerned with light microscopy and should be one of the first features examined when attempting to identify dinoflagellates.

An important feature in the identification of thecate dinoflagellates is the arrangement, number and shape of thecal plates (i.e., their tabulation). Different systems are in use for counting dinoflagellate plates and establishing their arrangement around the cell. Here we follow the Kofoidian system, which arranges dinoflagellate plates into 6 series: The apical, precingular,

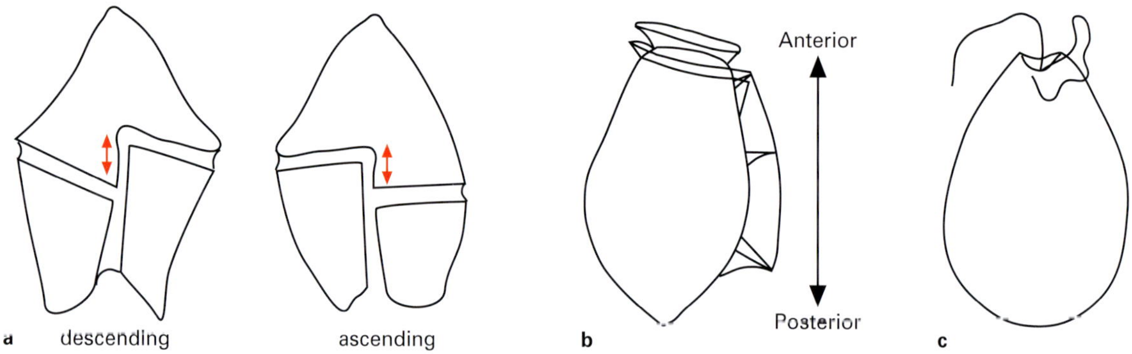

Fig. 8. a, Schematics of dinoflagellates with descending and ascending cingulum respectively. **b,** Dinoflagellates. A schematic drawing of a *Dinophysis* cell, in which the cingulum is located anteriorly. **c,** Schematic drawing of a *Prorocentrum* species in which there is no separate cingulum and sulcus.

cingular, postcingular, antapical and sulcal plates (Fig. 9). In addition to these 6 main series so called intercalary plates can also be present on epitheca and/ or hypotheca (Fig. 9b). If they are present, they are located between the apical and precingular plates or the postcingular and antapical plates and touch neither the apex nor the cingulum. Several other features of the theca are important as well, particularly the apical pore complex (APC), which is also useful in identification. This complex consists of a special pore plate in which the pore is located and a further plate, the canal or x plate. The shape of the pore can vary from simple slits to teardrop shapes. The APC can be very complex, e.g., in *Protoperidinium* species where the pore plate and canal plate can be surrounded by a 'list' or 'brim' as termed by Toriumi & Dodge (1993) (Fig. 9a). The tabulation of a thecate dinoflagellate usually characterises the genus, i.e., most species in the genus will have the same plate arrangement with few exceptions. To identify the species however, the shape of the plates also needs to be examined unless there are other obvious morphological features that help to identify the organism. We will provide tabulation patterns with relevant species descriptions of thecate dinoflagellates (Table 3). Note however, that different authors do not always agree in their interpretation of plate patterns. We will point out such discrepancies where necessary.

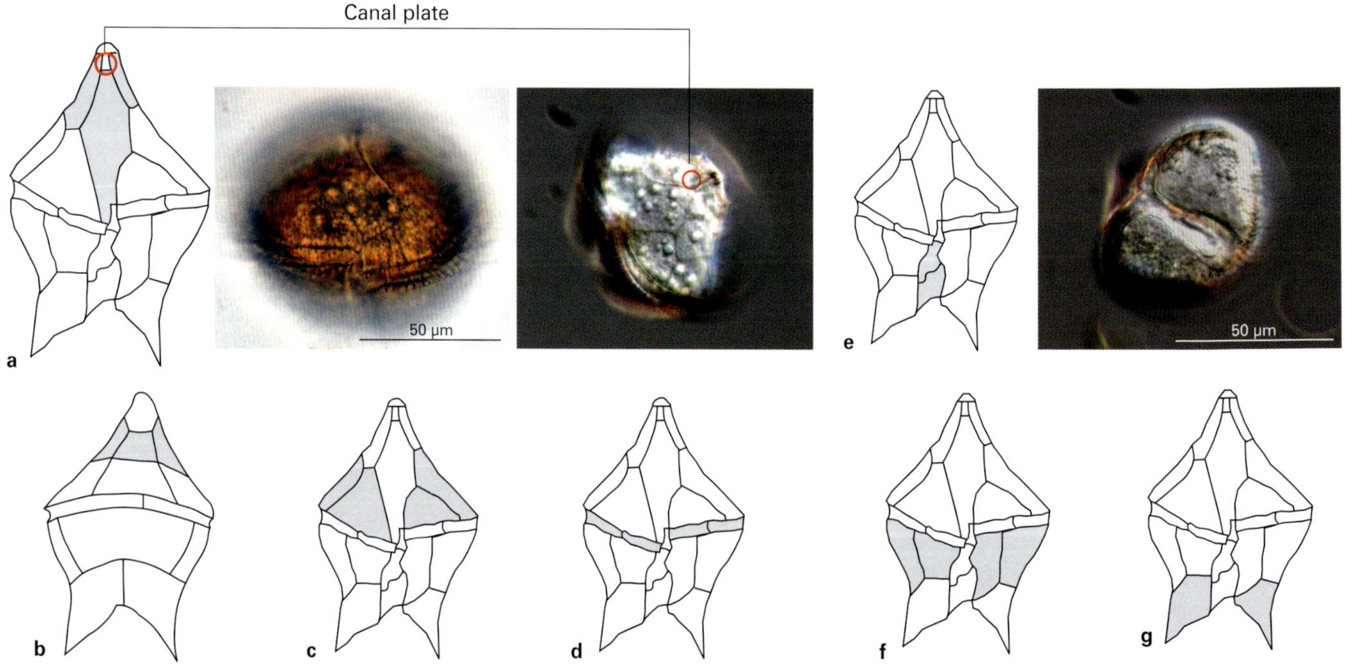

Fig. 9. Schematic of the dinoflagellate plates and their designation. **a,** apical plates; **b,** anterior intercalary plates; **c,** precingular plates; **d,** cingular plates; **e,** sulcal plates; **f,** postcingular plates; **g,** antapical plates

In some genera such as *Protoperidinium* and *Ceratium*, species can sometimes be identified on the basis of other morphological characteristics such as apical and antapical spines, but often the examination of thecal plates is essential. To give some examples, in *Alexandrium*, the shape of the first apical, sixth precingular and sulcal plates are important features, whereas in *Protoperidinium* the first apical and second anterior intercalary plates have to be examined. In the *Diplopsalis* group (a group of several thecate genera such as *Diplopsalis*, *Oblea* and *Preperidinium* resembling each other closely) the number of antapical plates is also important.

The examination of plate patterns is not an easy process and due to time constraints not always feasible in the routine identification of large numbers of samples. Plates are difficult to visualize using light microscopy. In many cases, e.g., *Protoperidinium* species the plates to be examined are located on both the ventral and dorsal side of the organism, which might necessitate the manipulation of the specimen on the slide. This requires some practice and can only be done once the sample has already been counted. For reliable identification epifluorescence microscopy in combination with the calcofluor stain (Fritz & Triemer 1985) should be used. If this is not available and the most important features on the dinoflagellate cell are not visible, the organism should be recorded at the genus level only.

Dinoflagellate reproduction

Dinoflagellates normally reproduce asexually by simple binary fission, but the exact mechanisms of division vary greatly between dinoflagellate groups. In *Alexandrium* and *Ceratium* for instance division is oblique with each of the daughter cell retaining half of the theca from the mother cell. Similarly in *Prorocentrum* and *Dinophysis* the cell expands laterally with a daughter inheriting one of the parent valves and forming a new valve inside the parent. This type of division is called desmoschisis. In other groups, e.g., *Protoperidinium*, two daughter cells are produced within the parent theca which is then absorbed with both daughter cells producing a complete new theca. This is known as elutheroschisis (Pfiester & Anderson 1987). Both, desmoschisis and elutheroschisis have been described as part of the complex life cycle of *Lingulodinium polyedrum* (Figueroa et al. 2005). In both types of reproductive pattern considerable changes in cell dimensions and shape are possible, e.g., in *Ceratium* species, which for an analyst unfamiliar with dinoflagellate life cycles can lead to species misidentifications (Figueroa 2008).

In some dinoflagellate species sexual reproduction has also been demonstrated. The fusion of two gametes can result in the formation of a motile planozygote. This is similar in morphology to the haploid vegetative cell but often larger. The fusion of two gametes to form the planozygote might be confused in some cases with cell division. Details such as the angle at which cells are attached to each other are important features distinguishing the two (Figueroa & Bravo 2005). The zygote can then start dividing again or form a cyst, the so-called hypnozygote, which is morphologically dissimilar from the vegetative cell (Marret & Zonneveld, 2003, Pfiester & Anderson 1987). The hypnozygote is very resistant to degradation and is thought to act as a resting stage that can survive adverse environmental conditions. These cysts sink to, and accumulate in the sediment where they can remain dormant for weeks or months forming a seedbed for a new population of vegetative cells. Often, for instance in the genus *Alexandrium,* the dormancy period is mandatory, i.e., during this period excystment is not possible. The exact mechanisms that trigger ex- and encystment in different dinoflagellate species are still not very well known, but for those species that have been extensively studied, several factors were shown to be important. Phosphate deficiency in the medium for instance triggered the formation of resting cysts in *Lingulodinium polyedrum* and *Alexandrium catenella* (Figueroa & Bravo 2005, Figueroa et al. 2005) and it has been argued by several authors, e.g., Wyatt & Jenkinson (1997) and Uchida (2001) that it is the actual population density that in some way triggers the formation of gametes and planozygotes. In addition dinoflagellate life cycles can be very complex with several types of cyst being produced by the same species (Figueroa & Bravo 2005). Although resting stages are predominantly a benthic feature, they can also occur in water samples, e.g., after stormy periods.

Table 3. Example of the standardized reporting scheme for dinoflagellate plate tabulation. In this case *Protoperidinium* plate tabulation is given as an example. Numbers refer to the numbers of plates in a given series and the signs designate the series itself, i.e. 6s denotes a dinoflagellate with 6 cingular (c) plates.

Apical pore complex		Apical (')	Anterior intercalary (a)	Precingular ('')	Cingular (c)	Sulcal (s)	Postcingular (''')	Posterior intercalary (p)	Antapical ('''')
Po	X								
+	+	4'	2-3a	7''	4c	6s	5'''	0p	2''''

3. Other planktonic organisms

Other flagellates

A great diversity of small flagellate species, belonging to a number of different taxonomic groups (e.g., cryptophytes, prasinophytes, coccolithophores, euglenophytes and raphidophytes; Fig. 10) occur in the oceans. Size ranges of the species in the different groups vary greatly and are not always a good criterion for identification. Raphidophytes for instance contain members usually smaller than 30 µm *(Heterosigma akashiwo)* while others exceed 100 µm *(Chattonella antiqua)*. Some groups such as the prasinophytes *(Micromonas pusilla)* and haptophytes also include picoplanktonic species, which are smaller than 3 µm.

These major groups vary in several important aspects: their pigment composition, presence or absence of an eyespot as well as the shape and number of chloroplasts. Raphidophytes for instance have many small chloroplasts whereas members of the Prasinophyceae, Cryptophyceae and Prymnesiophyceae (e.g., the foam forming *Phaeocystis pouchetii*) have only one or two.

Other characteristic features are the number and types of flagella and, related to the latter, types of motility. Flagella can be used for either pulling or pushing the cell forward (e.g., in the prymnesiophytes). In groups with two flagella sometimes only one flagellum is used in locomotion (e.g., raphidophytes). Therefore, it is often necessary to view live samples, when attempting to identify a species.

In chemically preserved samples, and when using light microscopy, flagellates can only be identified with great difficulty or not at all as cells might shrink and change their shape and/or flagella might be lost. We will provide some flagellate light micrographs, but tools such as Scanning electron microscopy (SEM) and Transmission electron microscopy (TEM) should be used for identification. (Descriptions of the major groups will be provided on page 173-180).

Ciliates

Ciliates are a very diverse group of organisms with about 7000 described species which are found in a vast range of habitats from marine pelagic and benthic species to freshwater ciliates and those living in soil.

Ciliates derive their name from the presence of hair like structures called cilia (Latin for 'eye lashes'). These are arranged in species specific patterns around the body and oral cavity or like in the ciliate *Paramecium* (a fresh water ciliate) are distributed more or less evenly all over the cell. In some species cilia are grouped in bundles forming so called cirri. The cilia are involved in locomotion, feeding and can also have a sensory capacity. A particular feature, that can also be useful in species identification, is the nuclear morphology (shape, number and size of the nuclei). In contrast to other organisms, ciliates have two types of nuclei, of which there can be several per cell. The macronucleus is involved in key metabolic processes such as protein synthesis, whereas the micronucleus contains two sets of chromosomes and is thought to be involved in sexual reproduction (for further information see Lee et al. 2000, a useful web resource is **http://www.liv.ac.uk/ciliate**). The structure of the nuclei is not visible in cells that have been preserved in Lugols iodine solution. However, they can be visualized after removal of the fixative, staining of the cells with DAPI and examination under the epifluorescence microscope (see methods section below).

Together with dinoflagellates ciliates form the main components of the microzooplankton. Many species are heterotrophic and some mixotrophic (e.g., *Myrionecta rubra* (Lohmann) Jankowski, 1976 and *Laboea strobila*, Lohmann, 1908 (see Stoecker et al. 2005). Food particles are usually taken up by phagocytosis. They are ingested via the mouth (cytostome) and become enveloped in food vacuoles within the cells. The contents are broken down via lysozomes. Traditionally, most free living ciliates were thought to be feeding on bacteria. However, it is now known that they can also ingest much larger food particles including some diatoms.

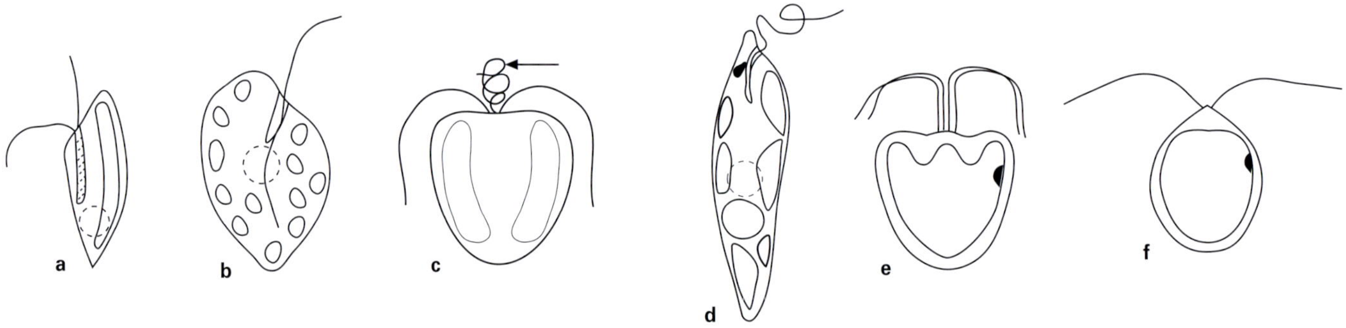

Fig. 10. Schematics of common flagellate groups. **a,** Cryptophyceae; **b,** Raphidophyceae; **c,** Prymnesiophyceae, with arrow pointing to the haptonema. **d-f,** Chlorophytes: **d,** Euglenophyceae; **e,** Prasinophyceae; **f,** Chlorophyceae. Black spots in d-f: eyespots; dotted lines = position of nucleus. e,f, redrawn from Throndsen (1997).

We here only provide images of some common species that frequently appear in phytoplankton samples and can be identified even in chemically preserved samples.

Notes on methodology

A wide range of techniques are available for the study of phytoplankton both with respect to different microscopy techniques and sample collection/ fixation and analysis methods. Increasingly molecular techniques are also used, although these are not yet practical for routine monitoring programmes with a high throughput of samples. Some microscopic techniques are also very complex and beyond the scope of this book. Here we restrict ourselves to the methods commonly used in phytoplankton monitoring. In the following section we will point out potential advantages and disadvantages of the different methods as all steps of the sample preparation and enumeration process can introduce artifacts that one needs to be aware of.

Sample fixation

Fixation with Lugol's solution is the most common method, although a variety of additional fixatives such as formalin and glutaraldehyde are also in use. The choice of fixative depends on a number of factors like the general purpose of the study (enumeration, biovolume calculations), the target organisms to be fixed and the envisaged period of storage of samples post analysis. Lugol for instance has the advantage of staining cells heavily and making them more conspicuous, e. g., in sediment rich samples. It is also recommended for the enumeration of microzooplankton species such as ciliates. Moreover, it also enhances the sinking of the particles in Utermöhl chambers (see below). However, Lugol fixation also has its disadvantages. The heavy staining obscures morphological characteristics, for instance the position of the nucleus, pyrenoids or other organelles. As Lugol also masks the organism's autofluorescence, samples preserved in this manner are also unsuitable for foodweb studies, e. g., the detection of mixotrophic species. A further problem is that Lugol can change the shape and size of the fixed organism. Both shrinkage and expansion of cells have been observed making biovolume calculations difficult (Gifford & Caron, 2000, Menden-Deuer et al. 2001, Zaraus & Irigoien 2008). The problem is compounded by the fact that not all species in a sample respond to a given fixative in the same way. Lugol also has a tendency to become depleted during long-term storage making continued monitoring of the stored samples and potentially addition of further fixative necessary. For long-term storage, formalin is a, therefore, better option. It is also preferable to Lugol when the use of epifluorescence microscopy is planned as formalin preserves or even enhances the cell's fluorescence patterns. Despite its disadvantages however, Lugol is relatively inexpensive and does not pose considerable health risks. It is also simple to prepare and apply to a sample and is, therefore, still the fixative of choice in many routine monitoring programmes, which do not require the identification of every single taxon to species level.

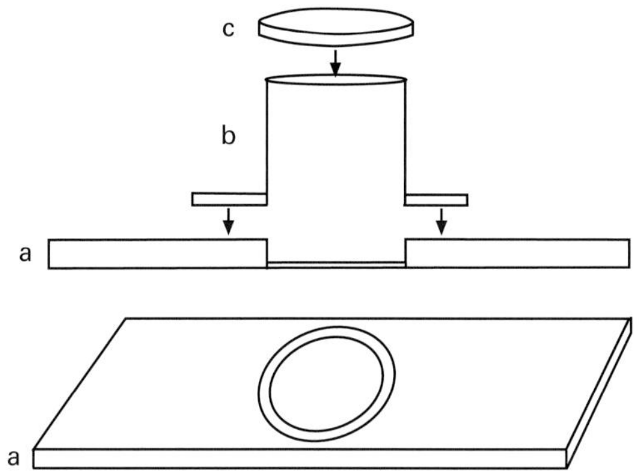

Fig. 11. Schematic drawing of an Utermöhl chamber. **a,** The base plate, containing the well in which the sample is concentrated. **b,** The Utermöhl chamber. **c,** The cover glass placed on top of the tower after it has been filled with the sample.

Sample analysis

The most commonly used method in the routine enumeration of phytoplankton is the so called Utermöhl method (Lund 1958, Utermöhl 1931), where raw samples are concentrated by simple sedimentation in Utermöhl chambers (Fig. 11a-c) and viewed under an inverted microscope. Sedimentation allows the enumeration of relative dilute samples, increasing the chance of capturing rare species. Utermöhl chambers are available in various volumes. The Utermöhl apparatus consists of a base plate (Fig. 11a) with a small well in which the sedimented cells are collected, and a cylinder which is placed on top of the base plate (Fig. 11b). The cylinder is aligned precisely with the collecting chamber, and the sample is then filled into the cylinder. Lastly, the cylinder is covered by a top plate to avoid sample evaporation (Fig. 11c). It is important to completely fill the chamber before closing it with the cover glass as otherwise the sample may leak. The sedimentation time varies, depending on the size of the cylinder and the size of the cells. For coastal samples chambers with a volume of 10, 25 or 50 ml are used, depending on the density of material in the original raw sample. For very dilute samples 100 ml chambers are also employed but these should be used with care as in these very large cylinders samples sometimes do not settle completely even after long periods of time.

Although the Utermöhl method is easy to use in principle, care needs to be taken when using the technique and during subsequent sample enumeration to achieve a consistent counting result. Before filling a chamber with the sample, the sample should be brought to room temperature to avoid excessive bubble formation which might hinder the sedimentation process. It is also essential that the sample is properly

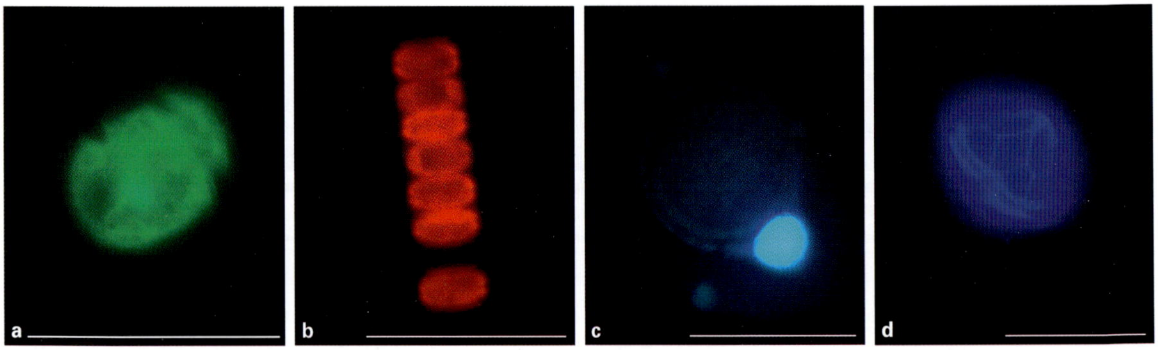

Fig. 12. a, Autofluorescence of the heterotrophic dinoflagellate *Gyrodinium calyptoglyphe*. **b,** Chlorophyll a autofluorescence of the diatom *Paralia sulcata*. **c,** *Gyrodinium spirale* with ingested diatom cell and DAPI stained nucleus. **d,** The dinoflagellate *Protoperidinium subinerme* Calcofluor stained thecal plates (the cell was fixed in formalin). Scale bars = 50 μm.

homogenized before addition to the Utermöhl chamber. This should be done by gently inverting the sample bottles many times. Violent shaking should be avoided as this might destroy delicate cells and break up diatom chains. Furthermore, the samples should be settled out in a vibration free place to ensure an even distribution of the cells on the slides and to avoid re-suspension of already settled cells.

This method enables the analyst to count larger sample volumes increasing the chances of finding less abundant cells even in sediment rich samples. However, it can also be very time consuming

Microscopy techniques

The most commonly used light microscopy techniques are brightfield, phase contrast and differential interference contrast (DIC). The latter two are particularly useful for colourless specimen, which provide little or no natural contrast. Both DIC and phase contrast are useful for the visualization of spines and other processes or the surface structure of thecae and frustules. These techniques are very valuable for purely taxonomic studies. For a more ecological approach a further technique, epifluorescence should also be used.

Epifluorescence microscopy

In a nutshell, in epifluorescence microscopy the particles in a sample are hit with light of a specified wavelength. This causes the temporary excitation of pigments in the sample to a higher energy level. When the energy level drops again the molecule emits light of a different wavelength, i.e. it fluoresces. This method has very wide applications in plankton research and indeed in all Life Sciences (Fig. 12). Chlorophyll for instance fluoresces red when hit with blue or UV light. This allows the enumeration of small autotrophic cells where not enough morphological features are visible for the identification of taxonomic groups.

Using appropriate fluorescent dyes it is also possible to visualize other cell organelles such as nuclei. A common, stain for DNA for instance is DAPI. When excited with UV light the stained material will fluoresce blue allowing the visualization of size and position of the nucleus in the cell. This is a very useful tool for the identification of ciliates as the number, position and morphology of the cell's nuclei are important diagnostic features. A further important substance is Fluorescent Brightener 28 (previously known as Calcofluor, Fritz & Triemer 1985). The fluorescent brightener attaches to cellulose. The thecal plates of dinoflagellates consist of cellulose, with the sutures between adjacent plates being somewhat thicker than the plates proper. Staining the cells with the brightener, therefore, allows the visualization of plate boundaries and, hence, an examination of the number and shape of the thecal plates of dinoflagellates. Fluorescent brightener is, therefore, a vital tool in the identification of dinoflagellates and the description of new species. Epifluorescence microscopy cannot easily be used on Lugol-fixed samples, as the iodine based stain obscures the chlorophyll fluorescence. However, the Lugol stain can be removed by adding a few drops of sodium thiosulphate to the settled sample. After post fixing the cells with formalin, epifluorescence normally works well.

Automated systems for particle recognition and counting

Many monitoring programmes and sampling campaigns generate a number of samples that have traditionally been counted manually often with sedimentation techniques such as Utermöhl or other sedimentation chambers. However, this manual sample preparation and analysis is extremely time consuming. Therefore, increasingly techniques are being developed which can count and identify particles of different sizes and shapes automatically. Such systems usually make use of two properties of the particles in a sample. The first is light scatter, in which a particle is hit with a light beam, usually a laser, and depending on the optical properties of the particle light will be scattered away from the particle in a characteristic pattern. The second property is fluorescence, in which particles, hit with a beam of light of a narrow wavelength range, emit light in a characteristic pattern depending on their chemical composition (e.g., the types of pigments present).

An example of an instrument using both physical properties to distinguish between and quantify different particles is the flow cytometer. Flow cytometers are often used for studies involving nano- and picoplankton. A drawback of flow cytometers is however that the size range of particles is limited but it allows a very high throughput of particles. Flow cytometers also work with a hardware (flow cell) that generates a laminar flow in which the particles are aligned and move past a light beam and sensor individually. The flow cell is expensive and requires a considerable degree of maintenance.

A more recent development is the FlowCAM, which combines some properties of conventional microscopy and flow cytometry. As with flow cytometry it uses light scatter and fluorescence properties of particles flowing through a flow chamber past a light source and sensor to discriminate between different types of particles, e.g., different phytoplankton species. In contrast to flow cytometry however, no unit for creating laminar flow conditions is needed, facilitating an easier set-up. The size range of objects that can be processed is also considerably larger than in flow cytometry (1 µm-3 mm compared to 0.5-20 µm) In addition, it is also possible to take digital images of the samples. The FlowCAM offers a range of set-ups from continuous imaging of the sample flowing through the flow chamber to more selective imaging using using fluorescence or scatter as a trigger. While the imaging of counted particles is an undoubted advantage, the process of fine tuning the equipment is very time consuming and the FlowCAM cannot easily be used in samples containing a lot of sediment.

A further recent development is the ZOOSCAN. This is similar to a conventional scanner but coupled with image recognition software for the identification and quantification of the scanned zooplankton samples (Grosjean et al. 2004).

However, although these new techniques offer exciting prospects, manual counting will remain the standard technique for the foreseeable future with the exception of flow cytometry, which is already the tool of choice for very small particles.

Ecological aspects concerning phytoplankton

Nutrient supply

To facilitate growth, phytoplankton requires light and carbon and a range of different inorganic nutrients. Whereas some of these are needed in considerable concentrations (macronutrients), e.g., nitrogen, phosphorus and silica (the latter, e.g., in diatoms and silicoflagellates). Some of the macronutrients like phosphorus and to a lesser degree nitrogen may be stored by the cells, others like silica cannot. Internal phosphorus pools allow up to five cell doublings, nitrogen pools up to three doublings when the relevant nutrient is depleted in the surrounding medium.

In addition to macronutrients, a suite of different chemical elements are required in smaller concentrations, for instance sulphur, potassium, sodium, calcium, magnesium and chlorine. These are termed micronutrients (Reynolds 2006). All are essential for the biosynthesis of cell metabolites (proteins, lipids, carbohydrates, nucleic acids and ATP, etc.) and, hence, biomass. In addition, micronutrients steer metabolic pathways as co-factors of enzymes. Examples of such micronutrients are iron, manganese, molybdenum, copper, cobalt, zinc, boron, selenium and vanadium. Some dinoflagellate species for instance have a special requirement for selenium (Mitrovic et al 2004). Another important micronutrient is iron (Veldhuis & de Baar 2005). This is particularly true in so called HNLC (High nitrogen low chlorophyll) areas where major nutrients such as nitrogen occur in sufficient concentrations to allow high primary production but where production rates are nevertheless very low. In these areas the addition of iron to the water will cause the rapid onset of a phytoplankton bloom. A deficiency of one of these micronutrients causes a decrease or complete cessation of growth.

Optimal nutrient conditions do not only depend on the absolute amounts of any one nutrient but also on the nutrient ratio. The ratio of the main chemical components C : N : P in phytoplankton cells is 106 : 16 : 1 (Redfield et al. 1968). This is called the Redfield Ratio and reflects a situation where nutrients in the ambient water are not limited. If the N : P ratio is higher, a phytoplankton bloom potentially will be phosphorus limited, if it is lower it will potentially be limited by nitrogen. For diatoms silica can be included in this ratio which is then 106 : 15 : 16 : 1 (C : Si : N : P). This ratio was first described for oligotrophic open ocean waters. In coastal nutrient enriched waters the ratio of carbon to nitogen and phosphorus can deviate considerably from the Redfield Ratio. For more details see, e.g., Reynolds (2006).

Primary production

The process of primary production is the transformation of inorganic carbon into organic substances (autotrophy). The necessary energy can be provided either by light (photosynthesis carried out by plants and bacteria) or by inorganic compounds (chemosynthesis in bacteria). The synthesized organic substances are used for the organism's basic metabolism, the build-up of new biomass and reproduction. Some metabolites also serve as defences against predators.

The biomass produced by autotrophic organisms forms the nutritional basis for all other trophic levels i.e., the consumers and detritivores of the marine food web. In addition to the phytoplankton, benthic micro- and macroalgae as well as some higher plants (e.g., seagrasses) are also involved in marine primary production. All benthic primary producers are limited to the shallow depths where light levels are sufficient and the depth where net primary production is still possible is termed the euphotic zone. By definition this reaches down to a depth where 1 % of the surface light is still available. The extent of this zone is negatively correlated with the concentration of particulate and dissolved substances in the water column. In other words, the more turbid the water the shallower is the euphotic zone.

The spatial distribution of phytoplankton biomass in the oceans is very patchy. Chemical factors such as macro- and micro nutrient supply but also physical factors such as light availability and temperature have a great impact on the primary production process.

Regions with a high productivity are typically coastal areas with high nutrient inputs by rivers, or the upwelling areas on the west coast of Africa and the Americas. In general, primary production in shallow areas on continental shelves is estimated to reach up to 200 g C $m^{-2} \cdot a^{-1}$. The production rates in shallow areas contribute almost a quarter of the total oceanic production although they occupy only 5 % of the area of the seas. Even in normally unproductive areas however so-called "hot spots" of productivity can occur where primary production can reach as high as 500 to 800 g C $m^{-2} \cdot a^{-1}$. One such hot spot is the Sargasso Sea, which is located in the western Atlantic in an area which is otherwise a desert in terms of nutrient supply. The Sargasso Sea has large standing stocks of seaweed of the genus *Sargassum* and extensive phytoplankton blooms also occur. Although the phytoplankton can grow in these low nutrient conditions, it is rapidly grazed by the microzooplankton (Lessard & Murrell 1996, 1998). The main reason for the high productivity of the Sargasso sea is the occurrence of eddies (cyclical current systems) that cause the regeneration of nutrients from deeper layers into the surface water (Ewart et al. 2008).

The recent development of satellite remote sensing techniques now allows a better estimate of primary production worldwide. The contribution of marine primary production to the global carbon cycle shows an average of 45 to 50 Pg $C \cdot a^{-1}$ (or 4,5 to $5 \cdot 10^{16}$ g $C \cdot a^{-1}$). According to the data from current literature given in Reynolds (2006) this corresponds almost to the contribution of terrestrial systems (56.4 Pg $C \cdot a^{-1}$).

Types of nutrition

All organisms on earth have three basic requirements:

1. They need a source of carbon (the basic building block for forming biomass).
2. They also require a source of hydrogen atoms (reducing equivalents). These are necessary in the processes of the conversion of different carbon compounds.
3. They need a form of energy to drive their metabolic processes.

There are fundamental differences between plants and animals in how they fulfil these requirements. In plants, including phytoplankton, the source of carbon comes from carbon dioxide (CO_2) and water (H_2O) is the source of the hydrogen atoms. In the well known process of photosynthesis these simple inorganic compounds are used to form complex sugars with the help of the photosynthetic pigments that give plants their characteristic colour (green, brown or red depending on the dominant pigment). Organisms such as plants which derive their energy from light are called phototrophs, whereas animals are chemotrophs. Likewise, using CO_2, an inorganic source of carbon, as only carbon source is referred to as autotrophy. Again this is the case in plants. Animals on the other hand are heterotrophs, which get their carbon from ingesting and breaking down organic carbon sources.

Marine autotrophs

In the plankton, diatoms are often the dominant group of autotrophs, but a range of small flagellates, e.g., raphidophytes, cryptophytes, haptophytes and euglenophytes (the latter also containing heterotrophic species) are also important. Although these organisms have photosynthetic pigments in chloroplasts and usually use CO_2 as carbon source, several species are also reported to take up dissolved organic substances to cover the carbon demand for example under bad light conditions (e.g., Tuchman et al. 2006, White 1974).

Marine heterotrophs

With respect to nutrition, the dinoflagellates (Dinophyceae) are one of the most interesting groups. Nearly 50 % do not have chloroplasts and are obligate heterotrophs (Lessard 1991). These either take up dissolved organic compounds or feed on particulate organic matter and some are parasitic (Jacobsen & Anderson 1986, Schnepf & Drebes 1988, Taylor et al. 2008).

A great diversity of organisms, from bacteria to diatoms and even zooplankton larvae, serve as food and a large number of feeding mechanisms have evolved for the capture of this diverse range of prey. These include feeding with a pallium (feeding veil), feeding with a peduncle and also direct engulfment. The latter was originally thought to be typical of athecate dinoflagellate species but has also been discovered in a number of thecate species (Hansen & Nielsen 1997, Stoecker et al. 1997).

This type of feeding can cause considerable deformation of the cells to the extent that they are barely recognizable.

A further nutritional strategy is kleptoplastidy where chloroplasts (plastids) of ingested phototrophic organisms remain functional temporarily within the predator cell (Janson 2004).

Mixotrophy

Mixotrophy is a further feeding strategy common among particularly the dinoflagellates (Stoecker 1999). Mixotrophic species are pigmented and, therefore, capable of photosynthesis but also take up organic matter by active ingestion of prey items. The number of species known to be capable of both photosynthesis and active feeding is rapidly increasing, e.g., through the increased use of epifluorescence microscopy. The genera *Dinophysis*, *Prorocentrum*, *Scrippsiella*, *Ceratium*, *Fragilidium*, *Gymnodinium* and others all contain mixotrophic species (Jacobsen & Anderson 1994, Jeong et al 2005, Skovgaard 1996, Stoecker et al. 1997).

Symbiosis

A special case is an organism, which can take advantage of photosynthetically active, endosymbionts. Even typical microzooplankton species like the ciliate *Myrionecta rubra* may become phototrophic due to endosymbiotic cryptophycean cells. *M. rubra* can form huge blooms during summer colouring surface waters intensely red. Many other species of ciliates are either mixotrophic or heterotrophic (Jeong et al. 2002).

Factors influencing photosynthesis

The photosynthetic rate of the primary producers depends on the amount of light, the water temperature and the availability of macro- and micronutrients in the water.

The light quality changes considerably when entering the water column with increasing depth. Long wavelengths (red and infrared) are absorbed near the surface. In clear water short wavelengths (blue) penetrate the deepest. However, in particle rich water the short wavelengths are absorbed by the chlorophyll of the phytoplankton as well as by detritus particles and dissolved organic substances. In these cases green light advances furthest down into the water column.

Chlorophyll, the most important photosynthetic pigment in all algal groups, has two absorption maxima, one in the blue and one in the red wavelength spectrum. All wavelengths intermediate (the so called green gap) cannot be used efficiently by chlorophyll. Hence, most of the algal groups have additional accessory pigments with absorption maxima in the green gap so that all available light energy can be used. Cyanobacteria and cryptophytes for example have phycocyanin and phycoerythrin, diatoms fucoxanthin and the autotrophic dinoflagellates diadinoxanthin as additional accessory pigments (Dring 1991).

As mentioned above the photosynthetic rate strongly depends on the amount of available light (i.e., photon flux rate) and follows a typical photosynthesis/light curve: With increasing photon flux rates the photosynthesis increases up to a maximum value above which the rate of photosysnthesis remains constant with increasing light supply. Still higher amounts of light can even cause a decrease an organism's photosynthetic rate. This is known as photoinhibition. The photosynthesis/light curves vary between taxonomic group and the adaptation state of the species.

The impact of chemical factors is described in the chapter "Nutrient supply", p. 25.

Food webs

The classic idea of a pelagic food chain is that the energy which is bound in the biomass of the phytoplankton (primary producers) is transferred via the zooplankton (primary and secondary consumers) to higher trophic levels such as fish. It has now become clear, however, that this simple view does not reflect the complexities of nature but that the sum of trophic interactions does instead form an intricate web. For instance, we now know that what used to be called phytoplankton is actually a very heterogeneous group of organisms with the truly autotrophic diatoms but also dinoflagellates, many of which are heterotrophic or mixotrophic (capable of photosynthesis as well as the uptake of organic material). The latter now form a separate compartment the so-called microzooplankton, which also includes other heterotrophic and mixotrophic microbes such as ciliates, but also less well studied groups such as the choanoflagellates. In the microzooplankton in particular foodweb interactions are very complex. Dinoflagellates and ciliates are capable of feeding on a considerable size range of organisms from bacteria to crustacean larvae (i.e., prey can be an order of magnitude larger than the predator). In this way the microzooplankton forms a link between the higher trophic levels and the microbial foodweb. It is now known that microzooplankton can have a larger effect than copepods in controlling phytoplankton biomass and might even be able to control diatom blooms (Irigoien et al. 2005).

The discovery of the importance of the picoplankton, a group of organisms even smaller than nano- and microplankton has further added to our knowledge of the complexity of marine foodwebs and to the rates of flux of anorganic and organic material in the oceans. In the open ocean picoplankton (mainly cyanobacteria) are the dominant primary producers (Le Gall et al. 2008). They also have a considerable role in the export of carbon from the surface to deeper waters (Richardson and Jackson 2007).

The discovery of the importance of these new trophic components and interactions has also led to a re-evaluation of the pathways and rates of transfer of energy in the marine foodweb. All trophic levels release dissolved organic matter (DOM) into the water. The largest proportion of these substances comes from the phytoplankton. The dissolved organic material is utilized as substrate by bacteria. These bacteria serve as food source for heterotrophic nanoflagellates (HNF) which again are the food source for microzooplankton. These are then incorporated by multicellular zooplankton (for example copepods). In this way energy is essentially recycled and returned to the classic food web. This cyclic process is called the "microbial loop" (Azam et al. 1983). However, the importance of the marine microbial component, particularly bacteria, was first recognized by Pomeroy (1974).

Annual succession of Phytoplankton

When nutrient supply is sufficient, primary production is generally a function of light, temperature, and carbon dioxide concentration (which is usually not a limiting factor in marine environments). Solar radiation penetrates the atmosphere and hits the sea surface where it is partially reflected depending on the angle of the sun. Therefore, in winter when the sun is low, more light is reflected off the surface, and only little light penetrates the water column. Hence, the euphotic zone, where light supply is sufficient for photosynthesis, is much shallower than in summer. Furthermore winter storms may cause high turbulence in coastal areas re-suspending large amounts of organic matter from the bottom into the water column further decreasing the amount of ambient light. However, bacterial decomposition is continuing, albeit with a decreased turnover at low water temperatures. Thus, the nutrients phosphorus, nitrogen and silicon are dissolved in their inorganic form in the water again. Additionally, in coastal areas nutrients enter the system via river run-off. As they are not incorporated into biomass because of the low primary production in winter the nutrient concentrations in the water column tend to rise.

With increasing angles of the sun more light reaches into the water column causing a deeper euphotic zone. Simultaneously the upper part of the water column becomes stratified due to the warming by the penetrating sun light. Hence, the so called mixed layer decreases reaching a point where it coincides with the euphotic zone. In this situation, conditions are usually favourable for the development of the spring phytoplankton bloom.

In coastal areas with sufficient supply of silicate, diatoms dominate the phytoplankton community. As silicate is needed for the construction of the frustule and diatoms have very short doubling times under optimal conditions, they have an advantage over other taxonomic groups in silica rich coastal waters. Later, silicate is often the limiting factor for the diatoms. This initial phytoplankton bloom is usually followed by a bloom of microzooplankton and larger zooplankton species such as crustacean larvae and copepods, which feed on the phytoplankton. Together with limiting nutrients zooplankton grazing contributes to the breakdown of the spring bloom. However, infection of the phytoplankton populations with viruses and parasites as well as species specific life cycles and sedimentation can also play a decisive role.

During summer the biomass of the phytoplankton community remains on relatively low levels because of nutrient limitation, although smaller peaks can appear. Dominant species are mainly dinoflagellates with both heterotrophic (e.g., *Protoperidinium* species) and mixotrophic species, for instance of the genus *Ceratium* (see chapter "Types of nutrition", p. 26).

In autumn remineralisation processes involving bacteria (and, therefore, the release of nutrients back into the water) exceed the nutrient uptake rates of the ambient phytoplankton populations and more nutrients become available. This triggers a second bloom of the phytoplankton community. As, especially in coastal areas, silicate is then available again, this bloom is normally once more dominated by diatoms. After the break down of the autumn bloom which is usually caused by light limitation the winter situation is reached again.

One factor not yet mentioned here is the formation of different types of resting stages for example by diatoms and dinoflagellates. Many planktonic species, including dinoflagellates and diatoms have complex life cycles including resting stages that are formed in unsuitable environmental conditions and sink to the sediment. When considering environmental controls of phytoplankton blooms the tolerance levels and requirements not only of the vegetative stages but also the resting stages need to be considered. This is particularly important as the resting stages can accumulate in the sediment, sometimes over many years. When conditions become favourable again they excyst i.e. they form a seedbed for a new bloom (Genovesi-Giunti et al. 2006, McQuoid & Godhe 2004).

The above description of phytoplankton succession patterns in coastal temperate waters, over the course of a year results from long-term observations and mean values but the onset, scale and composition of the spring and autumn blooms can vary considerably from year to year.

Control of phytoplankton biomass

Both phytoplankton and zooplankton population growth are caused by a complex interplay of factors such as light intensity, temperature and nutrient supply and predation (Cloern 1996). While it is true that the timing and magnitude of the spring bloom is often determined by physico-chemical factors this is not universally true. After warm winters for instance the survival rates of zooplankton are higher than in colder years. The spring population of zooplankton in relation to the phytoplankton such as diatoms is, therefore, larger and can suppress the magnitude or even delay the emergence of a phytoplankton bloom (Wiltshire & Manly 2004).

The control by abiotic factors is called "bottom up" because the phytoplankton biomass depends on a lower level (resources for the corresponding trophic level). In contrast to that, during summer the phytoplankton biomass is often controlled by the number of predators (zooplankton) present. If the ingestion rates of the zooplankton exceed the production rate of the phytoplankton the biomass of the primary producers is decreasing. This control is called "top down" because it is caused by the upper trophic levels, i. e., the presence of predators rather than resource availability. It is not only the traditional zooplankton grazing however, that can curtail a phytoplankton bloom. mass infection by viruses for instance have been shown to cause the collapse of *Phaeocystis* blooms (Jacobsen et al. 2007).

'Harmful algal blooms'

All unicellular algae that may have a deleterious effect on other aquatic species or on humans (including economic damage) are termed 'harmful algae'. However, this is not a systematic term, but encompasses a number of different algae taxa such as haptophytes, cyanobacteria, diatoms and dinoflagellates. Harmful algal blooms are a worldwide phenomenon with many harmful species also occurring in the North Sea (Elbrächter 1996). There are several different pathways via which harmful algae can exert a negative effect.

Mass blooms

Some microalgae can bloom in such cell densities that that they cause a discolouration of the water. Depending on the dominant pigments in the causative organism, these blooms can be reddish, brown or green (termed red, green or brown tides respectively). Mass blooms are harmful as they can lead to the formation of anoxic zones when the bloom decays and they also reduce the amenity value of public beaches, when masses of dying cells or mucus are washed up on the shore. A prime example for this problem is *Phaeocystis globosa*. In their life cycle *Phaeocystis* species pass through different stages, the most striking being a colonial stage with numerous non-motile cells embedded in buoyant mucilage. Mass accumulations of this mucilage on beaches are a well known phenomenon in the North Sea. Other bloom forming organisms are dinoflagellates such as *Akashiwo sanguinea* (red tide former) or *Karenia mikimotoi*. Diatoms such as *Asterionellopsis glacialis* (a brown tide former) are also known to form dense enough blooms to discolour the water. Not all mass occurrences have been demonstrated to have harmful effects. Examples are blooms of the dinoflagellate *Lepidodinium chlorophorum*.

Toxicity (Table 4)

Another harmful effect of major concern from both a public health and economic perspective is the production of toxins by microalgae. Almost all major phytoplankton groups, from diatoms *(Pseudo-nitzschia)* via a host of dinoflagellates (e. g., *Alexandrium*, *Protoceratium*, *Pfiesteria*) to other flagellates (e. g., *Prymnesium*, *Chrysochromulina*) can produce toxins

Table 4. Important toxic phytoplankton species confirmed to occur in the North Sea. *, There is usually more than one species of toxin; **, Krock et al. 2009.

	Toxin*						
Taxon	Saxitoxin	Okadaic acid	Domoic acid	Spirolides	Azaspiracids	Yessotoxin	Effect
Alexandrium tamarense	■						PSP
A. minutum	■						PSP
A. ostenfeldii				■			PSP, Spirolide toxicity
Dinophysis acuta		■					DSP
D. acuminata		■					DSP
D. norvegica		■					DSP
small thecate dinoflagellate**					■		
Protoceratium reticulatum						■	Impact on humans uncertain
Lingulodinium polyedrum						■	Impact on humans uncertain
Gonyaulax spinifera						■	Impact on humans uncertain
Pseudo-nitzschia pungens			■				ASP
P. multiseries			■				ASP
P. delicatissima							ASP
Haptophytes *(Prymnesium parvum)*							Haemolytic effect

that cause various diseases in humans and also fish kills. The effects in humans range from paralytic symptoms, the so called paralytic shellfish poisoning (PSP) via diarrhea (diarrhetic shellfish poisoning = DSP) to neurotoxic symptoms (NSP, ASP). These can in very severe cases even cause fatalities. Importantly, these algae do not necessarily need to occur in large cell numbers to have an adverse effect. Rather their toxins accumulate in organisms that feed on them, e.g., mussels and other commercially harvested shellfish causing toxicity in humans upon consumption. Their occurrence in commercially harvested shellfish also causes considerable economic losses as aquaculture businesses with contaminated stocks often have to be closed for considerable lengths of time, while the occurrence of toxic species persists.

Alien species

For years, the invasion of non-indigenous species into the North Sea and adjacent seas has been reported in the literature (Nehring 2003). Although these species are often reported as invaders, one has to distinguish between two completely different processes when discussing alien species. The first process is introduction, where an alien species is accidentally or purposefully introduced into an area as the result of human activities. The other is true invasion where an organism increases its range of distribution often as the result of changing environmental conditions. The latter is a normal process and a substantial number of 'invasions' are probably unsuccessful in the long-term because the biotic and abiotic conditions in the area receiving the new species are not favourable. The phenomenon of alien species is also known from planktonic ecosystems (Nehring 1998, but see Gomez 2008). Although natural transport by ocean currents may be responsible for the appearance of foreign phytoplankton in the North Sea, a major cause today is the introduction by ballast water (Gollasch et al. 2002) of ships and the settling via marine aquacultures (e.g., mussels). However, there is no universal agreement on the actual importance of ballast water as a vector for alien planktonic species (Smayda 2007).

In the North Sea phytoplankton, several non-indigenous species have been reported: *Odontella sinensis*, *Corethron criophilum*, *Stephanopyxis palmeriana*, *Pleurosigma planktonicum*, *Thalassiosira punctigera* and *Coscinodiscus wailesii*. Many of these alien species have not yet been demonstrated to have any adverse effects. However, problems might occur when the alien species is toxic or proliferates unchecked due to the absence of natural enemies (Occhipinti-Ambrogi 2007).

An example of an accidentally introduced species that has proliferated and has some harmful properties is *Coscinodiscus wailesii* (a large centric diatom with a diameter up to 500 µm). Some of the earliest records of this species in Europe were from the Celtic Sea in 1977 (Boalch & Harbour 1977), the English Channel (Rincé & Plaumier 1986) and the southern Irish Sea (Robinson et al. 1980). In the early 1980s *C. wailesii* also appeared in the German Bight in large numbers to become a conspicuous component of the ecosystem (Dürselen & Rick 1999, Wiltshire & Dürselen 2004). Conspicuous is the strong production of mucilage by this species in European waters which is not known from the pacific forms. This has caused considerable problems, e.g., by clogging fish nets during *Coscinodiscus wailesii* blooms.

Coscinodiscus wailesii in the North Sea can grow at a wide range of environmental conditions including temperatures ranging from < 0 °C to > 20 °C and salinities ranging from 24-35. The species occurs in the open ocean, in coastal areas and also in estuaries where both parameters vary considerably. In addition *C. wailesii* can also cope with a large range of nutrient concentrations. Wide environmental tolerances are a characteristic feature of many phytoplankton species however and might not by themselves serve as a complete explanation for the success of this species. However, *C. wailesii* has another advantage: its large size. This causes inefficient grazing by some herbivorous copepods (Roy et al. 1989). These prefer phytoplankton cells smaller than 100 µm while *C. wailesii* nearly always has a diameter above 350 µm. If *C. wailesii* is not suitable as food for copepods this can also have consequences for the upper trophic levels by facilitating blooms of this diatom. In contrast *Crangon crangon* for example feeds and digests the cells of *C. wailesii* efficiently. However, it was recently shown that not all copepods are incapable of handling *C. wailesi*. Jansen (2008) showed that *Temora longicornis* can handle and ingest *C. wailesii*, without diatom remains appearing in the fecal pellets. This illustrates that the potential effect of *C. wailesii* on the marine food web is not clear cut and might be difficult to detect. As with all other alien species it requires long-term observation before the real impact of such a species can be judged.

Many of the phytoplankton regarded as introduced to the North Sea, such as *Odontella sinensis*, are not currently a matter of concern. They have become a normal component of temperate planktonic communities without causing any apparent harm. Nevertheless, the possible immigration or introduction of new species ought to be monitored as the warming trends in the North Sea might increase the rate of successful establishment of new species including toxic microalgae (Nehring 1998). As the result of climate change the invasion or introduction of new species into the North Sea is becoming an increasingly important research topic.

How to use this book

Each taxon description is headed by an information bar for quick reference to some ecological and physiological characteristics of the species. This is followed by images of the organism, mainly from environmental samples, and a short description. For genera with many species we also include general schematics outlining some of the major features of the identification of species in this genus. The descriptions provide details not only on the traditional taxonomy of the organism but – where ever possible – also on how to identify the organism or to point out identification problems in Lugol-fixed samples. This book is dealing primarily with the identification of phytoplankton species and not their ecology. Therefore, we provide only very basic summaries of some particularly important topics that help the reader to put the taxonomic information into an ecological context. Relevant references to more detailed accounts will be provided throughout this book (see also Valiela 1995).

Legends for page headers

Seasonality for North Sea and adjacent waters

W	Winter	January-March
S	Spring	April-June
S	Summer	July-September
A	Autumn	October-December

	Normally no appearance
	Occasionally
	Moderate
	Peak occurrence

Bloom intensity and frequency for the North Sea

no	Not bloom forming
✿	Rarely in moderate abundance or biomass
✿✿	Frequently in moderate abundance or biomass
✿✿✿	Frequently in high abundance or biomass

Harmful events

	Physical damage to fish gills
	Water discolouration
	Anoxic event
	Slime forming
	Toxic

Bacteriastrum hyalinum Lauder, 1864

Family: Chaetocerotaceae

Season				Trophic mode	Shape	Harmful	Bloom	Resting stage
W	S	S	A	autotrophic	cylinder	no		yes

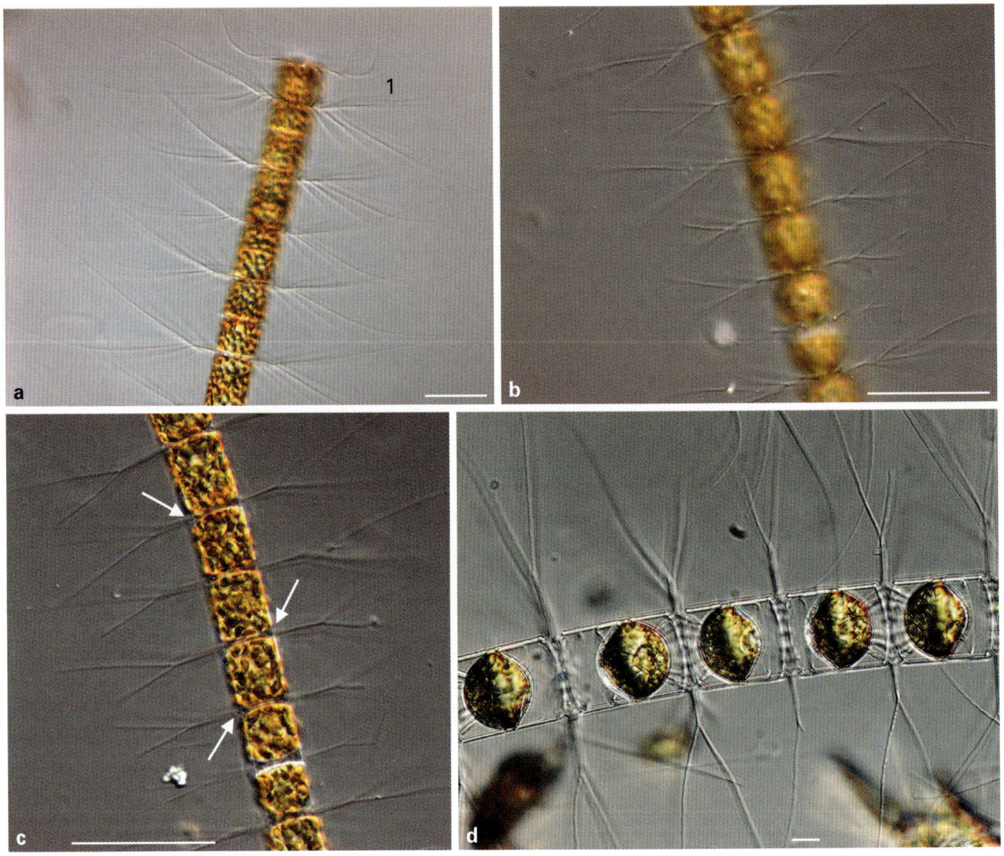

a, Chain of *Bacteriastrum hyalinum* showing both intercalary and terminal setae (1).
b, Chain with focus on the fused setae (arrows). **c,** The same chain with focus on the valves, showing the chloroplasts.
d, Live cells with resting spores. b,c, Scale bars= 50 µm; a,d, scale bar = 20 µm. Images taken with DIC optics.

Description

Cells cylindrical with a circular cross-section and united into chains by their 12-32 fused setae. Length of setae about twice the valve diameter. Distal. bifurcated ends of setae lying in a plane parallel to the chain axis. Terminal setae morphologically distinct from the fused setae and curved backwards (away from the chain), spirally undulate and not bifurcated. Chloroplasts numerous, small rounded.

Size

Diameter: 13-56 µm
Height (pervalvar axis): 20-80 µm

Distribution. *Bacteriastrum hyalinum* is a common species in temperate waters where it can form high biomasses especially in summer.

Similar species: *Bacteriastrum delicatulum* Cleve, 1897: It is also common in temperate waters. But this species is smaller (diameter 7-18 µm) than *B. hyalinum*, the cells are usually long cylindrical and have only 7-10 bristles. The branched ends lie in the valve plane.

Synonyms: none.

Literature: Boalch 1975, Drebes 1972, Hendey 1974.

Chaetoceros danicus Cleve, 1889 — Family: Chaetocerotaceae

Season				Trophic mode	Shape	Harmful	Bloom	Resting stage
W	S	S	A	autotrophic	elliptic cylinder	physical damage		no

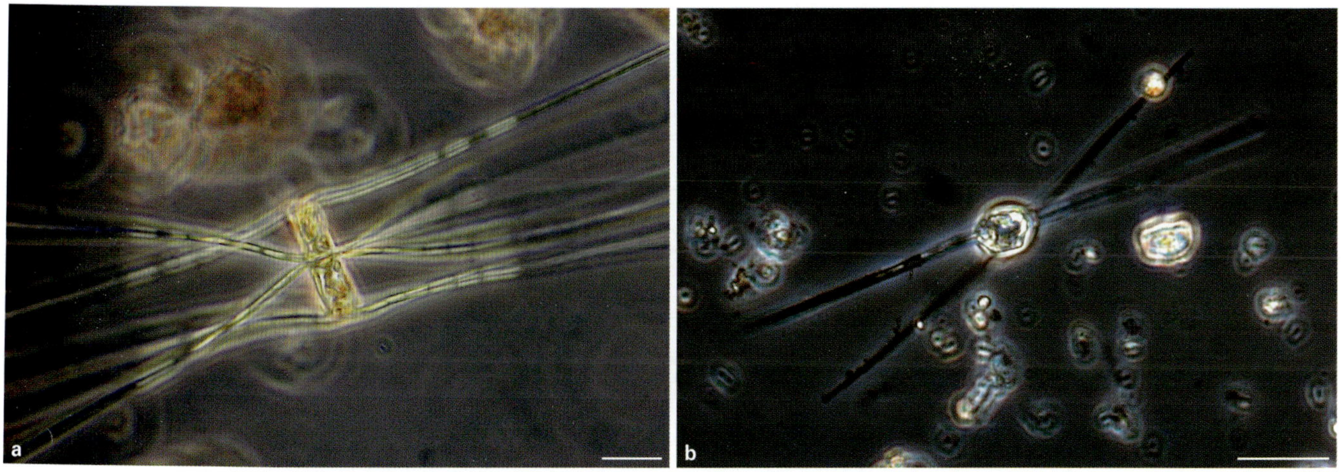

a, Phase contrast image of a pair of live cells of *Chaetoceros danicus* in girdle view. Scale bar = 25 µm.
b, Single Lugol-fixed cell in valve view. Scale bar = 20 µm.

Description

Cells solitary or connected into very short chains of two or three cells. In water mounts cells often seen in valve view showing the broadly elliptical to circular outline. .Valve surface flat, valve mantle relatively high with a distinct furrow towards the first girdle band. Aperture between adjacent cells narrow. Long, stiff setae originating at the valve margin, perpendicular to the pervalvar axis forming acute angles in valve view with the setae located diagonally opposite each other forming a straight line. Chloroplasts small, numerous, allso present in the setae.

Size

Large diameter (apical axis): 8-22 µm
Small diameter (transapical axis): 7-20 µm
Height (pervalvar axis): 8-30 µm

Distribution. The species has a worldwide distribution in coastal areas, occurring throughout the year and sometimes becoming moderately abundant.

Similar species In valve view cells appear similar to single cells of *Chaetoceros borealis*.

Synonyms: none.

Literature: Hendey 1974, Rines & Hargraves 1988

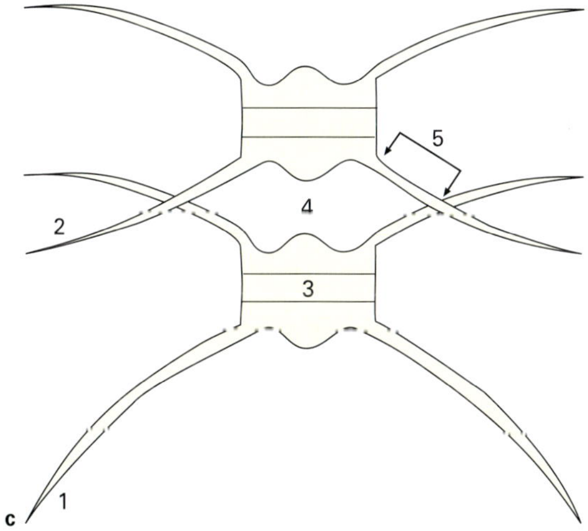

c, Schematic drawing of of a generalized *Chaetoceros* chain showing important diagnostic features: 1, Terminal setae; 2, Intercalary setae; 3, Girdle; 4, Aperture; 5, Basal part.

Chaetoceros borealis J. W. Bailey, 1854 — Family: Chaetocerotaceae

Season				Trophic mode	Shape	Harmful	Bloom	Resting stage
W	S	S	A	autotrophic	elliptic cylinder	physical damage	no	no

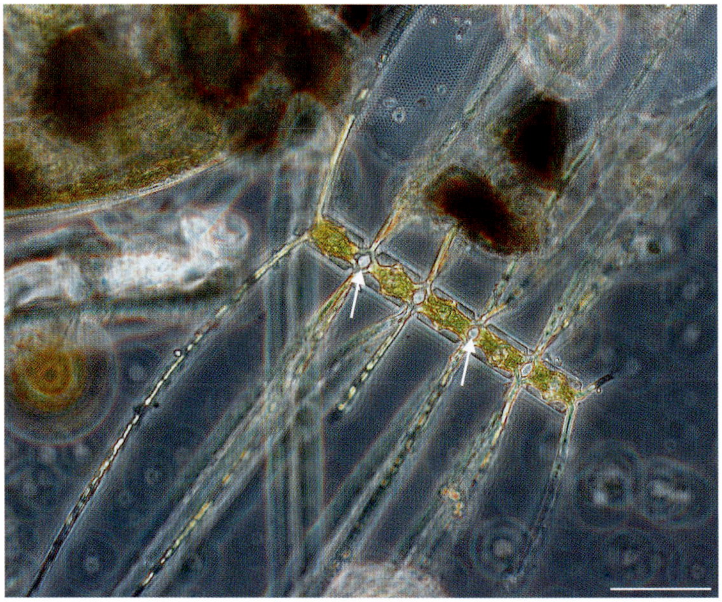

Phase contrast image of a live chain of *Chaetoceros borealis* with slightly plasmolysed cells. Arrows point to the apertures between cells. Scale bar = 50 µm.

Description

Cells with broadly elliptical cross-section forming straight, close-set chains. In girdle view cells rectangular, ratio of cell diameter to pervalvar axis variable. Valve face slightly convex and mantle high. Setae with short, thick basal part. At point of fusion of adjacent setae they curve as to become nearly perpendicular to the chain axis. Crossed setae of adjacent cells forming an acute angle which is nearly halved by the apical axis. Apertures between cells braod and elliptical to hexagonal (arrow). Chloroplasts small, numerous, occurring on main valve and in the setae

Size

Large diameter (apical axis): 12-46 µm
Height (pervalvar axis): 20-40 µm

Distribution. *Chaetoceros borealis* is probably a cosmopolitan species but is mainly found in temperate to cold waters. In North Sea plankton it is occasionally found in the plankton especially during spring.

Similar species: Chains of broader cells can be confused with *Chaetoceros densus*. The latter has a distinctly smaller aperture between cells and their setae do not have a basal part. Single small cells of *C. borealis* are similar to *Chaetoceros danicus* especially in valve view. Diagonally opposite setae of the latter form a straight line.

Synonyms: none.

Literature: Hendey 1974, Rines & Hargraves 1988.

Chaetoceros densus (Cleve) Cleve, 1901 — Family: Chaetocerotaceae

Season				Trophic mode	Shape	Harmful	Bloom	Resting stage
W	S	S	A	autotrophic	elliptic cylinder	physical damage	✿	no

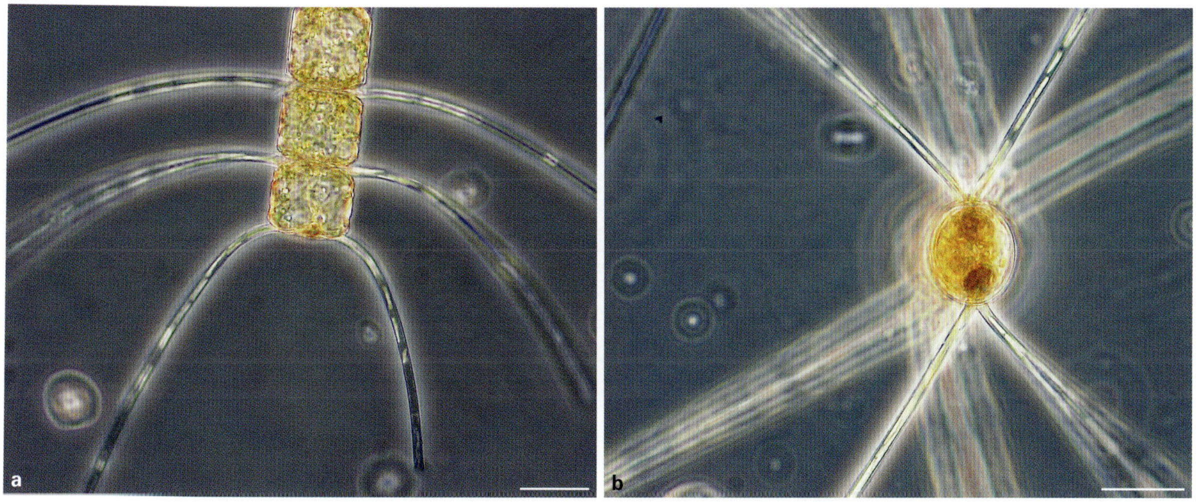

a, Live chain of *Chaetoceros densus* in girdle view. **b,** Chain in valve view. Phase contrast images. Scale bars = 25 µm.

Description

Cells rectangular in girdle view and broadly elliptical to nearly circular in valve view. Cells connected into straight chains with small apertures. Valves weakly convex with a moderately high mantle. Distinct furrow at the transition point from mantle to the first girdle band. Setae arising slightly inside the valve margin at the apical poles immediately crossing over with those of the adjacent cell. Setae rounded to rectangular in cross-section with tiny spines at the edges. Intercalary setae perpendicular to the chain axis or slightly curved towards the ends of the chain. Intercalary setae diverging at the same angle with respect to the apical axis. Chloroplasts numerous including within the strong setae.

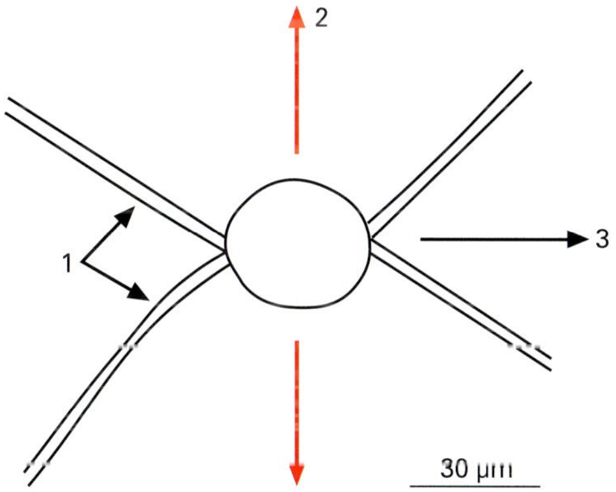

c, Schematic drawing of *Chaetoceros densus* in valve view showing the divergence angle between the two setae (1), direction of the transapical axis (2), and apical axis (3).

Size

Large diameter (apical axis): 10-55 µm
Small diameter (transapical axis): 8-46 µm
Height (pervalvar axis): 12-45 µm

Distribution. *Chaetoceros densus* is a cosmopolitan species that is occasionally found throughout the year becoming moderately abundant in late summer and autumn.

Similar species: *Chaetoceros eibenii*: It can be distinguished in valve view by the angle at which the setae diverge from the valve edge.

Synonym:
Chaetoceros borealis var. *densum* Cleve, 1897.

Literature: Hendey 1974, Rines & Hargraves 1988.

Chaetoceros eibenii (Grunow) Meunier, 1882 — Family: Chaetocerotaceae

Season				Trophic mode	Shape	Harmful	Bloom	Resting stage
W	S	S	A	autotrophic	elliptic cylinder	physical damage	✿	yes

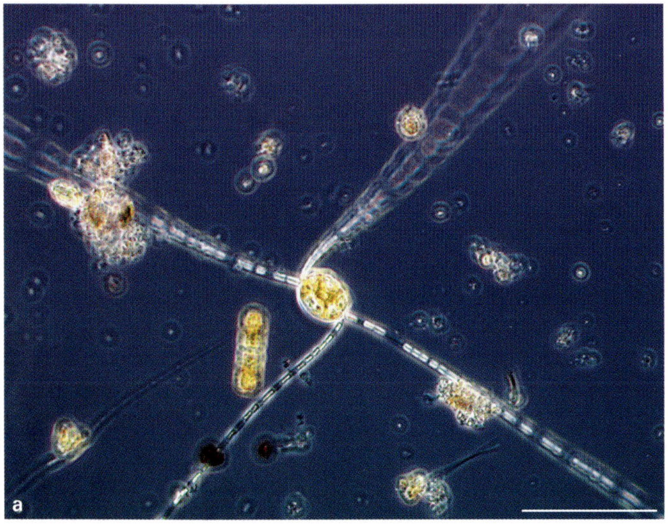

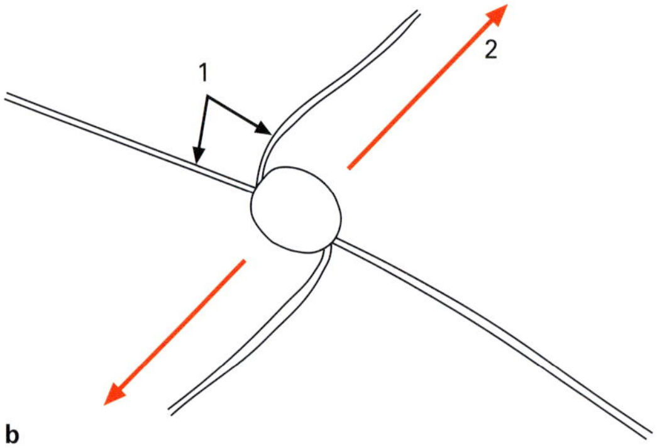

a, Phase contrast image of single cell of *Chaetoceros eibenii* in valve view. Scale bar = 50 µm.
b, Schematic drawing indicating the divergence angles of the setae (1) in relation to the transapical axis (2). One seta lies roughly on the apical axis while the second runs parallel to the transapical axis.

Description

Cells rectangular in girdle view and broadly elliptical in valve view. Cells connected in straight chains with larger elliptical to hexagonal apertures between adjacent valves. Valves flat to slightly concave with a short mantle and a small central process. Setae, with short basal part, arising slightly inside the valve margin. Setae quadrangular in cross-section and covered with small spines. One of the setae almost parallel to the apical axis, the other strongly diverging to become nearly parallel to the transapical axis. Chloroplasts small and numerous, also present within the strong setae.

Size

Large diameter (apical axis): 10-55 µm
Small diameter (transapical axis): 7-37 µm
Height (pervalvar axis): 8-30 µm

Distribution. *Chaetoceros eibenii* is distributed in coastal areas of warm to temperate waters around the world. In the North Sea it can be found from late spring to early autumn.

Similar species: *Chaetoceros eibenii* resembles *Chaetoceros densus*. However, in the latter the apertures between cells are usually smaller and in valve view the intercalary setae diverge at the same angle with respect to the apical axis.

Synonym:
Chaetoceros paradoxus var. *eibenii* (Grunow) Van Heurck, 1896.

Literature. Hendey et al. 1954, Rines & Hargraves 1988.

Chaetoceros* cf. *compressus Lauder, 1864

Family: Chaetocerotaceae

Season				Trophic mode	Shape	Harmful	Bloom	Resting stage
W	S	S	A	autotrophic	elliptic cylinder	no		yes

a, Central part of a live chain of *Chaetoceros* cf. *compressus*.
b, Terminal part of the chain showing both types of intercalary setae. Scale bars = 50 µm.

Description

Cells rectangular in girdle view and broadly elliptical in valve view, forming long straight chains which can be twisted about the chain axis. Valves flat or slightly convex in the centre with a low mantle. Two types of intercalary setae present. Intercalary setae regularly arranged along the chain and very delicate (sometimes difficult to sea with a light microscope). Additionally, occurrence of heavily silicified very long, intercalary setae curved and strongly directed towards the nearest end of the chain, distally almost parallel to the chain axis. Setae originating within the valve margin, having a short basal part. Apertures elliptical to hexagonal, moderately wide and slightly constricted in the centre. Chloroplasts numerous, small and plate-like.

Size

Large diameter (apical axis): 7-40 µm
Height (pervalvar axis): 12-30 µm

Distribution. *Chaetoceros* cf. *compressus* is a species of temperate to warm water regions which can be found throughout the year sometimes building moderate biomasses.

Similar species: *Chaetoceros compressus* can be distinguished from the similar species *Chaetoceros contortus* Schütt, 1895 by its resting spores, which are formed at one end of the vegetative cell, usually pressed against a vegetative valve. They have a lenticular shape and in contrast to *Chaetoceros contortus* possess a ring of fine hairs originating on the mantle of one valve.

Synonyms: none.

Literature: Hendey 1974, Rines 1999, Rines & Hargraves 1988, Shevchenko et al. 2006.

Chaetoceros lauderi Ralfs, 1864

Family: Chaetocerotaceae

Season				Trophic mode	Shape	Harmful	Bloom	Resting stage
W	S	S	A	autotrophic	elliptic cylinder	no		yes

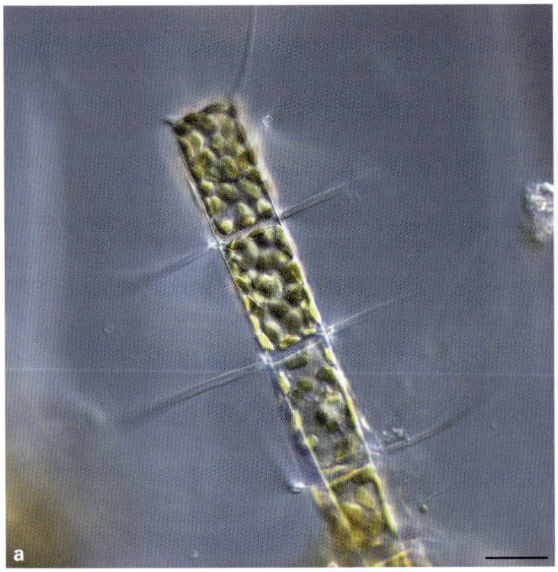

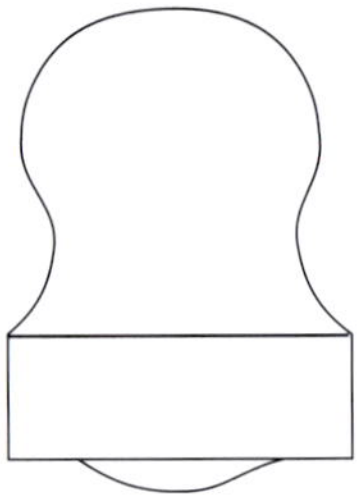

b

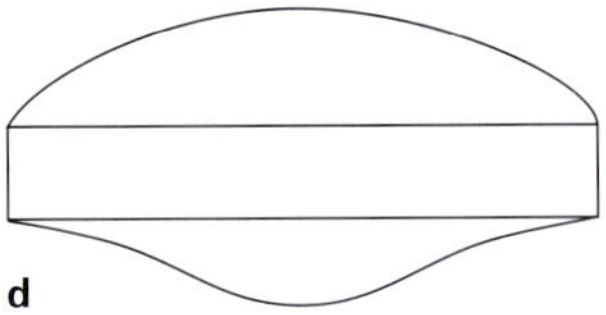

d

a, DIC image of a live chain of *Chaetoceros lauderi*. **b,** Schematic outline of a *C. lauderi* resting cyst. **c,** *C. teres* resting spore (Lugol-fixed). **d,** Schematic outline of a *C. teres* resting spore. Scale bars = 20 µm.

Description

Cells higher than wide in girdle view and elliptic in valve view. Apertures between valves very narrow, slit-like. Intercalary setae lying in a plane roughly perpendicular to the chain axis. The Terminal setae widely divergent. Many chloroplasts per cell.

Size

Large diameter (apical axis): 18-50 µm
Height (pervalvar axis): 18-60 µm

Distribution. *Chaetoceros lauderi* is a warm water to temperate species that occurs mainly in summer in the North Sea. In contrast, *C. teres* is a Northern cold to temperate form which is found in the North Sea mainly during late winter and spring when it can become moderately abundant.

Similar species: This species is very similar to *Chaetoceros teres*. Compared with *C. teres*, *C. lauderi* chains are slightly twisted. But the main distinguishing feature, is the resting spore which is highly vaulted and spiny in *C. lauderi* but smooth in *C. teres*. Unless resting spores can be seen the two species cannot be identified reliably.

Synonym: *Chaetoceros weissflogii* Schütt, 1905.

Literature: Rines & Hargraves 1988, Shevchenko et al. 2006.

Chaetoceros didymus Ehrenberg, 1845 — Family: Chaetocerotaceae

Season				Trophic mode	Shape	Harmful	Bloom	Resting stage
W	S	S	A	autotrophic	elliptic cylinder	no		yes

a, Live chain of *Chaetoceros didymus*; **b,** Resting stage of *C. didymus*.
DIC images. Scale bars = 20 μm.

Description

Cells rectangular in girdle view and with a narrowly elliptical cross-section building straight, usually long chains. Valve face concave with a protuberance in the centre (arrow). Mantle low. intercalary and terminal setae similar: long, coarse, straight and with a short basal part. Intercalary setae diverging at about 25-45 degrees to the chain axis; terminal setae almost parallel to the chain axis. Setae crossing over at the colony margin, four-sided in cross section, perforated with poroids and bearing spines and hair-like structures at their base. Labiate process centrally located. Resting spores smooth, paired, and held together by short thick curved setae. Primary valves with one or two undulations or a central protuberance. Secondary valves concave. Two large, plate-like chloroplasts per cell, each with a pyrenoid situated in the valve protruberances.

Size

Large diameter (apical axis): 10-40 μm
Height (pervalvar axis): 6-30 μm

Distribution. The species is distributed in temperate to warm water regions. It can be found throughout the year but is most abundant in summer and autumn.

Similar species: *Chaetoceros protuberans* H. S. Lauder, 1864 is a very similar species, which was previously regarded as a variety of *C. didymus* (*Chaetoceros didymus* var. *protuberans* (H. S. Lauder) H. H. Gran & K. Yendo, 1914). The latter lacks the short hair-like structures found at the base of the setae in *C. didymus*. Setae in *C. protuberans* are fused at a greater distance from the chain edge. *C. protuberans* is a warm water species and is occasionally found in the North Sea during summer.

Synonyms: none.

Literature: Hargraves 1979, Hernández-Becerril 1991, Rines & Hargraves 1988, Shevchenko et al. 2006, von Stosch et al. 1973.

Chaetoceros decipiens Cleve, 1873 Family: Chaetocerotaceae

Season				Trophic mode	Shape	Harmful	Bloom	Resting stage
W	S	S	A	autotrophic	elliptic cylinder	no	no	no

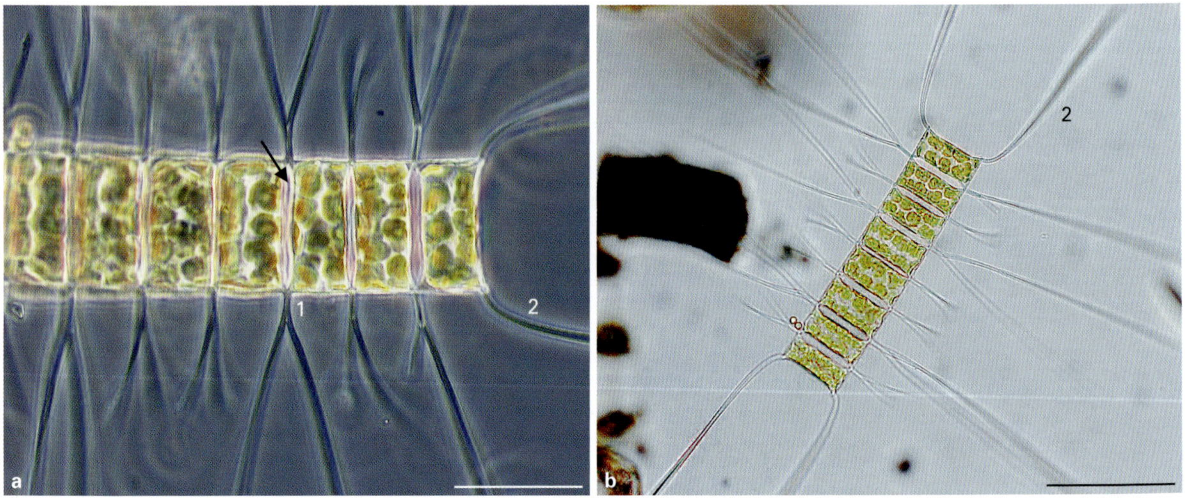

a, Phase contrast image of a live chain of *Chaetoceros decipiens* with the small apertures (arrow) and the partly fused setae at the valve margin (1). Scale bar = 50 µm.
b, Bright field image of a short chain of live cells showing the terminal setae (2). Scale bar = 100 µm.

Description

Cells narrowly elliptical in valve view and broadly rectangular in girdle view with concave valve face and high mantle. Chains long and straight. Adjacent cells touching with their valve poles. Apertures narrow. Setae long, thick and straight, without a basal part. Sibling intercalary setae fused proximally for a short distance (1), before diverging at an angle of 10-25 degrees from the colony axis. Terminal setae thicker and shorter than intercalary setae, initially broadly diverging from apical axis and curving parallel to the colony axis (2). Labiate process small, tube-like, eccentrically located and only found on the terminal valve. Setae polygonal in cross section (6-8 sides), perforated with poroids and bearing small spines. Several small chloroplasts per valve.

Size

Large diameter (apical axis): 10-85 µm
Height (pervalvar axis): 9-35 µm

Distribution. The species has a world-wide distribution and occurs in the plankton throughout the year, but mainly during spring.

Similar species: The fusion of adjacent setae at their base is characteristic for the species. *Chaetoceros lorenzianus* has a similar morphology but the setae are only fused where setae from adjacent valves cross. *Chaetoceros lorenzianus* is also a cyst former. *C. decipiens* is not.

Synonym:
Chaetoceros grunowii Schütt, 1895.

Literature: Rines & Hargraves 1988, Shevchenko et al. 2006.

Chaetoceros curvisetus Cleve, 1889 — Family: Chaetocerotaceae

Season				Trophic mode	Shape	Harmful	Bloom	Resting stage
W	S	S	A	autotrophic	elliptic cylinder	physical damage	✿✿	yes

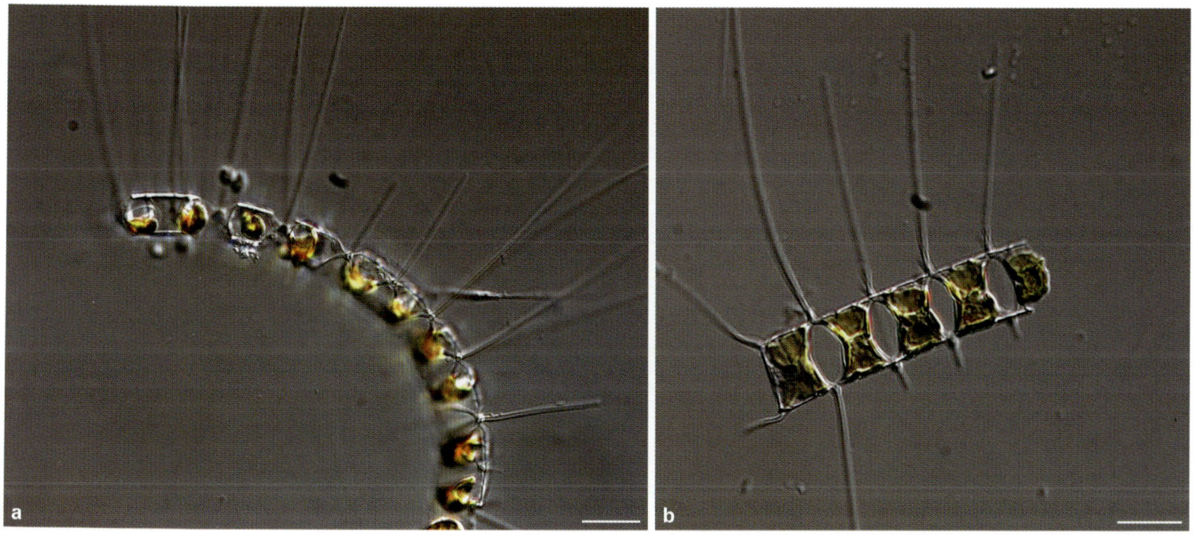

a, DIC of a curved chain of live *Chaetoceros curvisetus*.
b, Chain fragment in broad girdle view showing the characteristic aperture. Scale bars = 20 µm.

Description

Colonies curved and relatively long. Cells rectangular in girdle view and a low mantle. Apertures between adjacent cells wide and elliptical, sometimes with a central notch.
Intercalary and terminal setae similar, long, thin and with a short basal part. Setae cross over at the colony margin, curving in the same direction towards the outside of the coiled colony and almost perpendicularly to the colony axis. Setae circular in cross-section, perforated by small poroids and bearing spiral rows of spines. One chloroplast present. Labiate process a centrally located tube-like structure, only discernable with SEM
Resting spores are spiny, with broadly convex primary valve and almost flat secondary valve

Size

Large diameter (apical axis): 7-30 µm

Distribution. *Chaetoceros curvisetus* is a cosmopolitan species, mainly in temperate to warm waters.

Similar species: See *Chaetoceros debilis.*

Synonyms: none.

Literature: Rines & Hargraves 1988, Shevchenko et al. 2006.

Chaetoceros debilis Cleve, 1894 — Family: Chaetocerotaceae

Season				Trophic mode	Shape	Harmful	Bloom	Resting stage
W	S	S	A	autotrophic	elliptic cylinder	physical damage	✿✿	yes

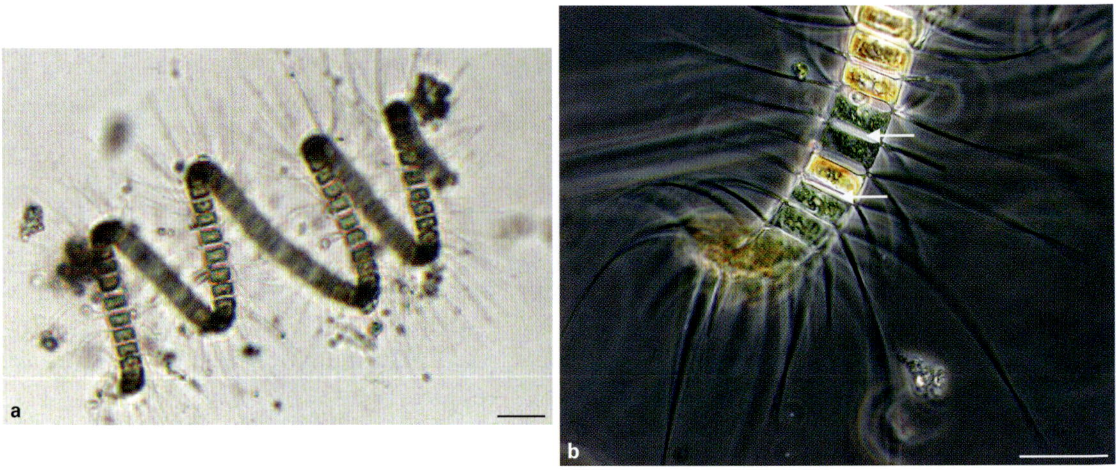

a, Bright field image of a Lugol-fixed spiral chain of *Chaetoceros debilis*.
b, Phase contrast image of a live chain showing the outwardly directed setae and apertures between adjacent cells (arrows). Scale bars = 25 µm.

Description

Chains long and spirally twisted. Cells rectangular in girdle view with a low mantle and elliptic in valve view. Valves of adjacent cells not touching. Apertures between cells narrow and often very slightly constricted in their centre. Setae long, thin and with a short basal part. Terminal setae thicker than intercalary setae, perpendicular to the colony axis. Intercalary setae diverging at an angle of 30-70 degrees from the colony axis, crossing slightly outside the colony margin, and pointing in one direction towards the outside of the spiral chain. Setae circular in cross-section and perforated with poroids. One chloroplast per cell. Resting spores: Primary valve with two long spines, directed towards the corners of the parent cell. Both primary and secondary valves vaulted and with undulations.

Size

Large diameter (apical axis): 8-40 µm
Small diameter (transapical axis): 5-25 µm
Height (pervalvar axis): 6-20 µm

Distribution. *Chaetoceros debilis* is a cosmopolitan species mainly in cooler to temperate waters. It can be found from spring to autumn sometimes forming blooms.

Similar species: *Chaetoceros curvisetus* Cleve, 1889 also forms spiral chains. However, these spirals have a wider diameter and the apertures between adjacent cells are larger, and broadly oval.

Synonym:
Chaetoceros vermiculatus Schütt, 1895.

Literature: Rines & Hargraves 1988, Shevchenko et al. 2006.

Chaetoceros diadema (Ehrenberg) Gran, 1897 Family: Chaetocerotaceae

Season				Trophic mode	Shape	Harmful	Bloom	Resting stage
W	S	S	A	autotrophic	elliptic cylinder	no		yes

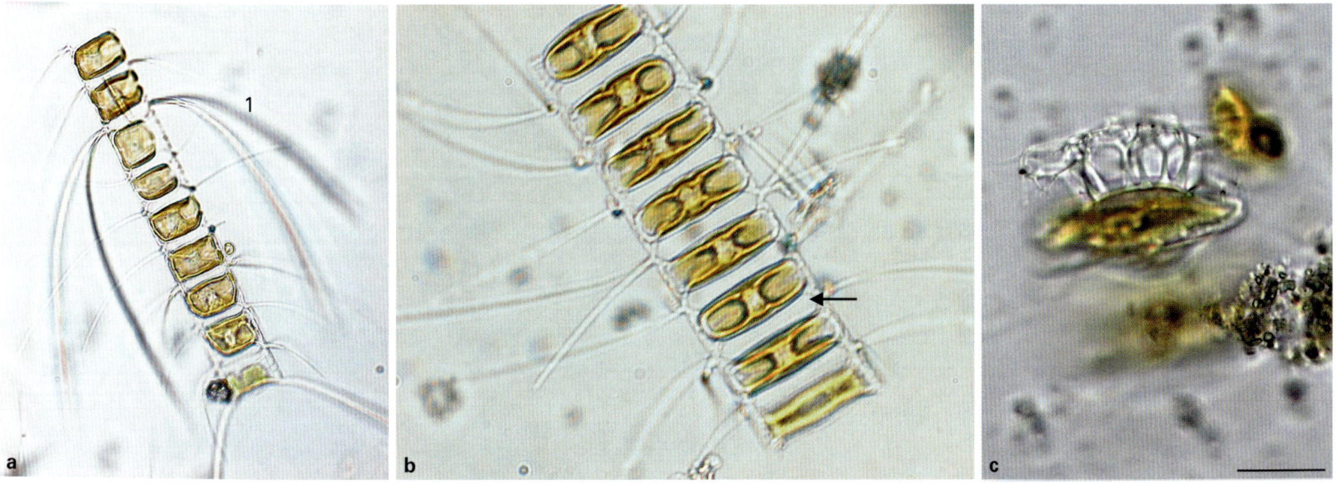

a, Bright field image of a live chain of *Chaetoceros diadema* with normal intercalary setae, a pair of special (thickened) intercalary setae (1) and the terminal setae. Scale bar = 50 µm.
b, Bright field image of a live chain with the typical apertures between adjacent cells.
c, Resting spore. b, c, scale bars = 20 µm.

Description

Cells joined into chains, sometimes slightly twisted around the pervalvar axis. Cells rectangular in girdle view with slightly rounded margins. Valves elliptical with a flat valve face except a convex central section. Adjacent valves not touching. Setae originating slightly inside the valve margin. Basal part short to medium sized. Setae from adjacent valve fused close to the valve margin before diverging. Intercalary setae sometimes but not always perpendicular to the chain axis. Terminal setae diverging at a wide angle. Sometimes two thicker setae pairs turned towards each other present within the chain (1). Apertures between adjacent cells wide and elliptical with a slight constriction in the centre (arrow). One plate like chloroplast present. Resting spores with evenly vaulted primary valves and with 2-12 dichotomously branched spines. Secondary valve smooth and inflated.

Size

Large diameter (apical axis): 10-50 µm
Height (pervalvar axis): 5-25 µm

Distribution. *Chaetoceros diadema* is a cosmopolitan species which occurs in coastal areas throughout the year mainly during spring and early summer.

Similar species: *Chaetoceros debilis* has a similar frustule shape and shape of apertures. But this species is usually smaller, forms spirally twisted chains and the setae are all directed to one side of the colony.

Synonyms:
Syndendrium diadema Ehrenberg, 1854,
Chaetoceros distans var. *subsecundus* Grunow ex Van Heurck, 1880,
Chaetoceros groenlandicus Cleve, 1896,
Chaetoceros paradoxus Peragallo, 1897,

Literature: French III & Hargraves 1985; Hargraves 1972, 1979; Rines & Hargraves 1988; Shevchenko et al. 2006.

Chaetoceros socialis Lauder, 1864 — Family: Chaetocerotaceae

Season				Trophic mode	Shape	Harmful	Bloom	Resting stage
W	S	S	A	autotrophic	elliptic cylinder	physical damage	✿✿✿	yes

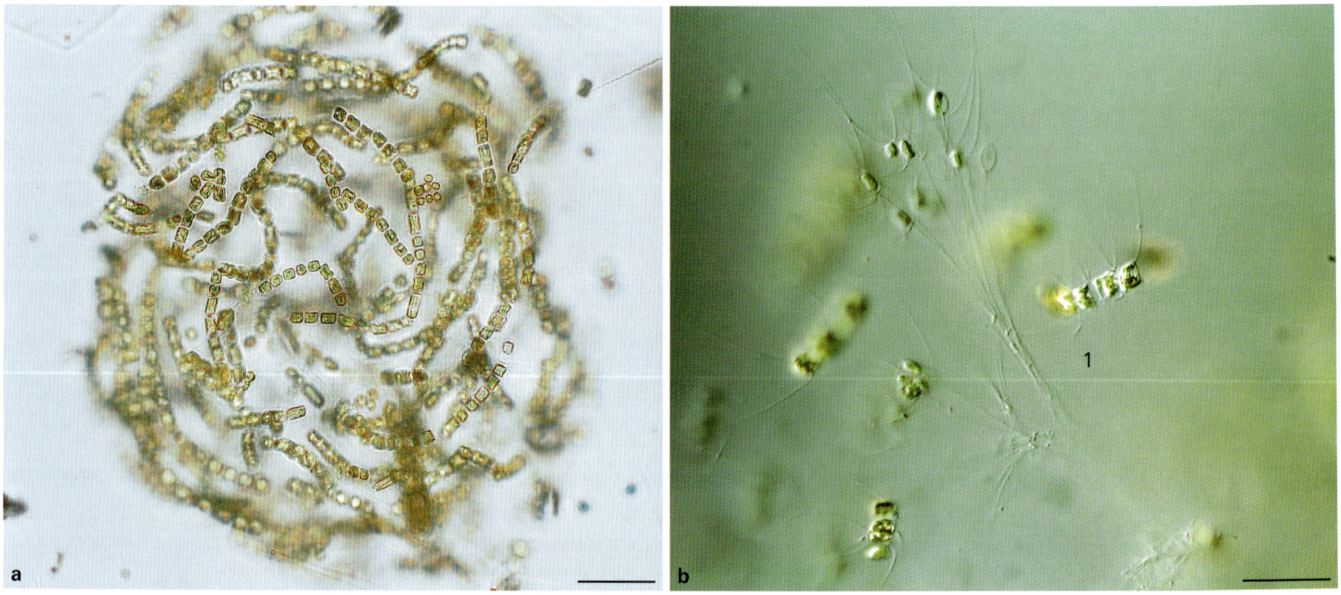

a, Bright field image of a large *Chaetoceros socialis* colony.
b, *Chaetoceros socialis* f. *radians*; image courtesy of Lars Edler. Scale bars = 50 µm.

Description

Cells small, cylindrical with an elliptical cross-section and united into small flexible chains. Adjacent cells in a chain not usually touching each other. Setae arising from within the valve margin. Three short setae per cell. The fourth very long and connecting with the long setae from adjacent cells and chains to form colonies that can contain hundreds of cells.
Important: Live, intact colonies usually spherical with the long setae pointing towards the interior of the colony. After fixation with Lugol colony shape often becoming irregular or disintegrating entirely. One plate-like chloroplast per cell, located in the girdle area.

Note

Formerly regarded as a separate species, *Chaetoceros radians* Schütt, 1895 is now considered a form of *C. socialis*: *Chaetoceros socialis* f. *radians* (F. Schütt) A. I. Proshkina-Lavrenko, 1963. Colonies are also rounded but sometimes showing two relatively small aggregates of short chains located opposite each other. These are connected radially by their long setae (1). When only fragments of colonies are found in fixed samples the form can unequivocally be distinguished from the species only by the resting spores which are spiny in the form and smooth in the species.

Size

Large diameter (apical axis): 2-15 µm
Small diameter (transapical axis): 2-10 µm
Height (pervalvar axis): 2-12 µm

Distribution. *Chaetoceros socialis* probably has a worldwide distribution in coastal waters with its centre of distribution in colder water where it can form blooms. In the North Sea it occurs from spring to autumn being most abundant in late spring.

Similar species: None.

Synonyms: none for the species.
For forma:
Chaetoceros radians Schütt, 1895,
Chaetoceros socialis f. *vernalis* A. I. Proshkina-Lavrenko, 1953.

Literature: Rines & Hargraves 1988, Shevchenko et al. 2006.

Chaetoceros subtilis Cleve, 1896

Family: Chaetocerotaceae

Season				Trophic mode	Shape	Harmful	Bloom	Resting stage
W	S	S	A	autotrophic	elliptic cylinder	no	no	yes

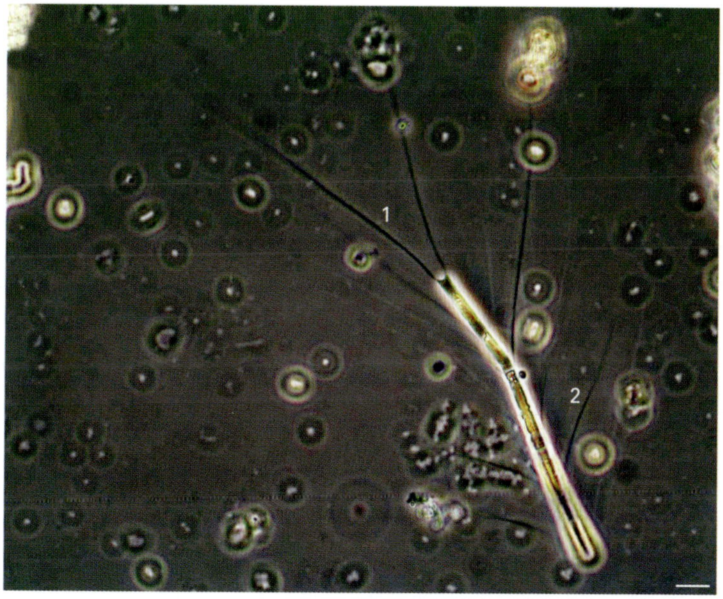

Phase contrast image of a short live chain of *Chaetoceros subtilis* showing terminal (1) and intercalary setae (2). They are similar in morphology and point towards one end of the chain. Scale bar = 10 µm. Image courtesy of Regina Hansen.

Description

Cells rectangular in girdle view with an elliptical cross-section in valve view occurring either joined into short chains or solitary. Epivalve slightly convex and hypovalve concave (i. e. a heterovalvate species) with adjacent valves fitting together tightly without apertures. Setae originating at the valve edge, straight and all directed towards one end of the chain. Terminal setae somewhat thicker and much longer than the intercalary ones. Resting spores with unequally vaulted valves covered with spines. One plate-like chloroplast per cell.

Size

Large diameter (apical axis): 2-14 µm
Height (pervalvar axis): 4-25 µm

Distribution. The species is mainly distributed in brackish water.

Similar species: none.

Synonyms: none.

Literature: Rines & Hargraves 1988, Shevchenko et al. 2006.

Coscinodiscus asteromphalus Ehrenberg, 1844 Family: Coscinodiscaceae

Season				Trophic mode	Shape	Harmful	Bloom	Resting stage
W	S	S	A	autotrophic	cylinder	water discolouration	✿✿	no

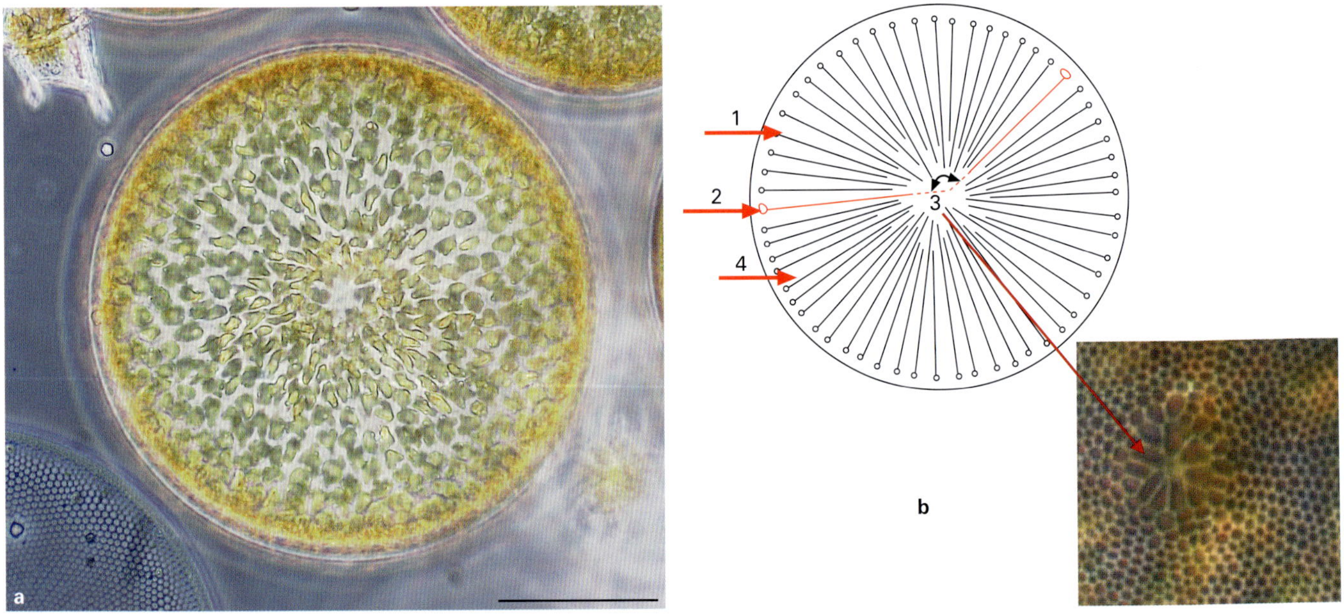

a, Phase contrast image of *Coscinodiscus asteromphalus* in valve view. Scale bar= 100 µm.
b, Schematic drawing indicating important features in the genus *Coscinodiscus*:
1, Location of marginal ring of labiate processes; 2, The two larger labiates (macrorimoportulae);
3, Central hyaline (non-areolated) area. Instead of a hyaline area there can be a rosette of a few large areolae;
4, Areolation pattern, with the lines representing radial rows of areolae.
Schematic adapted from Hasle & Syvertsen 1997.

Description

Cells discoid in girdle view with slightly convex valves and a relatively long pervalvar axis. Central valve area somewhat depressed. Coarse structure of polygonal areolae showing primary radial and secondary spiral pattern with a large rosette in the centre. Areolae nearly of the same size (3-5 in 10 µm), with smaller ones only close to the valve margin. One ring of labiate processes visible in the light microscope and situated close to the margin; the two larger labiate processes (macrorimoportulae) can be found at an angle of about 120-135° apart. Chloroplasts numerous, small and plate-like.

Size

Diameter: 80-400 µm
Height (pervalvar axis): 40-200 µm

Distribution. *Coscinodiscus asteromphalus* is a cosmopolitan species which can form blooms.

Similar species: In size and areolae pattern *Coscinodiscus asteromphalus* is similar to *Coscinodiscus argus* Ehrenberg, 1839 and *Coscinodiscus centralis* Ehrenberg, 1844. The marginal processes of *C. asteromphalus* are closer to the margin and the central rosette is different in structure.

Synonyms:

Coscinodiscus radiatus var. *asteromphalus* (Ehrenberg) Ehrenberg, 1854,
Coscinodiscus asteromphalus var. *conspicua* Grunow, 1883,
Coscinodiscus asteromphalus var. *genuina* Grunow, 1884.

Literature: Brooks 1975c, Hasle & Lange 1992, Hendey 1974, Werner 1971.

Coscinodiscus concinnus W. Smith, 1856 Family: Coscinodiscaceae

Season				Trophic mode	Shape	Harmful	Bloom	Resting stage
W	S	S	A	autotrophic	cylinder	anoxia	✿✿	no

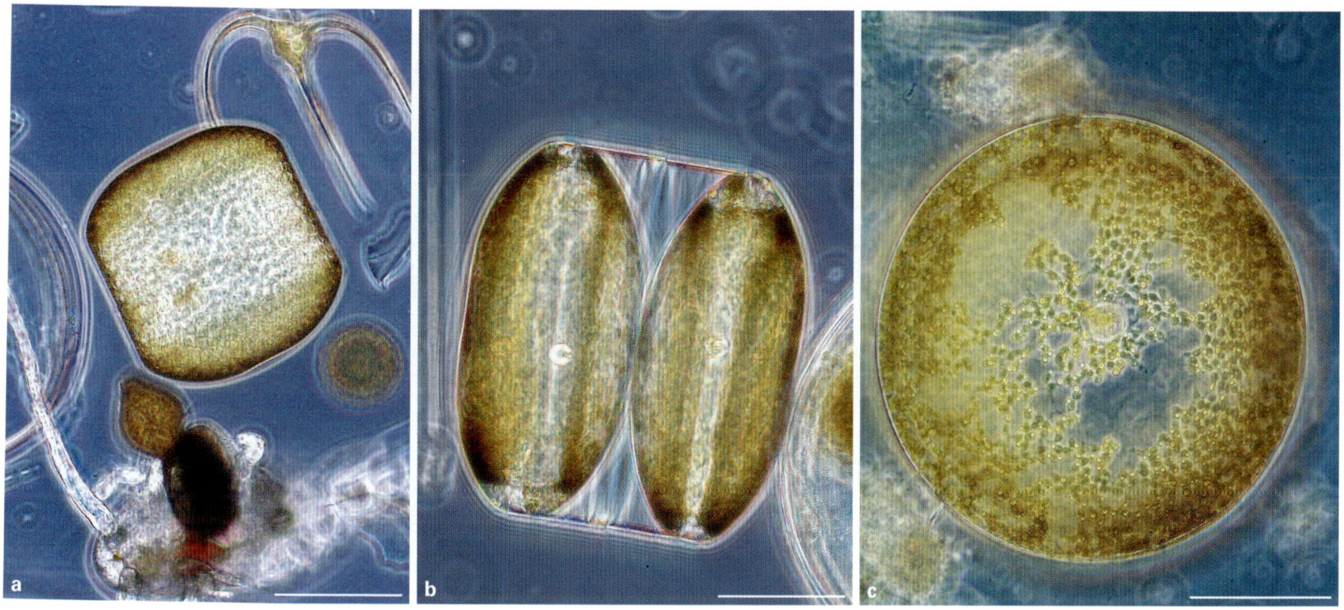

a, Live drum-shaped cell of *Coscinodiscus concinnus* in girdle view. Scale bar = 100 µm.
b, Girdle view of a more discoid live cell in division. Scale bar = 50 µm.
c, Live cell in valve view; image courtesy of Regina Hansen. Scale bar = 100 µm.
All images taken in phase contrast optics.

Description

Cells appearing in two morphological forms (dimorph): Cells either discoid or drum shaped in girdle view with thin-walled, convex valve face. Central valve face flat or slightly concave and with a large rosette or hyaline region. Small areolae (7-9 in 10 µm) arranged into bundles of rows that are bordered by distinct hyaline lines running from the single ring of marginal labiate processes to the valve centre. Two larger marginal labiate processes spaced 135° apart. Girdle band consisting of several rings. Chloroplasts numerous small, rounded, and plate-like.

Size

Diameter: 110-500 µm
Height (pervalvar axis): 60-500 µm

Distribution. The species can either be found oceanic or in coastal areas mainly in colder water. It can dominate the plankton in late winter or spring with high biomasses.

Similar species: *Coscinodiscus concinnus* is distinguished from *C. centralis* by its longer pervalvar axis and the larger distance between valve margin and processes.

Synonyms: none.

Literature: Boalch 1971, Brooks 1975b.

Coscinodiscus granii Gough, 1905

Family: Coscinodiscaceae

Season				Trophic mode	Shape	Harmful	Bloom	Resting stage
W	S	S	A	autotrophic	cylinder	no	no	no

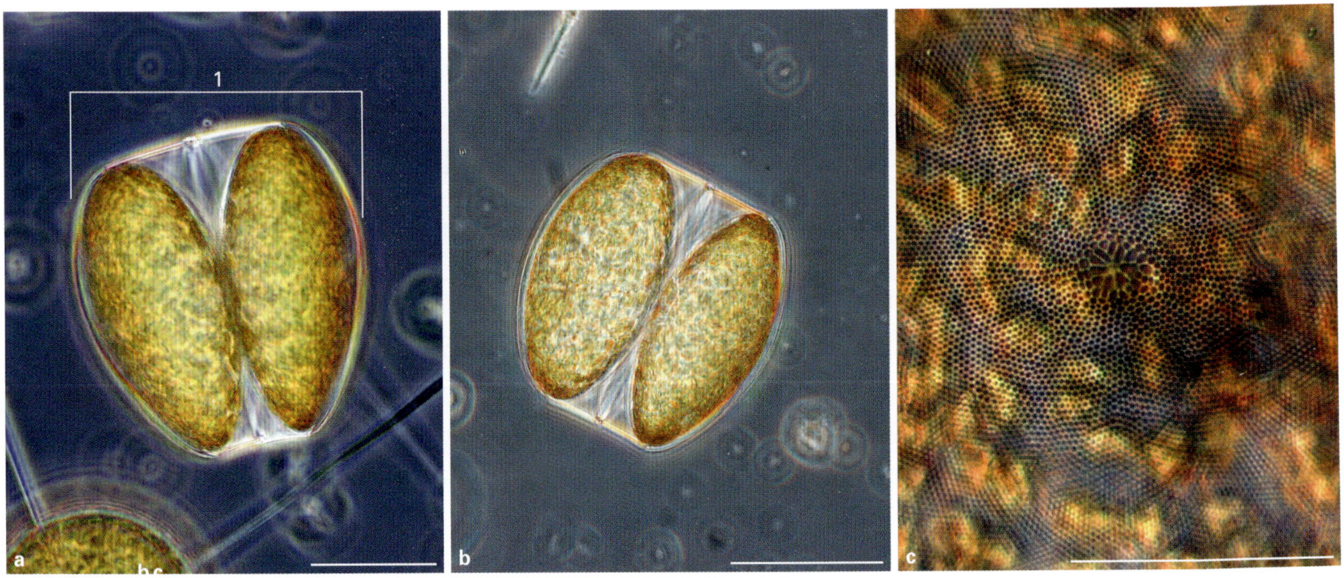

a, b, Phase contrast images of *Coscinodiscus granii* in girdle view. Scale bars = 50 µm (a) and 100 µm (b).
c, Detail of the central rosette. Scale bar = 50 µm.

Description

Cells wedge-shaped in girdle view depending on the orientation of the cell. Largest valve convexity not in the centre but near the highest part of the cell margin (1). Valve face with a more or less pronounced central rosette of larger areolae and a radial pattern of areolae (8-11 in 10 µm) decreasing in size from the centre to the margin. One ring of marginal processes with two larger processes 120-135 degrees apart. Chloroplasts numerous, discoid.

Size

Diameter: 40-200 µm
Height (pervalvar axis): 30-180 µm at widest part

Distribution. *Coscinodiscus granii* is a cosmopolitan species with a wide temperature tolerance. It can be found throughout the year in low to moderate abundance.

Similar species: The height difference of the valve margins can be very small. This species can, therefore, be confused with *Coscinodiscus concinnus* when seen in girdle view.

Synonyms: none.

Literature: Boalch 1971, Brooks 1975a, Hendey 1974.

Coscinodiscus radiatus Ehrenberg, 1840

Family: Coscinodiscaceae

Season				Trophic mode	Shape	Harmful	Bloom	Resting stage
W	S	S	A	autotrophic	cylinder	no	no	no

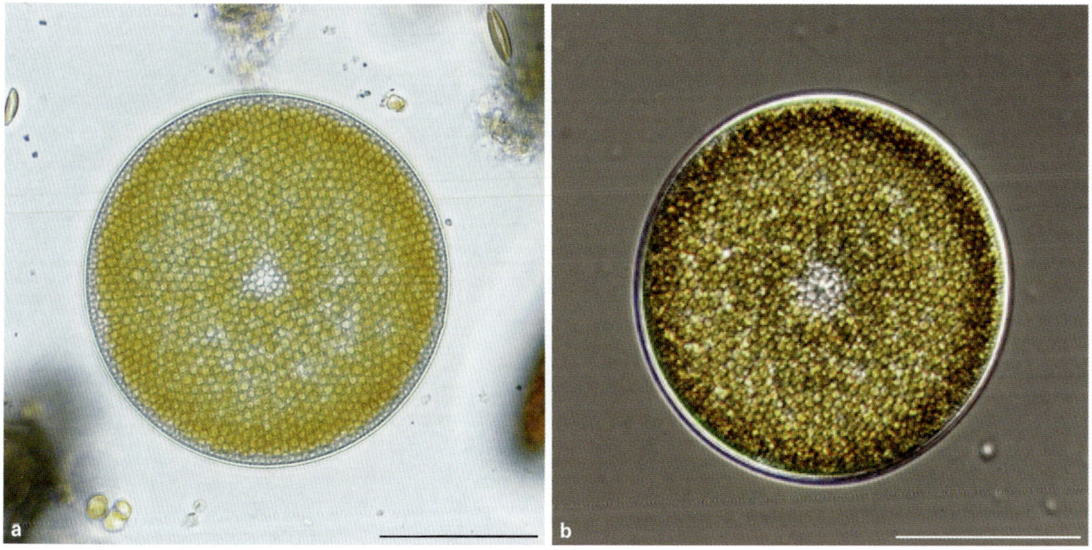

a, Bright field image of *Coscinodiscus radiatus* in valve view. Scale bar = 50 µm.
b, Phase contrast image of *C. radiatus* in valve view. Scale bar = 20 µm.

Description

One of the smaller *Coscinodiscus* species. Cells box-shaped in girdle view with the valves considerably wider than high. Large areolae forming distinct rows radiating from the valve centre, with a central rosette of slightly larger areolae. Secondary rows can appear as spirals. Areolae coarse compared to other *Coscinodiscus* species (2-4 in 10 µm) and easily seen with the light microscope. Areolae becoming smaller towards the valve margin. Chloroplasts numerous, small and rounded.

Size

Diameter: 30-180 µm
Height (pervalvar axis): 15-60 µm

Distribution. *Coscinodiscus radiatus* is a cosmopolitan species which is frequently found in plankton especially in oceanic areas.

Similar species: *Coscinodiscus marginatus*: When seen in valve view this species can be distinguished from *C. radiatus* by having more irregular rows of areolae.

Synonym:
Coscinodiscus borealis Ehrenberg, 1862.

Literature: Hasle & Sims 1986, Hernández Becerril 2000.

Coscinodiscus wailesii Gran & Angst, 1931 — Family: Coscinodiscaceae

Season				Trophic mode	Shape	Harmful	Bloom	Resting stage
W	S	S	A	autotrophic	cylinder	slime/foam	✿✿✿	no

a, Girdle view of *Coscinodiscus wailesii*.
b, Valve view of *C. wailesii*, showing the fine valve areolation and central hyaline area (arrow).
c, Valve centre of a live cell, showing the central hyaline area and areolation patterns
All are phase contrast images. Scale bars = 100 µm.

Description

Cells often as high as wide in girdle view. Valve face flat or appearing to have a slightly undulating valve margin, depending on the focal plane. In valve view a prominent hyaline area is visible in the centre with wide interstriae radiating from it, resulting in an irregular fasciculation, 5-6 valve areolae in 10 µm. Areolae small but cribra are still visible with the light microscope. Two marginal rings of processes. The outer ring containing the two larger processes, 120-180 degrees apart. Chloroplast numerous, shape irregular.

Size

Diameter: 250-500 µm
Height (pervalvar axis): 120-500 µm

Distribution. The species was introduced into European waters from the Pacific during the 1970s. It has wide temperature and salinity tolerances. In the North Sea *Coscinodiscus wailesii* can bloom from late winter to autumn. It also appears in the western Baltic Sea.

Similar species: Because of its size, the typical rectangular form in girdle view, the fine pattern of areolae and the form of chloroplasts the species is easily identified.

Synonyms: none.

Literature: Fernandes et al. 2001.

Actinoptychus senarius (Ehrenberg) Ehrenberg, 1843 — Family: Heliopeltaceae

Season				Trophic mode	Shape	Harmful	Bloom	Resting stage
W	S	S	A	autotrophic	cylinder	no	no	no

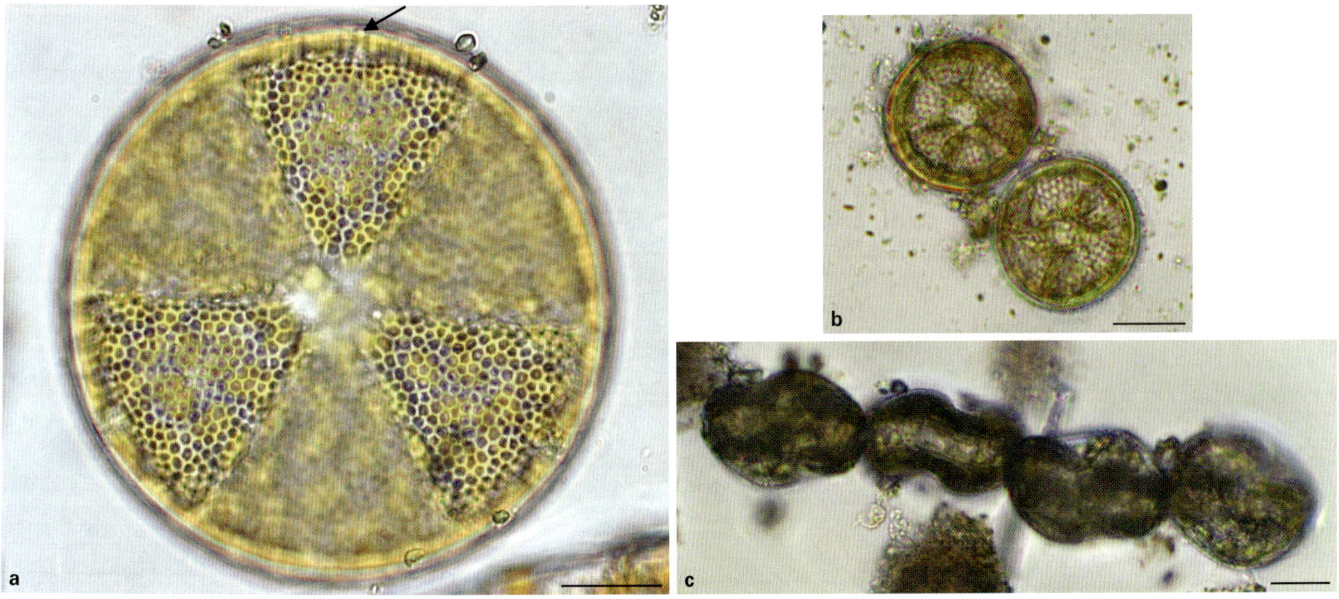

a, Live single cell of *Actinoptychus senarius* in valve view focused on the raised sectors (arrow pointing to one of the labiate processes).
b, Two small Lugol-fixed pair of cells in valve view.
c, Four Lugol-fixed, attached cells in girdle view.
All images in bright field. Scale bars = 20 µm.

Description

Species easily identified by its hexagonal central, non-areolated area and division of valve face into usually six sectors with alternating raised and depressed areas. Each raised sector with a marginal labiate process (arrow). Areolation coarse and easily visible in water mounts. Chloroplasts numerous and large.

Size

Diameter: 20-150 µm
Height (pervalvar axis): 10-70 µm

Distribution. *Actinoptychus senarius* is a cosmopolitan species and can be abundant in cold to temperate coastal waters.

Similar species: Because of the six alternating areas on their valve face this species cannot be confused with other species.

Synonyms:
Actinocyclus senarius Ehrenberg, 1838,
Actinocyclus undulatus J. W. Bailey, 1842,
Actinocyclus undulatus Kützing, 1844,
Actinoptychus undulatus (J. W. Bailey) Ralfs, 1861,
Actinoptychus undulatus Bailey ex Hustedt, 1927.

Literature: Hasle & Syvertsen 1997.

Actinocyclus octonarius Ehrenberg, 1838 — Family: Hemidiscaceae

Season				Trophic mode	Shape	Harmful	Bloom	Resting stage
W	S	S	A	autotrophic	cylinder	no	no	no

DIC image of an empty frustule of *Actinocyclus octonarius* from a permanent slide mounted in Naphrax. Scale bar = 50 µm.

Description

Cells drum shaped with a flat valve face and pronounced, abruptly curving mantle. Cells solitary. Valves circular in valve view. Central annulus highly variable (hyaline to completely areolate valve centres observed in clonal cultures). Areolae on the valve face arranged in sectors, separated from each other by complete radial rows of areolae. Towards the distal end of these rows a single opening of the labiate process is visible. Density of areolation considerably higher near the mantle than on the valve face. Numerous chloroplasts around the cell margin. Pseudonodulus (a large pore in the valve) near the valve margin large, but sometimes difficult to see, unless in empty valves.

Size

Diameter: 50-300 µm
Height (pervalvar axis): 20-120 µm

Distribution. The species has a world-wide distribution in plankton communities mainly in coastal areas.

Similar species: In routine analyses the species can be confused with other centric diatoms. The valve structures (central annulus, areolation and processes) are important for species identification. *Actinocyclus octonarius* differs from *Roperia tesselata* (Roper) Grunow ex Pelletan, 1889, another species with a pseudonodulus, in the areolation pattern. In *R. tesselata* areolation is linear in the valve centre but more irregular towards the margin.

Synonym:
Actinocyclus ehrenbergii Ralfs, 1861.

Literature: Hasle & Syvertsen 1997; Hendey 1964, 1974.

Podosira stelligera (J. W. Bailey) Mann, 1907 — Family: Hyalodiscaceae

Season				Trophic mode	Shape	Harmful	Bloom	Resting stage
W	S	S	A	autotrophic	spheroid	no	no	no

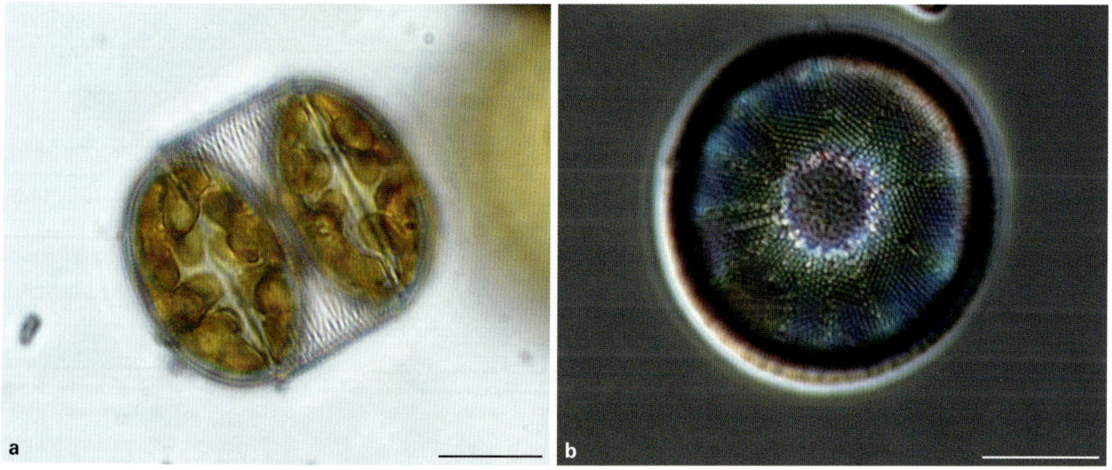

a, Bright field image of a pair of live cells of *Podosira stelligera* connected by common girdle bands.
b, Frustule in valve view, showing the central non-areolated area (permanent slide). Scale bars = 20 µm.

Description

Cells lenticular in girdle view with strongly convex valves, a circular cross-section and without a marked cingulum. Valve area divided into separated radial sectors which are clearly areolated in contrast to the indistinctly structured central area. A pair of cells before division cylindrically stretched because of numerous clearly visible girdle bands. Small rounded chloroplasts strongly pigmented and arranged star shaped. Nucleus located centrally in the centre of the cell.

Size

Diameter: 10-100 µm
Height (pervalvar axis): 12-130 µm

Distribution. *Podosira stelligera* has a worldwide distribution in the littoral zone. After turbulence it regularly occurs in the plankton.

Similar species: In shape this species is similar to *Hyalodiscus scoticus* (Kützing) Grunow, 1879 which also forms a pair of cells connected by a common girdle band before division. Usually *H. scoticus* is smaller. In valve view *P. stelligera* shows a distinct fasciculate areolation with a small clear central area in contrast to *H. scoticus* which has a primary radial and two secondary crossing areolation patterns and a large clear central area.

Synonyms:
Hyalodiscus stelliger J. W. Bailey, 1854,
Podosira maculata W. Smith, 1856,
Melosira maculata Lagerst., 1876.

Literature: Hendey 1974.

Melosira moniliformis (O. F. Müller) C. Agardh, 1824

Family: Melosiraceae

Season				Trophic mode	Shape	Harmful	Bloom	Resting stage
W	S	S	A	autotrophic	cylinder	no	no	yes

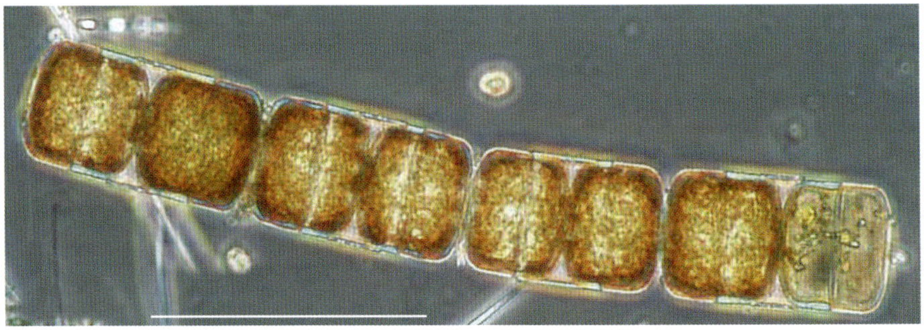

Phase contrast image of a chain of *Melosira moniliformis*. Scale bar = 200 µm. Image courtesy of Regina Hansen.

Description

Melosira moniliformis cells usually wider than high in girdle view and united into beadlike chains by mucilage pads. Linking spines not well developed. Cells often connected in twos or threes by their girdles. Valve face flat with small spines or granules. Mantle parallel to the pervalvar axis with a slight curvature towards the valve face. Labiate processes found over the entire valve surface, but concentrated on the valve face and the mantle edge. Chloroplasts irregularly lobed and discoid, lying in the peripheral cytoplasm. Fine areolae structure only observable in light microscopy as a tiny point pattern. Auxospore formation has been observed.

Size

Diameter: 25-70 µm
Length (pervalvar axis): 30-100 µm

Distribution. *Melosira moniliformis* is a typical littoral form that can occasionally be found in the plankton especially during late winter and early spring.

Similar species: none.

Synonyms:
Conferva moniliformis O. F. Müller, 1783,
Lysigonium moniliforme (O. F. Müller) Link, 1820,
Lysigonium moniliforme (O. F. Müller) Trevisan, 1848,
Melosira borreri var. *moniliformis* (O. F. Müller) A. Grunow, 1878.

Literature: Crawford 1977.

Stephanopyxis turris (Greville) Ralfs ex Pritchard, 1861 — Family: Stephanopyxidaceae

Season				Trophic mode	Shape	Harmful	Bloom	Resting stage
W	S	S	A	autotrophic	cylinder	no	✿	yes

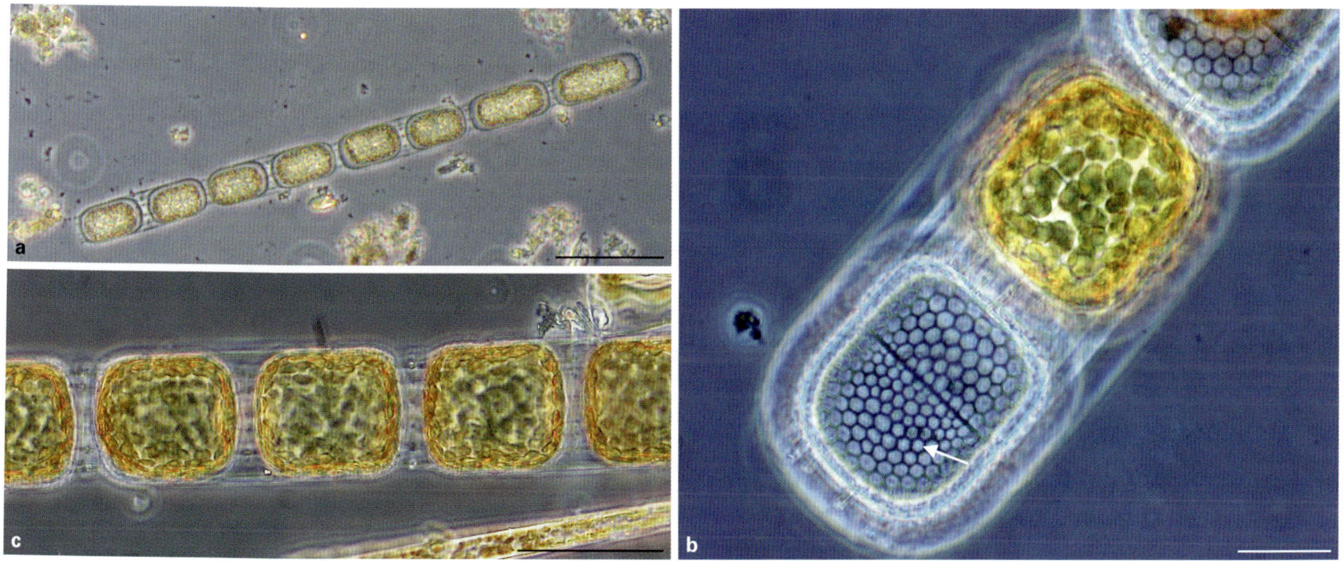

a, A Lugol-fixed chain of *Stephanopyxis turris*. Scale bar = 100 µm.
b, Live cells of the species showing detail of areolation (arrow) and processes. Scale bar = 20 µm.
c, Part of a live chain. Scale bar = 50 µm.
All images taken with phase contrast optics.

Description

Individual, vegetative cells of *Stephanopyxis turris* usually slightly higher than wide (with a high valve mantle). Cells coarsely areolated with the large hexagonal areolae. Areolae similar in size across the valve and visible with light microscopy. Adjacent cells joined into short, straight chains by the external parts of the long, hollow labiate process located in a marginal ring on the junction between valve face and valve mantle. A less conspicuous second ring of labiate processes also present. Chloroplasts. numerous small and rounded. *S. turris* is known to produce resting spores.

Size

Diameter (valvar axis): 20-90 µm
Height (pervalvar axis): 20-100 µm

Distribution. *Stephanopyxis turris* can be found sporadically from spring to autumn in temperate to warm waters sometimes building higher biomasses in late spring or early summer.

Similar species: The species is similar to *Stephanopyxis palmeriana* (Greville) Grunow, 1884 which has areolae of decreasing size from the centre of the valves to the margin. The morphology of *S. turris* chains also resembles those of *Skeletonema costatum* which, however, are much smaller.

Synonym:
Creswellia turris Greville & Arnott, 1857.

Literature: Hendey 1974, Round 1973

Paralia sulcata (Ehrenberg) Cleve, 1873

Family: Paraliaceae

Season				Trophic mode	Shape	Harmful	Bloom	Resting stage
W	S	S	A	autotrophic	cylinder	no	no	yes

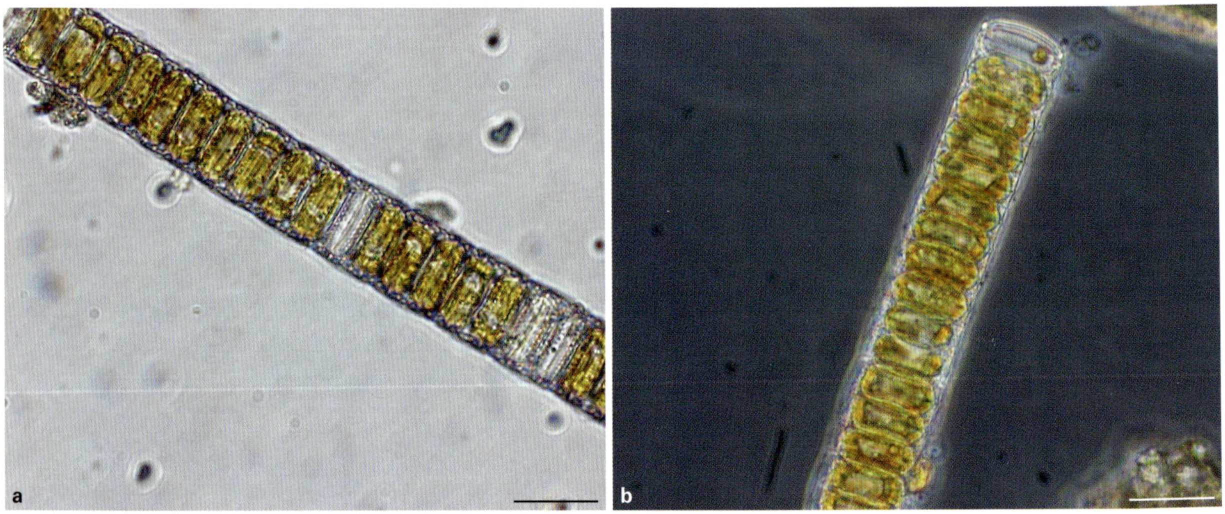

a, Bright field image of a live chain of *Paralia sulcata*.
b, Live chain in phase contrast.
Scale bars = 50 µm.

Description

Cells discoid with circular cross-section, joined into straight close-set chains. Sibling valves linked by marginal, interlocking, spatulate spines (spatulate = broad and flat distally with a narrow tapering base), their tops slit parallel to the valve margin. A ring of well developed (heavily silicified) labiate processes occurring at the extreme edge of the inside of the valve mantle. The slit-like internal openings located parallel to the mantle edge. Fewer than 10 discoid small chloroplasts per cell.

Size

Diameter: 8-80 µm
Height (pervalvar axis): 3-15 µm

Distribution. *Paralia sulcata* is a cosmopolitan littoral form which frequently appears in the plankton. Especially in late winter and early spring it can be the dominating species in plankton samples.

Similar species: In chain form the species is easy to identify. A single cell in valve view can be confused in light microscopy with other centric diatoms of similar size, e.g., *Cyclotella* species. *P. sulcata* shows a ring of radial ridges (which do not reach the margin) in the flat part of the valve face. In contrast, the radial structures on the undulated valves of *Cyclotella* species are running to (or close to) the valve margin.

Synonyms:
Paralia marina (W. Smith) Heiberg, 1863,
Gallionella sulcata Ehrenberg, 1839,
Melosira sulcata (Ehrenberg) Kützing, 1844,
Orthoseira marina W. Smith, 1856.

Literature: Crawford 1979, Sawai et al. 2005.

Lauderia annulata Cleve, 1873 — Family: Lauderiaceae

Season				Trophic mode	Shape	Harmful	Bloom	Resting stage
W	S	S	A	autotrophic	cylinder	no	✿✿	no

a, Part of a live chain of *Lauderia annulata* focussed on the chloroplasts close to the valve.
b, The same chain focussed on the central plasma-strand (see arrow).
c, DIC image of a short live chain showing the external tubes of the strutted processes at the valve margin (1).
a, b, Bright field images. Scale bars = 50 µm (a, b), 20 µm (c).

Description

Cells compact and cylindrical with a circular cross-section, forming straight, tight but fragile chains, easily breaking up into small sections or single cells. Valves weakly convex with a slight depression in the centre and numerous small strutted processes on the valve face and in the marginal zone. The external tubes arising from the marginal strutted processes visible in light microscopy (1). One labiate process and some occluded processes situated at the valve margin. Numerous strutted processes scattered on valve face and margin. Chloroplasts numerous, small, lobed and plate-like.

Size

Diameter: 15-75 µm
Height (pervalvar axis): 20-95 µm

Distribution. *Lauderia annulata* can be found in coastal waters mainly in spring but also in summer (still abundant around Helgoland in July).

Similar species: The shape of *Lauderia annulata* cells in girdle view closely resembles that of *Detonula pumila*. Valves of adjacent cells in *Detonula pumila* are not directly attached to each other. In valve view *Lauderia annulata* is very difficult to identify using light microscopy during routine analyses. In valve view cells can closely resemble single cells of *Thalassiosira* species.

Synonym:
Lauderia borealis Gran, 1900.

Literature: Hendey 1974, Kaczmarska et al. 2006, Syvertsen & Hasle 1982.

Detonula confervacea (Cleve) Gran, 1900   Family: Skeletonemaceae

Season				Trophic mode	Shape	Harmful	Bloom	Resting stage
W	S	S	A	autotrophic	cylinder	no	no	yes

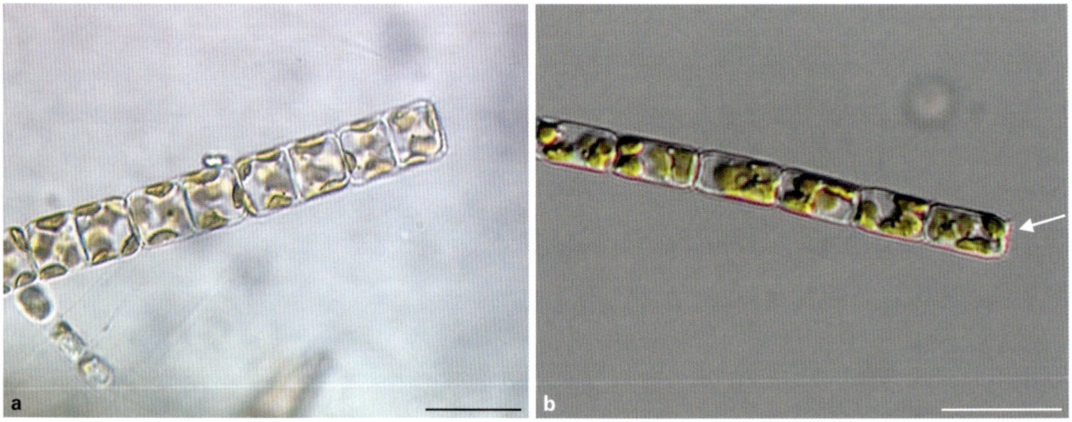

a, Bright field image of a live chain of *Detonula confervacea*. Scale bar = 20 µm.
b, DIC image of a chain of live cells. The external tubes of the small strutted processes are just discernable (arrow).

Description

Cells small cylindrical with a circular cross-section forming tight chains. Valve face flat with a marginal rim of small spines (strutted processes) linked with those in adjacent cells. The fine connecting thread from the strutted process in the valve centre usually not visible in light microscope. One labiate process situated at the valve margin. Chloroplasts small, rounded and plate-like.

Size

Diameter: 6-20 µm
Height (pervalvar axis): 15-30 µm

Distribution. *Detonula confervacea* has a preference for cold water and is found sporadically during winter in coastal waters of the Northern Seas.

Similar species: none.

Synonyms:
Lauderia confervacea Cleve, 1896,
Detonula cystifera Gran, 1900.

Literature: Kaczmarska et al. 2006, Syvertsen 1979.

Detonula pumila (Castracane) Gran, 1900 — Family: Skeletonemaceae

Season				Trophic mode	Shape	Harmful	Bloom	Resting stage
W	S	S	A	autotrophic	cylinder	no		not known

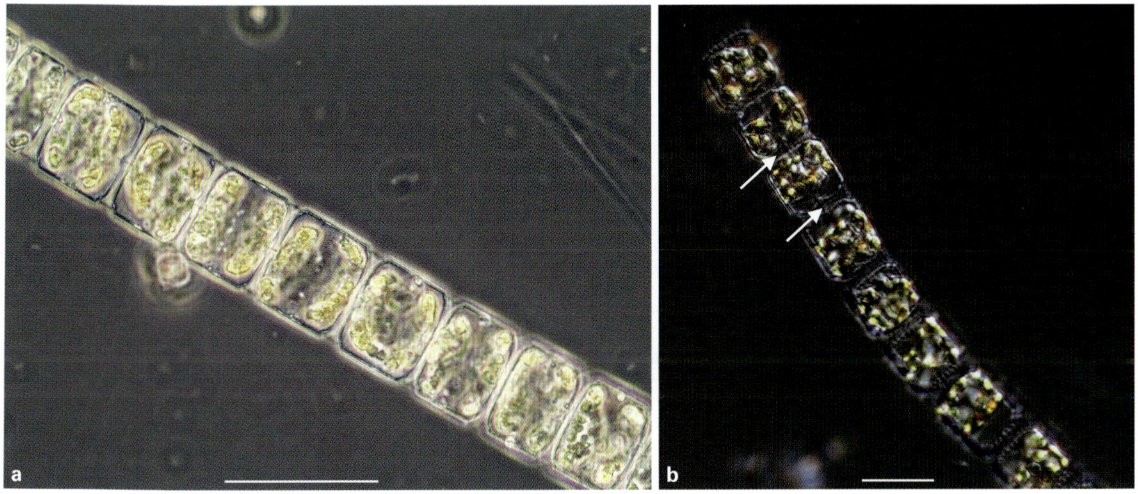

a, Phase contrast image of a formalin-fixed chain of the diatom *Detonula pumila*; image courtesy of Regina Hansen. Scale bar = 50 µm.
b, Live chain, showing the connecting threads (arrows) between adjacent cells. Scale bar = 20 µm.

Description

Cells cylindrical with a circular cross-section and forming straight chains. Valves flat or slightly convex, but always slightly indented in the centre and with one ring of marginal strutted processes from which delicate threads emerge. Threads overlapping with those of adjacent cells. Distinct thread from central strutted process connecting adjacent cells. Each valve with one marginal labiate process. Chloroplasts numerous and star-shaped.

Size

Diameter: 15-45 µm
Height (pervalvar axis): 15-120 µm

Distribution. *Detonula pumila* can be found in coastal waters during summer and autumn mainly in warmer waters.

Similar species: Chains of *Detonula pumila* resemble those of *Lauderia annulata*. In chains of the latter, valves of adjacent cells are directly in contact with each other. In *Detonula pumila* chains there is a small space between cells.

Synonyms:
Lauderia pumila Castracane, 1886,
Thalassiosira condensata Cleve, 1900,
Schroederella delicatula Pavillard, 1913,
Schroederella schroederi (Bergon) Pavillard, 1925.

Literature: Hasle 1973b.

Skeletonema Greville

Family: Skeletonemaceae

Season				Trophic mode	Shape	Harmful	Bloom	Resting stage
W	S	S	A	autotrophic	cylinder + 2 half ellipses	water discolouration	✿✿✿	yes

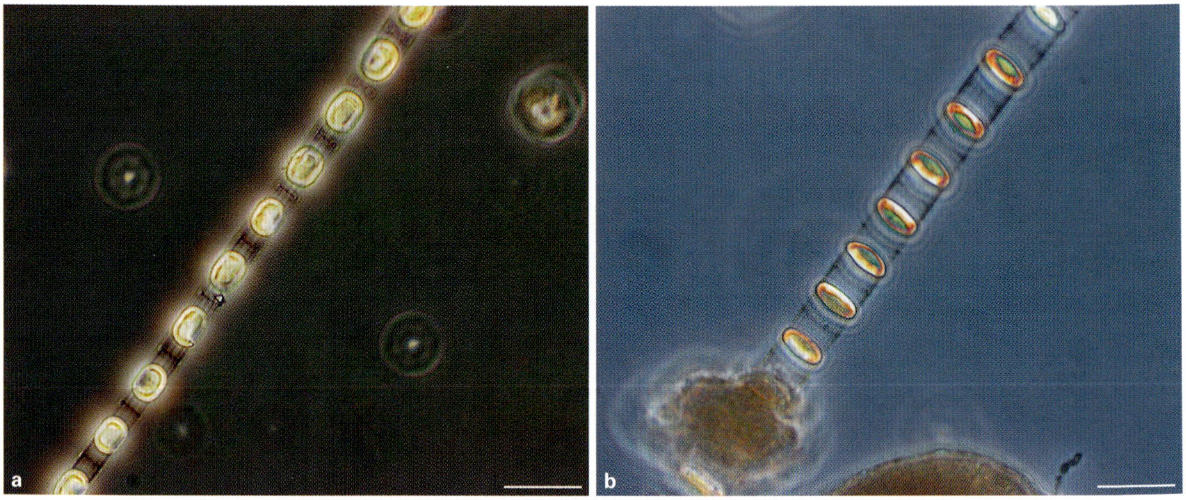

a, b, Phase contrast images of live chains of *Skeletonema costatum* sensu lato with different cell proportions. Scale bars = 20 µm.

Description

Shape lenticular, elliptical or long cylindrical in girdle view. Valves convex with a marginal ring of strutted processes connecting cells into long straight chains. Cells containing two plate-like chloroplasts.

Note

Recent molecular research has established that the species previously as *Skeletonema costatum* (Greville) Cleve, 1878 is actually several separate species with distinct biogeography. These distinctions cannot be made in routine-monitoring. These species are distinguished on the basis of the arrangement and shape of their valve processes, their numbers of chloroplasts and the pattern of interlinkage of cells. The *Skeletonema* species around Helgoland (and other parts of the North Sea) is probably not *S. costatum*. What we are describing here is *S. costatum* 'sensu lato', i.e., the images shown here would normally have been assigned the name *Skeletonema costatum*, although this will require revision, follwoing the findings of recent studies.

Size

Diameter: 3-20 µm
Height (pervalvar axis): 2-30 µm

Distribution. The *Skeletonema* species found in Helgoland appears throughout the year and can form blooms in the autumn.

Similar species: The genus *Skeletonema* was recently divided into several species closely resembling each other. For a reliable identification scanning and transmission electron microscopy should be used.

Synonyms: none.

Literature: Hasle 1973a, Round 1973, Sarno et al. 2005, Zingone et al. 2005.

Porosira glacialis (Grunow) E. Jørgensen, 1905

Family: Thalassiosiraceae

Season				Trophic mode	Shape	Harmful	Bloom	Resting stage
W	S	S	A	autotrophic	cylinder	no	✿	yes

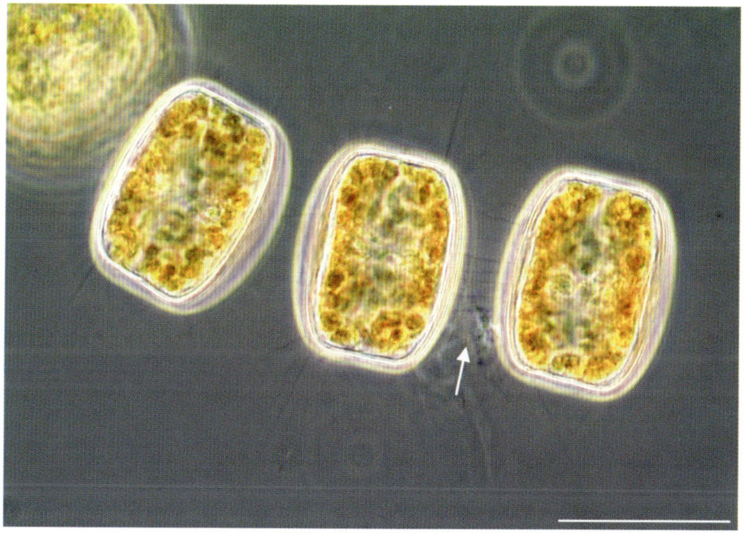

Phase contrast image of a short chain of live *Porosira glacialis* in girdle view showing the central connecting threads. Scale bar = 50 µm. Image courtesy of Regina Hansen.

Description

Cells discoid or narrowly cylindrical in girdle view. Valves slightly concave with rounded margins and radial areolation. One large labiate process near the margin and numerous strutted processes situated on the valve face. Processes not distributed in an obvious pattern as in *Thalassiosira* species. Numerous delicate threads connecting cells into loose chains that easily disintegrate in fixed samples, leaving only single cells. Chloroplasts numerous, small and plate-like.

Size

Diameter: 30-70 µm
Height (pervalvar axis): 25-60 µm

Distribution. The species is distributed in cold water to temperate regions. In the North Sea it is mainly found during winter and spring.

Similar species: Large single cells in valve view can be confused with *Coscinodiscus* species, smaller ones with *Thalassiosira* species. In girdle view, single cells look similar to *Lauderia annulata*, but the latter can be recognized by the conspicuous plasma strand in the centre of the cell.

Synonyms:
Podosira hormoides var. *glacialis* Grunow, 1884,
Podosira glacialis (Grunow) Cleve, 1896,
Lauderia glacialis (Grunow) Gran, 1900,
Porosira antarctica O. G. Kozlova, 1962.

Literature: Hasle 1973; Kaczmarska et al. 2006; Villareal & Fryxell 1983a, 1990.

Thalassiosira angulata (Gregory) Hasle, 1978 Family: Thalassiosiraceae

Season				Trophic mode	Shape	Harmful	Bloom	Resting stage
W	S	S	A	autotrophic	cylinder	no		no

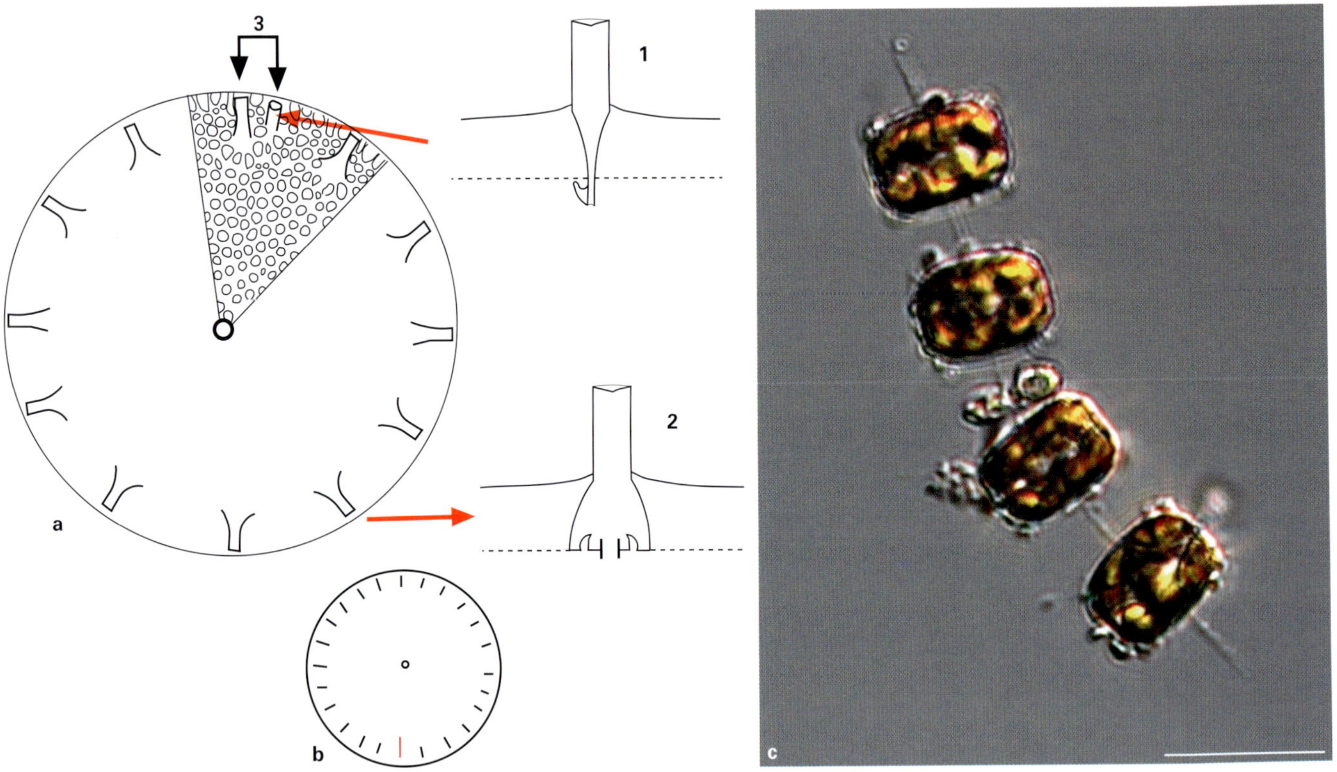

a, Schematic drawing of the principal characteristics of the diatom genus *Thalassiosira*, viewing the cell in valve view: 1, Labiate process; 2, Strutted process; 3, distance of labiate in relation to strutted process. **b,** Schematic drawing of *T. angulata* in valve view indicating the position of the strutted processes and the labiate process (shown in red). Schematic adapted from Hasle & Syvertsen 1997. **c,** DIC image of a short chain of *Thalassiosira angulata*. Scale bar = 20 µm.

Description

Pervalvar axis usually shorter than the diameter in girdle view. Valve face flat and mantle smoothly curved. The single central connecting thread distinctly longer than pervalvar axis. External tubes of the marginal strutted processes often readily seen in light microscopy. Valve view revealing hexagonal areolae (8-18 in 10 µm) in curved tangential rows, occasionally in straight rows or in sectors. No distinct central areola present. Marginal processes widely spaced. Labiate process large and always located closely to one of the marginal strutted processes. Transition between valve mantle and first girdle band sharp. Chloroplasts numerous small and rounded.

Size

Diameter: 12-40 µm
Height (pervalvar axis): 8-22 µm

Distribution. *Thalassiosira angulata* is distributed in the North Atlantic Ocean and can be found throughout the year mainly in cold to temperate water. In the German Bight it is most abundant and appears in autumn and winter.

Similar species: none.

Synonyms:
Orthoseira angulata Gregory, 1857,
Thalassiosira decipiens (Grunow) E. G. Jørgensen, 1905.

Literature: Hasle 1978, Hoppenrath et al. 2007, Mills & Kaczmarska 2006.

Thalassiosira anguste-lineata (A. Schmidt) G. Fryxell & Hasle, 1977

Family: Thalassiosiraceae

Season				Trophic mode	Shape	Harmful	Bloom	Resting stage
W	S	S	A	autotrophic	cylinder	no	no	no

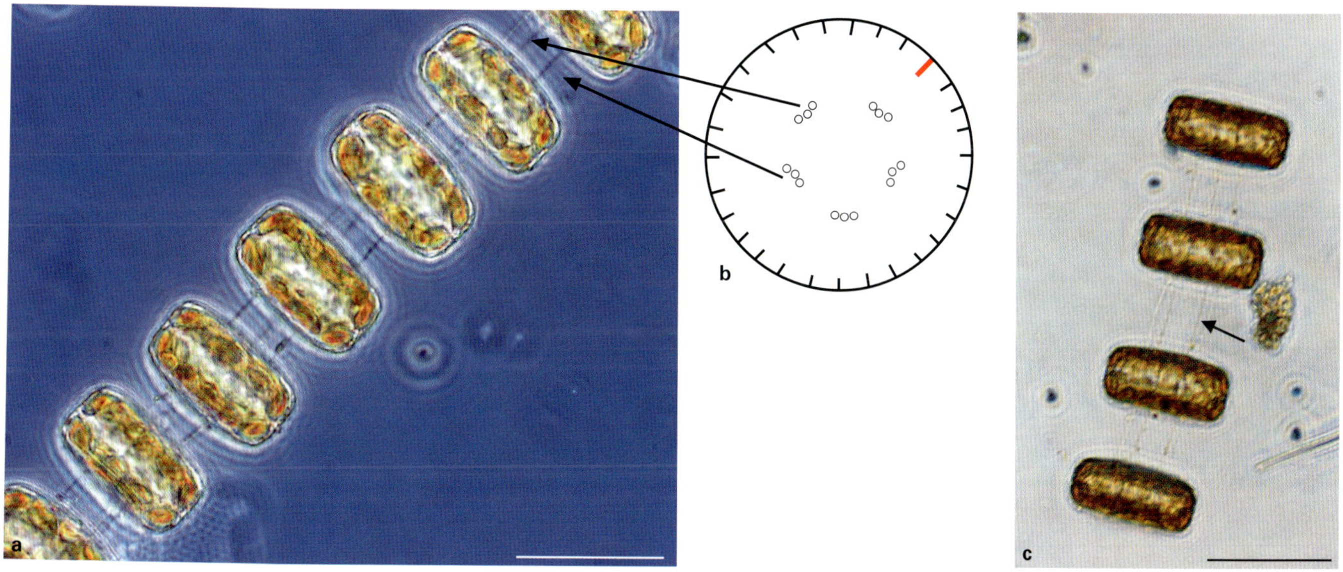

a, Live chain of *Thalassiosira anguste-lineata* with several connecting threads between the cells. **b,** Schematic of cell in valve view showing the location of the 'central' strutted processes. **c,** Bright field image of *T. anguste lineata*. Scale bars = 50 µm.

Description

Cells usually wider than high in girdle view. Valve face flat and valve mantle shallow. In valve view: Strutted processes not arranged as a central cluster but as several clusters of almost linearly arranged processes surrounding the valve centre. Threads issuing from individual processes in each cluster uniting to form thicker threads linking adjacent cells. Connecting threads distinct and visible in light microscopy (in girdle view). One ring of marginal strutted processes present, their external tubes narrowing slightly towards their distal end. A single labiate process lying just inside this ring and halfway between two processes. 3-6 marginal processes found within 10 µm. Areolation variable. Chloroplasts numerous small rounded pads.

Size

Diameter: 15-80 µm
Height (pervalvar axis): 10-40 µm

Distribution. *Thalassiosira anguste-lineata* is a cosmopolitan species which can frequently be found in the plankton mainly during spring.

Similar species: *Thalassiosira mediterranea* (Schröder) Hasle, 1990 is similar in girdle view, i.e. with several connecting threads, but it is a much smaller species.

Synonyms:
Coscinodiscus anguste-lineatus A. Schmidt, 1878,
Coscinodiscus polychordus Gran, 1897,
Thalassiosira polychorda (Gran) E. Jorgensen, 1899,
Coscinosira polychorda (Gran) Gran, 1900.

Literature: Fryxell & Hasle 1977, Hoppenrath et al. 2007, Kaczmarska et al. 2006.

Thalassiosira eccentrica (Ehrenberg) Cleve, 1904 — Family: Thalassiosiraceae

Season				Trophic mode	Shape	Harmful	Bloom	Resting stage
W	S	S	A	autotrophic	cylinder	no	no	no

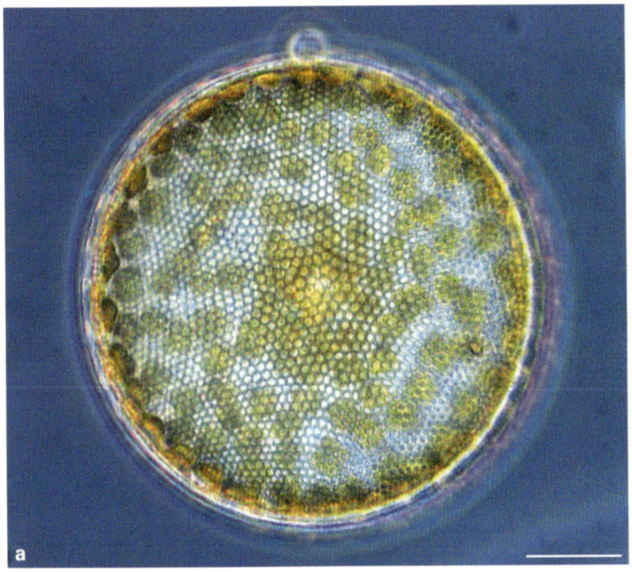

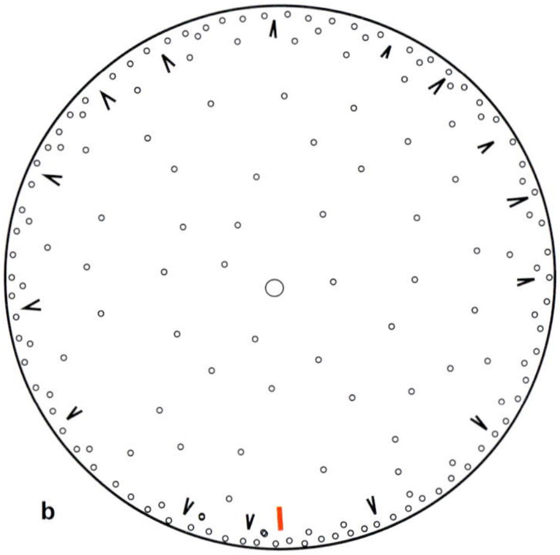

a, Single live cell of *Thalassiosira eccentrica* in valve view showing the typical areolation.
b, Schematic drawing showing the type and arrangement of processes (labiate process indicated in red).
Schematic adapted from Hasle & Syvertsen 1997. Scale bars = 20 µm.

Description

Valve face of *Thalassiosira eccentrica* flat with a rounded mantle. Connecting thread about twice the length of the cell height. In valve view, areolae in tangential rows, slightly fasciculate and decreasing in size from the centre (3-5 in 10 µm) to the margin (5-9 in 10 µm). Single central strutted process adjacent to a central areola and surrounded by seven areolae. Strutted processes scattered over the valve face and in two marginal rings. One ring of pointed spines and the single labiate process located inside the marginal rings of strutted processes. Chloroplasts typically brown, numerous and small.

Size

Diameter: 15-110 µm

Distribution. According to Hasle (1976) *Thalassiosira eccentrica* is nearly cosmopolitan except in the Antarctic, Subantarctic and Arctic zones. Based on available records, it might also be a coastal species. It is found mainly during spring and early summer but also in the autumn.

Similar species: None, although as in many *Thalassiosira* species *T. eccentrica* might also be difficult to identify in Lugol-fixed samples, particularly when cell contents such as chloroplasts obscure the aerolation pattern.

Synonyms:
Coscinodiscus eccentricus Ehrenberg, 1841,
Coscinodiscus labyrinthus Roper, 1858.

Literature: Fryxell & Hasle 1972, Hasle 1976, Hoppenrath et al. 2007, Kaczmarska et al. 2006.

Thalassiosira hendeyi Hasle & G. Fryxell, 1977 — Family: Thalassiosiraceae

Season				Trophic mode	Shape	Harmful	Bloom	Resting stage
W	S	S	A	autotrophic	cylinder	no	no	no

a, DIC image of a live cell of *Thalassiosira hendeyi* from net plankton. Scale bar = 20 µm.
b, Bright field image of a live cell in girdle view with the threads emerging from the marginal strutted processes discernable. Scale bar = 50 µm. Image b courtesy of Hanne Halliger.

Description

Cells rectangular in girdle view, with a slightly shorter pervalvar axis than the diameter and a slightly undulating marginal ridge. Valve face with two labiate processes (with long external tubes, see arrows) and three rings of small strutted processes. Strutted processes not visible in the light microscope . Areolae (5-6 in 10 µm) arranged in tangential rows. Cells normally solitary but sometimes forming short chains. Chloroplasts numerous, small and rounded.

Size

Diameter: 40-110 µm
Height (pervalvar axis): 30-90 µm

Distribution. This species is considered a warm water to temperate species. It has been reported, e.g., from the North Sea (Helgoland and Sylt). It is found mainly in spring and autumn.

Similar species: In *Thalassiosira punctigera* longer processes are also visible in girdle view. These are occluded processes in contrast to *T. hendeyi.* However, this difference cannot be seen using light microscopy.

Synonym:
Coscinodiscus hustedtii Müller-Melchers, 1953.

Literature. Hoppenrath et al. 2007, Kaczmarska et al. 2006.

Thalassiosira minima Gaarder, 1951

Family: Thalassiosiraceae

Season				Trophic mode	Shape	Harmful	Bloom	Resting stage
W	S	S	A	autotrophic	cylinder	no	✿	no

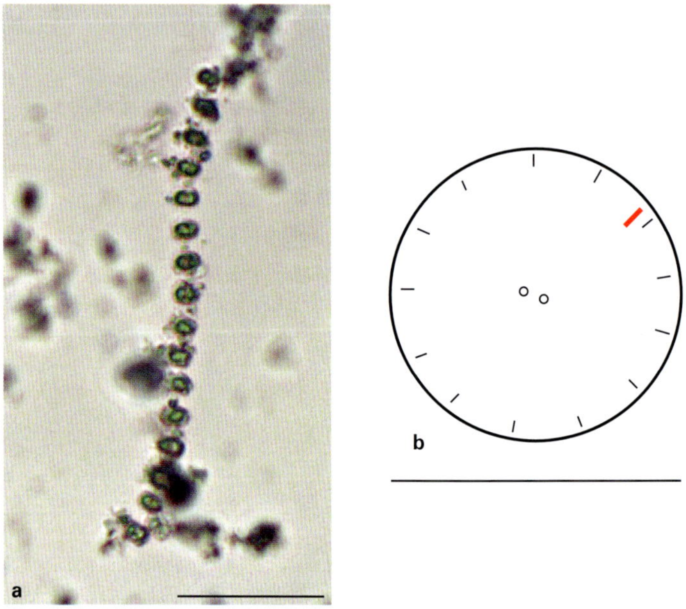

a, Bright field image of a Lugol-fixed chain of *Thalassiosira minima* with attached detritus particles. Scale bar = 50 µm. **b,** Schematic drawing showing the arrangement of processes on the valve face (labiate process marked in red). Schematic adapted from Hasle & Syvertsen 1997. Scale bar = 10 µm.

Description

Cells rectangular in girdle view slightly depressed in the centre of the valves. Diameter up to twice the length of the pervalvar axis. Valve face exhibiting one to three central strutted processes (typically two). The single labiate process situated close to one marginal strutted process. Fine pattern of areolae (30-40 in 10 µm) barely seen in the light microscope. Cells connected into chains with the connecting thread twice as long as the pervalvar axis. Chloroplasts few, rounded.

Size

Diameter: 5-15 µm
Height (pervalvar axis): 3-10 µm

Distribution. *Thalassiosira minima* is a cosmopolitan species except of polar regions. It can be found throughout the year. The highest abundance in the German Bight can be observed during spring.

Similar species: In general valve outline, *Thalassiosira minima* resembles very small specimen of *T. nordenskioeldii*, when seen in girdle view. But the two species differ clearly when seen in valve view (e.g., 2-3 central strutted processes in *T. minima* as opposed to one in *T. nordenskioeldii*).

Synonyms:
Coscinosira floridana I. C. G. Cooper, 1958,
Thalassiosira floridana (I. C. G. Cooper) Hasle, 1972.

Literature: Hasle 1980, Hoppenrath et al. 2007, Kaczmarska et al. 2006.

Thalassiosira nordenskioeldii Cleve, 1873 Family: Thalassiosiraceae

Season				Trophic mode	Shape	Harmful	Bloom	Resting stage
W	S	S	A	autotrophic	cylinder	no	✿✿✿	yes

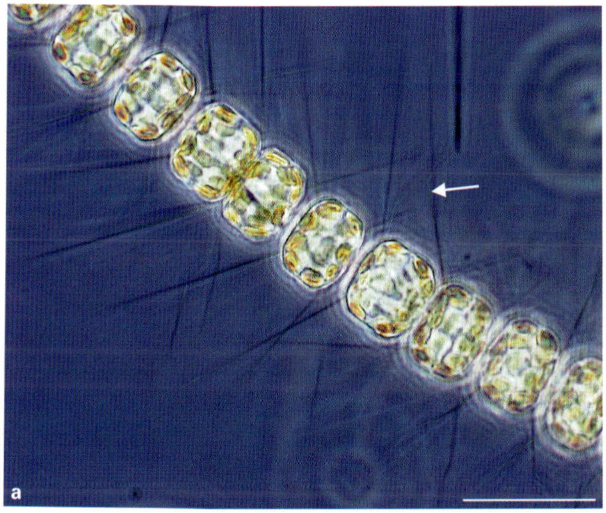
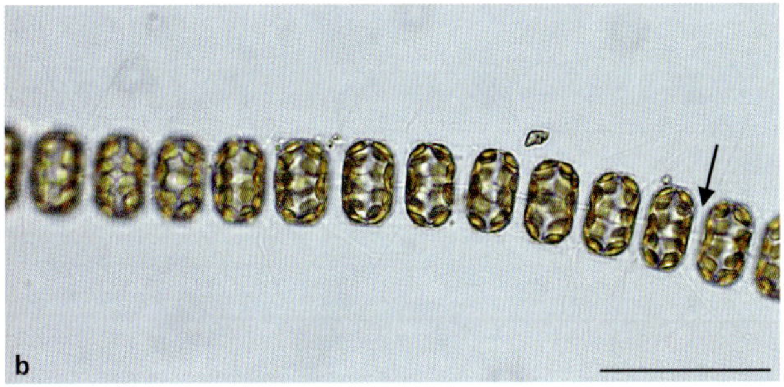
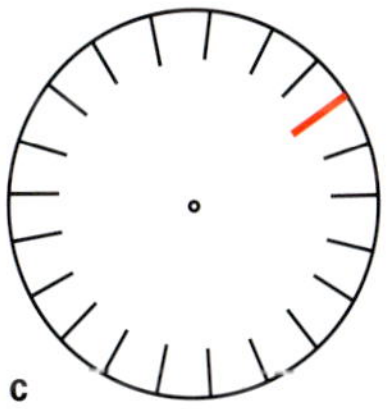

a, Phase contrast image of live chain of *Thalassiosira nordenskioeldii* in girdle view, showing the external tubes of the strutted processes (arrow).
b, Bright field image of a live chain, showing the short central connecting thread (arrow).
c, Schematic drawing of valve face indicating the type and position of processes.
Schematic adapted from Hasle & Syvertsen 1997. Scale bars = 20 µm.

Description

Cells octagonal in girdle view with a pronounced concavity around the valve centre and a high, oblique valve mantle. Valve face with one ring of marginal strutted processes with long, distinctive external tubes (three marginal processes within 10 µm), one central strutted process and a single labiate process with a variable position within this ring of strutted processes. Connecting thread between two valves the same length or shorter than the pervalvar axis. Areolation in radial rows, delicate but visible in light microscopy. Cells with numerous small plate-like chloroplasts.

Size

Diameter: 10-50 µm
Height (pervalvar axis): 5-33 µm

Distribution. *Thalassiosira nordenskioeldii* is a cold to temperate species in the Northern hemisphere. In the North Sea the species mainly appears during spring when it can form blooms.

Similar species: This species has a similar valve outline in girdle view to *T. aestivalis* Gran, 1931 but in the latter species the connecting threads are usually longer than in *T. nordenskioeldii*.

Synonyms: none.

Literature: Durbin 1974, 1977, 1978; Hoppenrath et al. 2007; Kaczmarska et al. 2006; Syvertsen 1979.

Thalassiosira punctigera (Castracane) Hasle, 1983

Family: Thalassiosiraceae

Season				Trophic mode	Shape	Harmful	Bloom	Resting stage
W	S	S	A	autotrophic	cylinder	no	✿	no

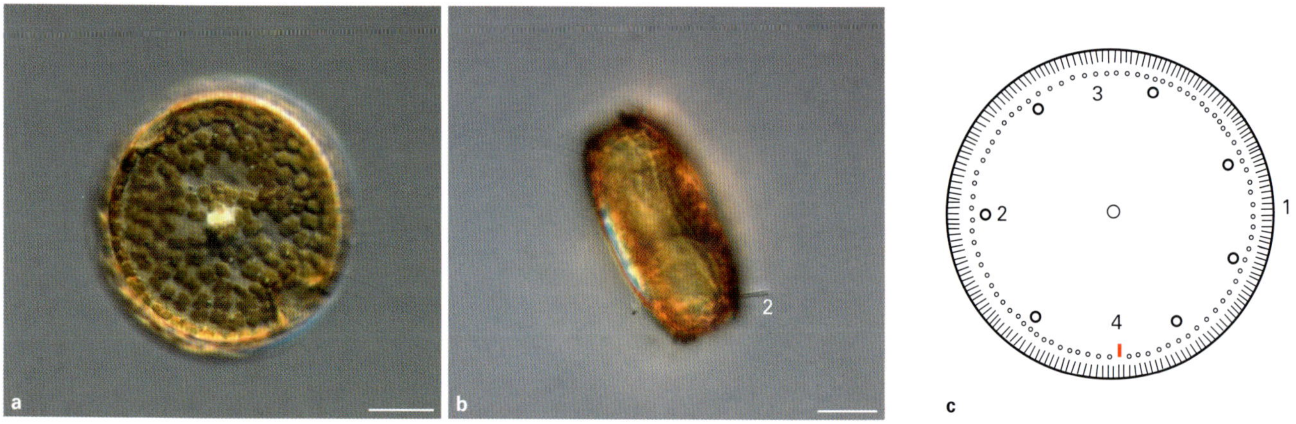

a, b, DIC images of *Thalassiosira punctigera*. **a,** Valve view of live cell showing the external tubes of the strutted processes; **b,** Girdle view of the same cell with the occluded process visible (2). **c,** Schematic drawing of the valve face showing the different processes (1, ribbed margin; 2, occluded process; 3, strutted processes; 4, labiate process). Schematic adapted from Hasle & Syvertsen 1997. Scale bars = 20 µm.

Description

Cells solitary or in very short chains connected by a thin connecting thread. Valve face convex in girdle view with a ribbed valve margin. Three types of processes discernable in valve view: Long tubes (occluded processes) large, variable in number and widely spaced. A ring of densely spaced strutted processes with short external tubes located close to the valve margin (four to five processes in 10 µm). One large labiate process with a long external tube. Chloroplasts numerous and small. Nucleus located centrally.

Size

Diameter: 40-185 µm
Height (pervalvar axis): 20-120 µm

Distribution. *Thalassiosira punctigera* is now a cosmopolitan species. Until the 1970s is was frequently found in the Pacific area and rarely in the North Atlantic. During the 70s it spread around European waters and is today found occasionally throughout the year, but most commonly during spring and early summer.

Similar species: In girdle view the occluded processes and the clearly visible central plasma strand make this a distinctive species. In valve view a typically dense layer of chloroplasts at the cell margin and a dark spot in the middle, indicating the strong central plasma strand are typical and easily seen in water mounts.

Synonyms:
Ethmodiscus punctiger Castracane, 1886,
Ethmodiscus japonicus Castracane,
Thalassiosira japonica Kisselev, 1935,
Thalassiosira angstii (Gran) Makarova, 1970.

Literature: Chepurnov et al. 2006, Hasle 1983, Hoppenrath et al. 2007.

Thalassiosira constricta Gaarder, 1938 — Family: Thalassiosiraceae

Season				Trophic mode	Shape	Harmful	Bloom	Resting stage
W	S	S	A	autotrophic	cylinder	no	✿	no

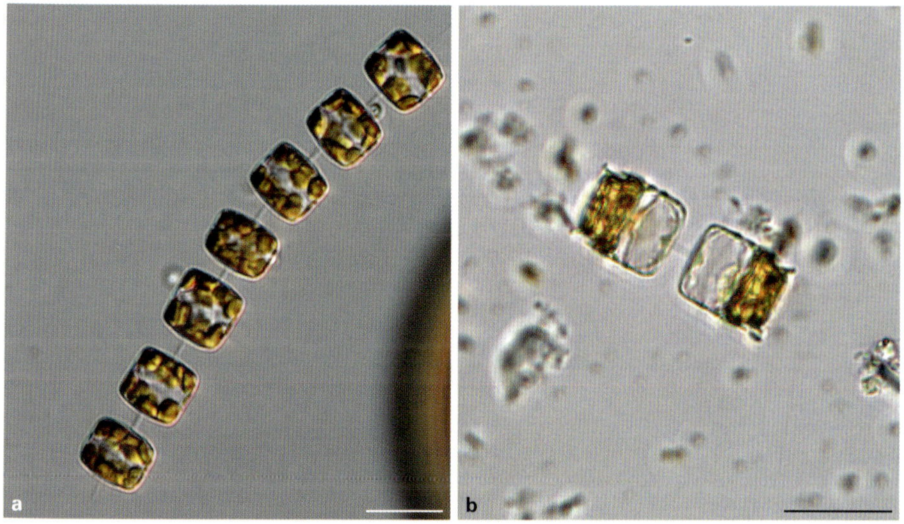

a, Live chain of *Thalassiosira constricta*. **b,** Lugol-fixed cells with resting spores. Scale bars = 20 µm.

Description

Cells with a similar height and width or higher than wide. Cells usually occurring in chains. Valve face flat with slightly rounded margins. Valve face with a cluster of central strutted processes and one ring of strutted processes. One labiate process located within the marginal ring of strutted processes, half-way between two strutted processes. Three to five marginal strutted processes in 10 µm.

Size

Large diameter (apical axis): 12-32 µm

Distribution. The species has been reported from the Arctic and sub-arctic, the Skagerrak and North Sea proper (Helgoland).

Similar species: *Thalassiosira gravida, Thalassiosira hyalina*.

Synonyms: none

Literature: Heimdahl 1971, 1974.

Thalassiosira rotula Meunier, 1910

Family: Thalassiosiraceae

Season				Trophic mode	Shape	Harmful	Bloom	Resting stage
W	S	S	A	autotrophic	cylinder	no	✿✿	no

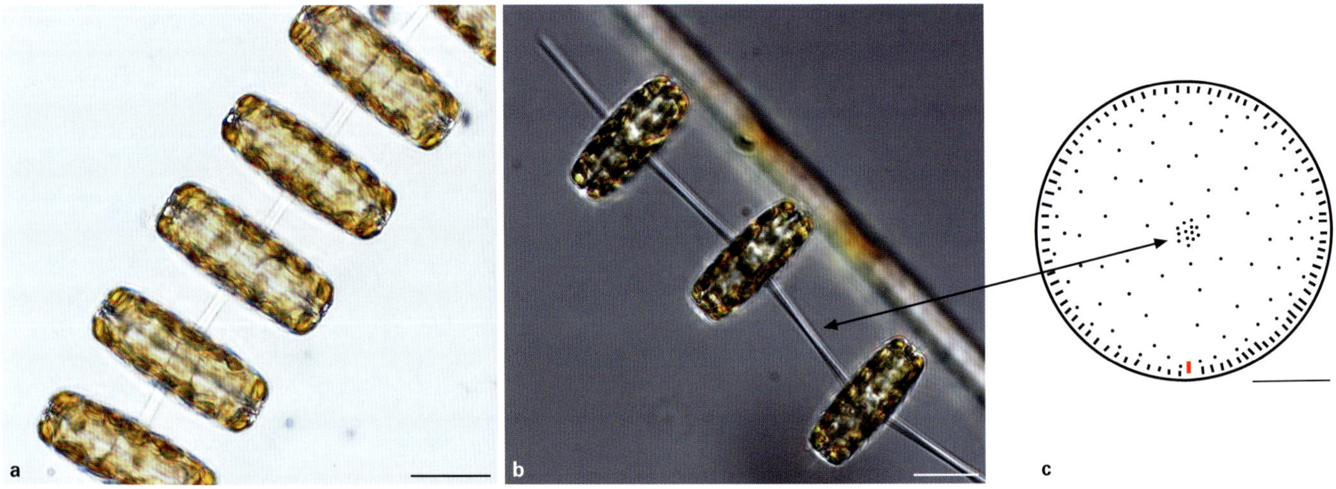

a, Bright field image of a live chain of *Thalassiosira rotula*.
b, DIC image of a short chain with the arrow pointing to the thick bundle of connecting threads. Scale bars = 20 µm.
c, Schematic drawing showing the position and arrangement of the ring of marginal strutted processes, central cluster of processes and the labiate process (in red). Schematic adapted from Hasle & Syvertsen 1997. Scale bar = 10 µm.

Description

Cells considerably wider than high in girdle view and rectangular in outline. Short, thick connecting threads easily discernable in girdle view. Base of the connecting threads appearing thickened as the thread actually consists of a bundle of strutted processes emerging from the centre of the valve face. Further strutted processes scattered all over the valve surface. Valve face with radial ribs, mantle with areolae. One ring of marginal strutted processes (12-15 processes in 10 µm). One labiate process near the valve margin. Cells with numerous small chloroplasts.

Size

Diameter: 8-60 µm
Height (pervalvar axis): 5-20 µm

Distribution. *Thalassiosira rotula* is considered cosmopolitan but it is very difficult to distinguish from *Thalassiosira gravida* Cleve, 1896. It can be found frequently during spring and summer often building high biomasses.

Similar species: *Thalassiosira gravida* is a very similar species. These species are not easy to distinguish in water mounts. The valve face in *T. rotula* shows radial ribs while a radial areolation is found in *T. gravida*. However, *T. rotula* is also distinguished from *T. gravida* on the basis that one of its girdle bands is irregularly thickened, which is discernible with light microscopy when viewing the cells in girdle view.

Synonyms: none.

Literature: Hoppenrath et al. 2007, Kaczmarska et al. 2006, Syvertsen 1977.

Dactyliosolen fragilissimus (Bergon) G. R. Hasle, 1997 — Family: Rhizosoleniaceae

Season				Trophic mode	Shape	Harmful	Bloom	Resting stage
W	S	S	A	autotrophic	cylinder	no	no	no

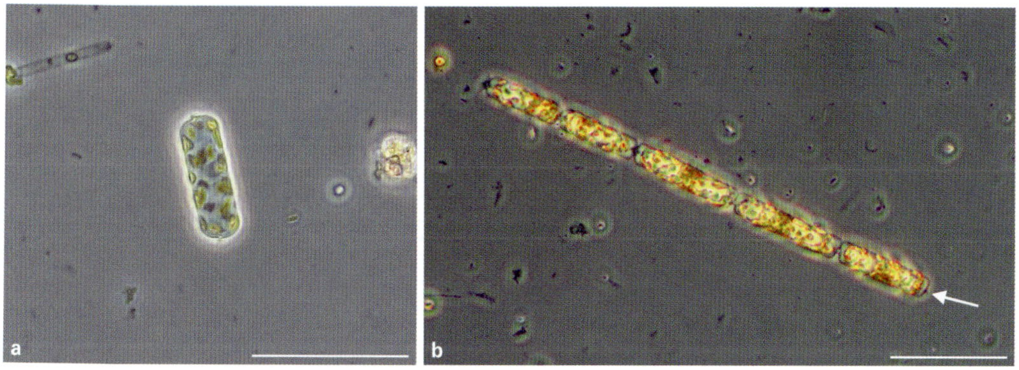

a, Single live cell of *Dactyliosolen fragilissimus*.
b, Short chain with the process on the valve face discernable (arrow).
Both are phase contrast images. Scale bars = 50 µm.

Description

Cells long and cylindrical with a circular cross-section forming loosely connected fragile chains. Cells several times longer than wide in girdle view. Girdle segments smooth and difficult to discern in light microscopy. Valves weakly convex with a small oblique spine near the centre of the valve face (see arrow in Figure b). Chloroplasts numerous, small and plate-like. Nucleus located close to the cell wall.

Size

Diameter: 8-70 µm
Length (pervalvar axis): 40-300 µm

Distribution. *Dactyliosolen fragilissimus* is a cosmopolitan and occurs sporadically in coastal waters throughout the year.

Similar species: Because of the spine in the centre of the valve face, this species is quite distinct but it has a very delicate valve and can, therefore, be overlooked.

Synonym: *Rhizosolenia fragilissima* Bergon, 1903.

Literature: Hasle 1975, Hasle & Syvertsen 1997.

Guinardia delicatula (Cleve) G. R. Hasle, 1997 — Family: Rhizosoleniaceae

Season				Trophic mode	Shape	Harmful	Bloom	Resting stage
W	S	S	A	autotrophic	cylinder	no	✿✿✿	no

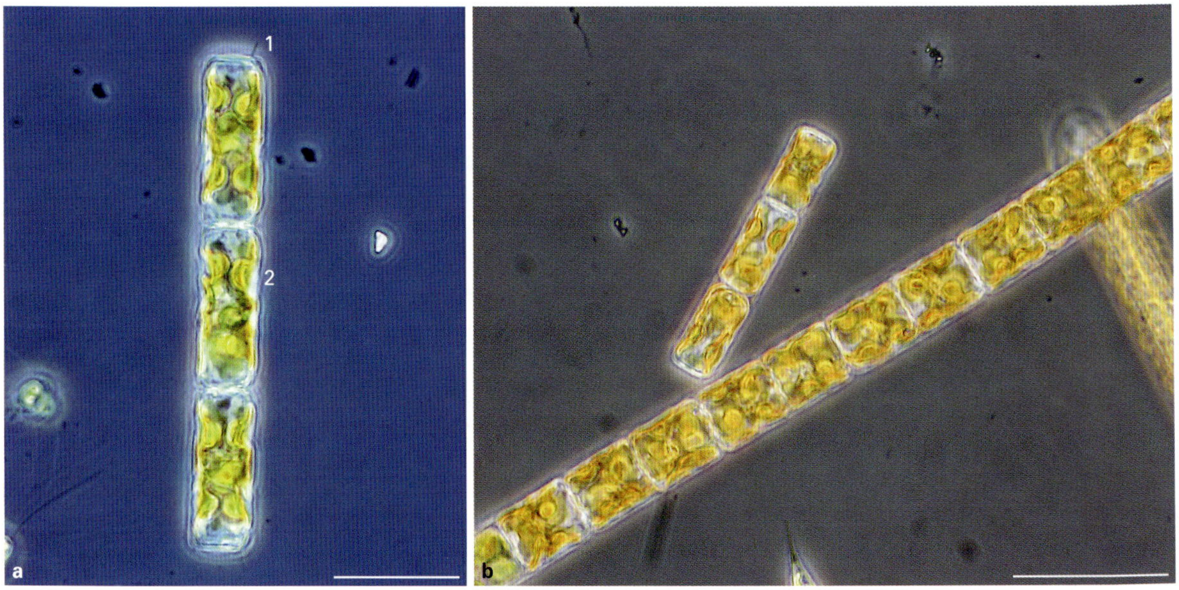

a, b, Phase contrast images of live chains of *Guinardia delicatula*: 1. The marginal spine; 2. The chloroplast and pyrenoid. Scale bars = 50 µm.

Description

Cells cylindrical. Valve face flat but slightly rounded at the margins. A spine (1) arising from the valve margin, directed obliquely to the pervalvar axis and fitting into a depression on the adjacent valve. Girdle bands not distinct. Cells forming close-set straight chains. Cells containing some large irregularly lobed and strongly pigmented chloroplasts each with a pyrenoid (2).

Size

Diameter: 8-40 µm
Length (pervalvar axis): 20-70 µm

Distribution. *Guinardia delicatula* occurs in coastal waters becoming most abundant in late spring and summer. In recent years this species has started appearing earlier in the year and is commonly present until the autumn.

Similar species: Small cells of *Guinardia delicatula* resemble *Leptocylindrus danicus*. But the chloroplasts of the latter are smaller. A clear sign for *Guinardia delicatula* is the conspicuous marginal spine on the valves.

Synonym:
Rhizosolenia delicatula Cleve, 1900.

Literature: Hasle 1975, Hasle & Syvertsen 1997.

Guinardia striata (Stolterfoth) G. R. Hasle, 1997 Family: Rhizosoleniaceae

Season				Trophic mode	Shape	Harmful	Bloom	Resting stage
W	S	S	A	autotrophic	cylinder	no	✿	no

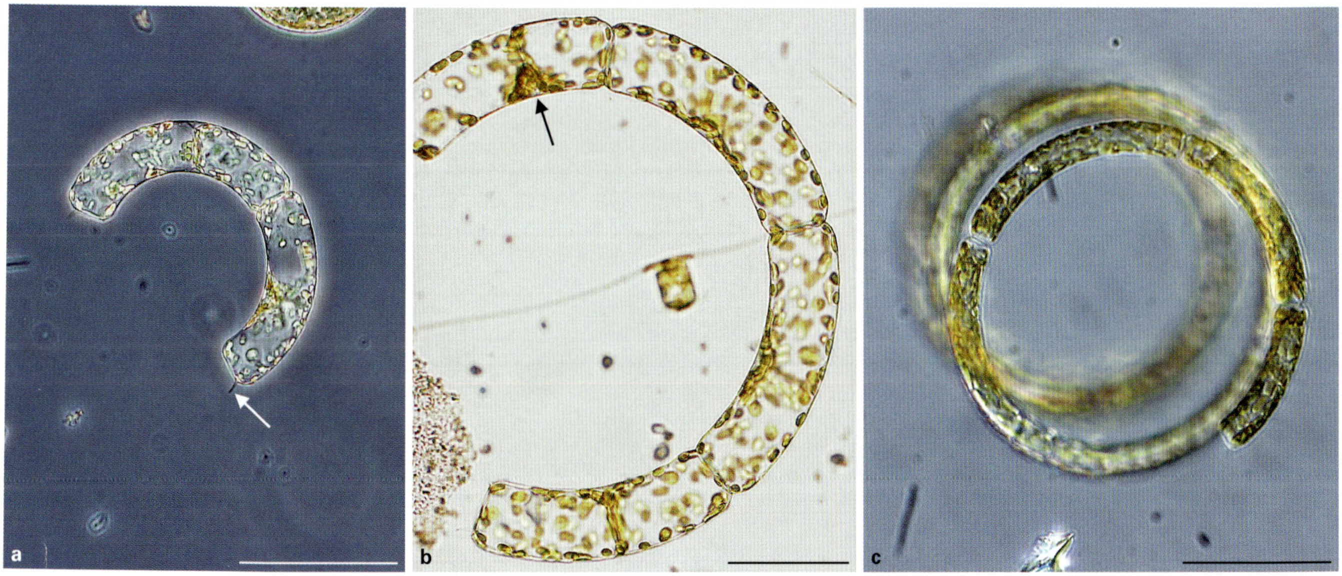

a, Phase contrast image of two live cells of *Guinardia striata* (arrow pointing to the prominent spine). **b,** Bright field image of a live chain. **c,** Phase contrast image of a long, curved chain. a, b, Scale bars = 50 µm; c, Scale bar = 100 µm.

Description

Cells large with a curved pervalvar axis resulting in close-set spiralling chains. Valve face flat but slightly rounded at the margins. A stout marginal spine (a, arrow) arising from the valve margin, fitting into a depression on the adjacent valve. Cells containing numerous small elliptic chloroplasts. Nucleus situated centrally, close to the cell wall (b, arrow).

Size

Diameter: 10-50 µm
Length (pervalvar axis): 30-300 µm

Distribution. *Guinardia striata* is found in coastal waters mainly in summer and autumn, sometimes becoming moderately abundant.

Similar species: none.

Synonyms:
Eucampia striata Stolterfoth, 1879,
Rhizosolenia stolterfothii H. Peragallo, 1888.

Literature: Hasle 1975, Hasle & Syvertsen 1997.

Guinardia flaccida (Castracane) H. Peragallo, 1892 Family: Rhizosoleniaceae

Season				Trophic mode	Shape	Harmful	Bloom	Resting stage
W	S	S	A	autotrophic	cylinder	no		no

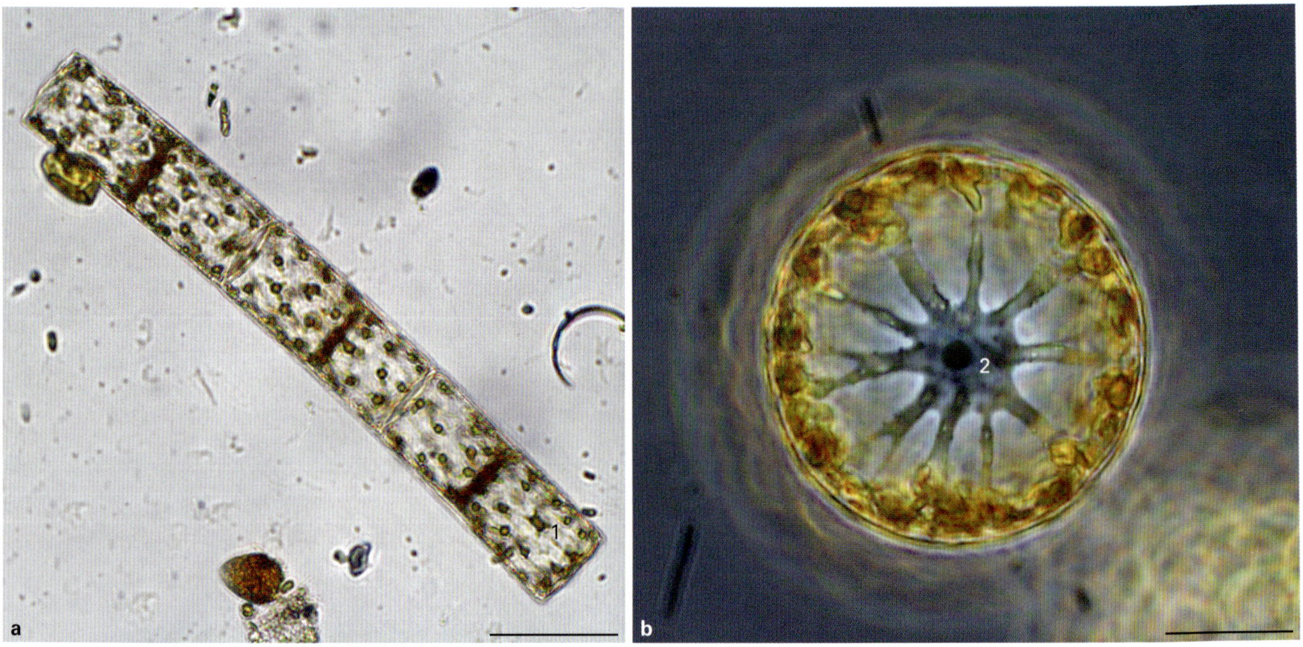

a, Bright field image of *Guinardia flaccida* in girdle view. Scale bar = 50 µm.
b, Phase contrast image of *G. flaccida* in valve view. Scale bar = 20 µm.

Description

Cells cylindrical and forming straight, usually short, chains. Diameter larger than in *Guinardia delicatula*. Valve face appears undulating in girdle view. A rudimentary, inconspicuous process (probably with an internal labiate structure) located on the valve face. Distinct girdle bands visible in light microscopy. Cells containing numerous large, star shaped chloroplasts (1), also visible in Lugol-fixed samples. Nucleus located centrally. In valve view several plasma strands radiating from this central area (2).

Size

Diameter: 25-95 µm
Length (pervalvar axis): 45-160 µm

Distribution. The species is found in coastal waters throughout the year, becoming most abundant in late spring and early summer.

Similar species: This species is quite distinct.

Synonyms:
Rhizosolenia flaccida Castracane, 1886,
Rhizosolenia castracanei Cleve, 1889,
Henseniella baltica Schütt, 1894,
Guinardia baltica Schütt, 1896.

Literature: Hendey 1974.

Proboscia alata (Brightwell) Sundström, 1986 — Family: Rhizosoleniaceae

Season				Trophic mode	Shape	Harmful	Bloom	Resting stage
W	S	S	A	autotrophic	cylinder	no	no	yes

Phase contrast image of part of a single live cell of *Proboscia alata* showing the blunt valve extension (arrow). Scale bar = 100 µm.

Description

Cells long cylindrical and often slightly curved with circular cross-section. Seen in lateral view, valves tapering into a proboscis with a blunt tip. Adjacent cells sometimes united into short chains, with the proboscis on one valve coming to rest in a depression on the valve of the adjacent cell. Girdle bands in two columns with pores and areolae. Chloroplasts numerous and small.

Size

Diameter: 3-15 µm
Length (pervalvar axis): up to 700 µm

Distribution. *Proboscia alata* has a world-wide distribution.

Similar species: The species could be confused with *Rhizosolenia hebetata* which, however, has a distinctly blunt process.

Synonyms:
Rhizosolenia alata Brightwell,1858,
Rhizosolenia alata f. *gracillima* (Cleve) Gran, 1905,
Rhizosolenia alata f. *genuina* Gran, 1908.

Literature: Jordan et al. 1991, Sundström 1986, Sunesen & Sar 2007, Van De Meene & Pickett-Heaps 2002.

Rhizosolenia imbricata Brightwell, 1858 — Family: Rhizosoleniaceae

Season				Trophic mode	Shape	Harmful	Bloom	Resting stage
W	S	S	A	autotrophic	cylinder	no		no

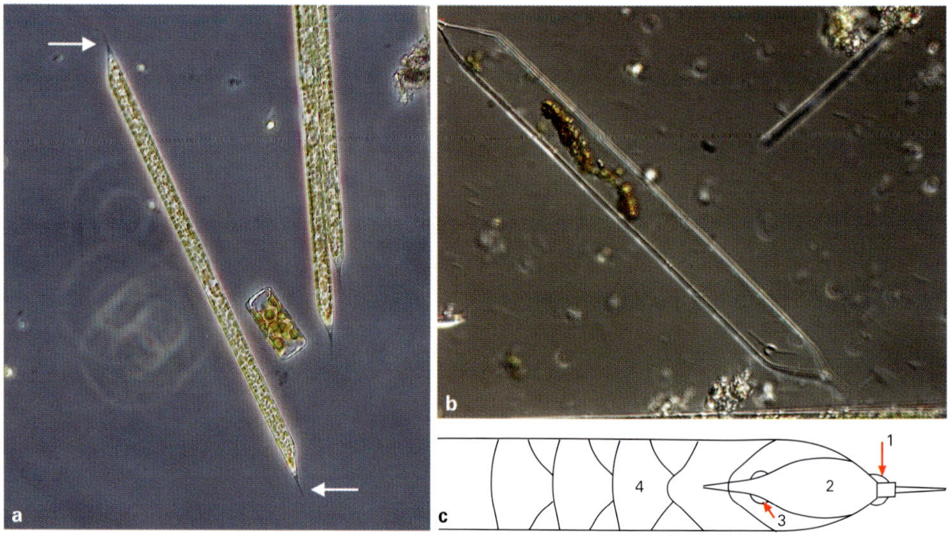

a, Phase contrast image of a single live cells of *Rhizosolenia imbricata*, here together with *Guinardia delicatula*. Scale bar = 100 µm. **b,** Empty valve showing the contiguous area and clasper. **c,** General schematic of *Rhizosolenia* indicating major morphological features: 1, Otarium; 2, Contiguous area; 3, Clasper; 4, Girdle segment.

Description

Cells long and cylindrical, with an extremely long pervalvar axis and a slightly elliptical cross-section and pointed wedge-shaped valves. Cells either solitary or linked into chains. In lateral view valve ends obliquely conical with an angle of about 45° ending with a short pointed process (see arrow). Process swollen at the base then narrowing abruptly. Otaria extending along swollen part of process. Cell wall clearly structured. Girdle segments arranged in two dorso-ventral columns and recognizable in light microscopy in empty frustules. Cells containing numerous small rounded chloroplasts. Nucleus situated centrally in the cell usually marked by a accumulation of chloroplasts.

Note

Cells with a dorso-ventral symmetry due to the pointed wedge shaped outline and morphology of the processes. The ventral side is defined as that having the valve depression holding the process of an adjacent valve after chain formation (see below). Two small wing-like membranous costae (called otaria) occur opposite each other at this swollen basal part. When a chain is formed the process comes to lie in the contiguous area, essentially an imprint of the process structure on the adjacent cell. The position where the otaria of the adjacent cell are lying, two membranous claspers can be found which link cells together.

Size

Large diameter (apical axis): 5-40 µm
Length (pervalvar axis): up to 650 µm

Distribution. The species is widely distributed and appears from spring to autumn especially in coastal areas. *Rhizosolenia imbricata* is most common in warmer water during summer time when blooms of the species might occur.

Similar species: In shape *Rhizosolenia imbricata* is similar to *Rhizosolenia styliformis*. But the latter species is normally longer and can be distinguished clearly by the pattern of the girdle segments (see *R. styliformis*) and the shape of the contiguous area (usually only visible in empty valves).

Synonyms:
Rhizosolenia shrubsolei Cleve, 1881,
Rhizosolenia imbricata var. *striata* (Greville) Grunow, 1882,
Rhizosolenia imbricata var. *shrubsolei* (Cleve) Schröder, 1906.

Literature: Sundström 1986, Sunesen & Sar 2007.

Rhizosolenia styliformis Brightwell, 1858

Family: Rhizosoleniaceae

Season				Trophic mode	Shape	Harmful	Bloom	Resting stage
W	S	S	A	autotrophic	cylinder	no		yes

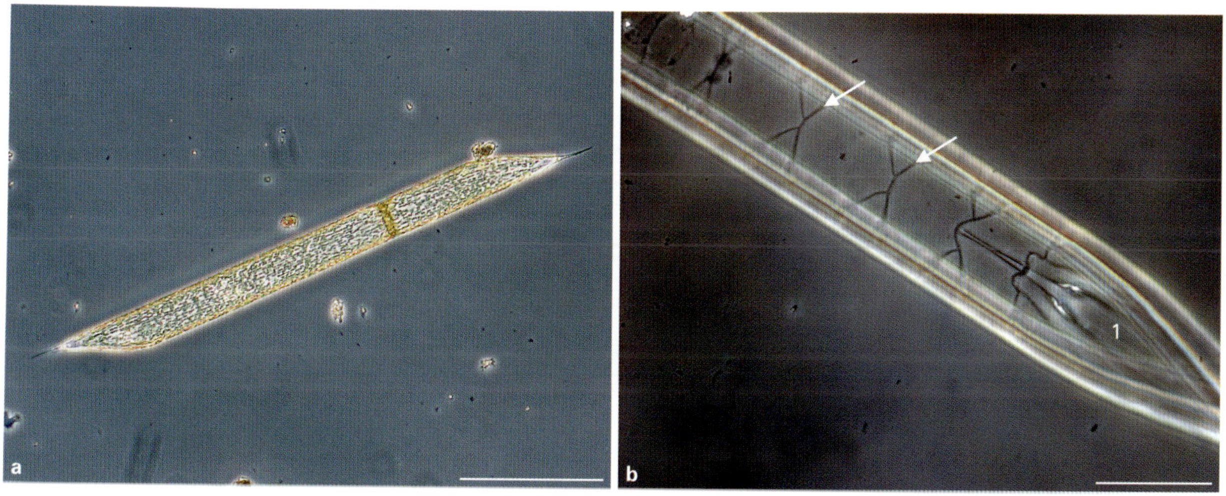

a, Phase contrast image of a single live cell of *Rhizosolenia styliformis*. Scale bar = 200 μm.
b, Phase contrast image of part of an empty cell with girdle band pattern (arrows) and contiguous area (1). Scale bar = 50 μm.

Description

Cells long cylindrical with a circular cross-section, solitary or linked into chains. In lateral view valves obliquely conical with an angle of about 45° ending with a short pointed process. In ventral view valves gradually narrowing towards the process. Otaria sometimes ending below the process base and extending a short distance along its basal part. Corresponding imprint of the sister cell (contiguous area) and arrangement of the girdle segments visible (1) in empty cells in light microscopy (see arrows). Cells with numerous small rounded chloroplasts. Nucleus situated more or less centrally in the cell and usually masked by an accumulation of chloroplasts.

Size

Diameter: 20-100 μm
Length (pervalvar axis): up to 1500 μm

Distribution. The species is found mainly in oceanic plankton from spring to autumn, sometimes forming blooms. In coastal plankton it is only found sporadically.

Similar species: The morphology of *Rhizosolenia styliformis* is very similar to *Rhizosolenia imbricata* but the former is normally much larger and can clearly be distinguished by the pattern of the girdle segments. The morphology of the process is also very different with the contiguous area in *R. styliformis* much narrower and more elongate than in *R. imbricata*.

Synonyms:
Rhizosolenia styliformis var. *polydactyla* van Heurck, 1909,
Rhizosolenia styliformis var. *longispina* Hustedt, 1914.

Literature: Hendey 1974, Sundström 1986.

Rhizosolenia hebetata* f. *semispina (Hensen) Gran, 1905 — Family: Rhizosoleniaceae

Season				Trophic mode	Shape	Harmful	Bloom	Resting stage
W	S	S	A	autotrophic	cylinder	no		no

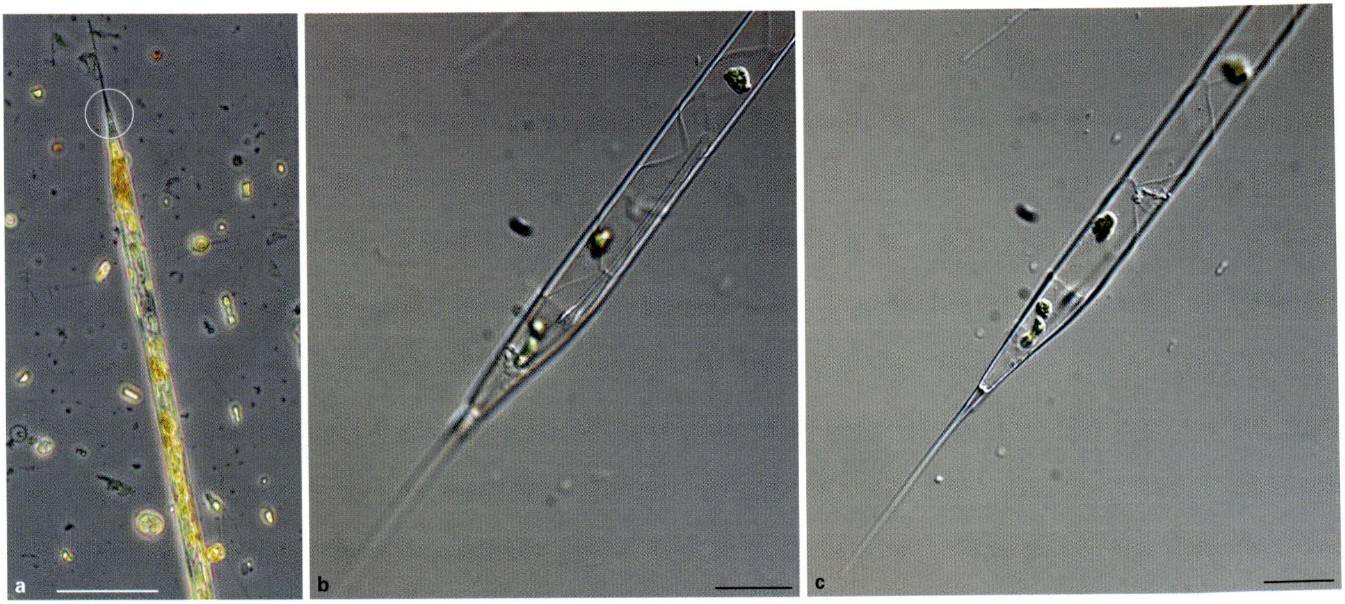

a, Phase contrast image of part of a single Lugol-fixed cell of *Rhizosolenia hebetata* f. *semispina* with the pointed valve process (see circle). Scale bar = 50 µm.
b, c, DIC images of the valve end of a live cell of *R. hebetata* f. *semispina*:
b, Focus on the contiguous area; **c,** Focus on the process and otarium.

Description

Cells long cylindrical with a circular cross-section and pointed wedge-shaped valves. In lateral view valves obliquely conical with an angle of about 20-25°, gradually narrowing towards the process. Otaria pointed and extending a few micrometres along the basal part of the process. The species has two morpho-types. In *Rhizosolenia hebetata* J. W. Bailey, 1856 valves pointed and terminating in a short blunt process. In *R. hebetata* f. *semispina* process longer and tapering towards a pointed tip (1). Moreover, dimorphic cells can be found showing one valve each of both types. Small chloroplasts numerous, elliptic and grainy.

Size

Diameter: 15-40 µm (*R. hebetata* f. *semispina* 5-15 µm)
Length (pervalvar axis): up to 750 µm

Distribution. This species is a cold water oceanic form and can be found during winter and spring sometimes dominating the phytoplankton community at low biomasses. The "*semispina*" form mainly appears in spring.

Similar species: This species (with blunt process) is similar in shape to *Proboscia alata*. In ventral view the "semispina" form resembles *Rhizosolenia setigera*. However, the ratio between length of the process and cell length is significantly smaller for *R. hebetata* f. *semispina*. In lateral view it could be confused with *Rhizosolenia imbricata*. But for this species the angle of the oblique valve is larger, and the process is shorter.

Synonyms:
Rhizosolenia semispina Hensen, 1887,
Rhizosolenia styliformis var. *semispina* (Hensen) G. Karsten, 1905,
and for the species:
Rhizosolenia hebetata f. *hiemalis* Gran, 1904.

Literature: Sundström 1986.

Rhizosolenia setigera Brightwell, 1858

Family: Rhizosoleniaceae

Season				Trophic mode	Shape	Harmful	Bloom	Resting stage
W	S	S	A	autotrophic	cylinder	no		yes

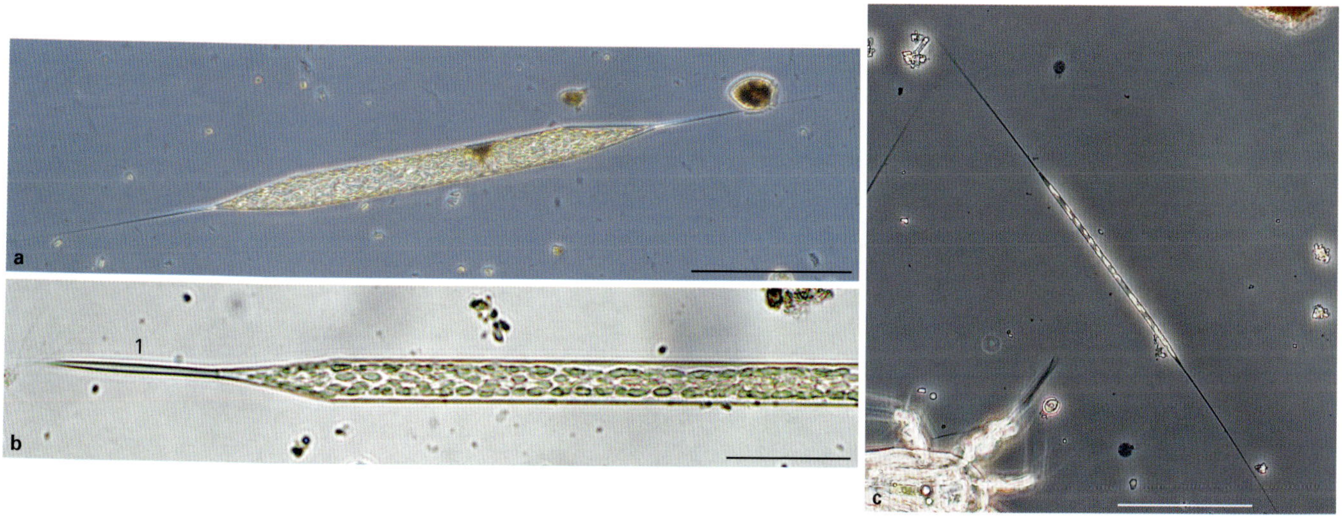

a, Phase contrast image of a single Lugol-fixed cell of *Rhizosolenia setigera*. Scale bar = 200 µm.
b, Bright field image of part of a Lugol-fixed cell of *Rhizosolenia setigera* f. *pungens* with the broadened central part of the process (1). Scale bar = 50 µm.
c, Phase contrast image of an extremely small live cell of *R. setigera*. Scale bar = 100 µm.

Description

Cells long cylindrical with a circular cross-section and conical valves (slightly curved), running into a long process, slightly broadened at the basis and gradually tapering towards the end. Otaria lacking. Mainly found as single cells, rarely forming short chains. Cells containing numerous small elliptic chloroplasts. Nucleus situated centrally close to the cell wall.

Note

Formerly considered as a separate species *Rhizosolonia pun gens* Cleve-Euler, 1937 is now regarded as *Rhizosolenia setigera* f. *pungens* (Cleve-Euler) Brunel, 1962. The processes of this form are clearly broadened in their central part (1).

Size

Diameter: 5-50 µm
Length (pervalvar axis): up to 600 µm

Distribution. The species is eurytherm and euryhaline and for that reason can be found in coastal plankton throughout the year. The main distribution is in cooler water during spring.

Similar species: none.

Synonyms (for the species):
Rhizosolenia japonica Castracane, 1886,
Rhizosolenia hensenii Schütt, 1900.

Literature: Sundström 1986, Sunesen & Sar 2007, Van de Meene & Picket Heaps 2004.

Neocalyptrella robusta (G. Norman ex Ralfs) D. U. Hernández-Becerril & M. E. Meave del Castillo, 1997

Family: Rhizosoleniaceae

Season				Trophic mode	Shape	Harmful	Bloom	Resting stage
W	S	S	A	autotrophic	elliptic cylinder	no	no	no

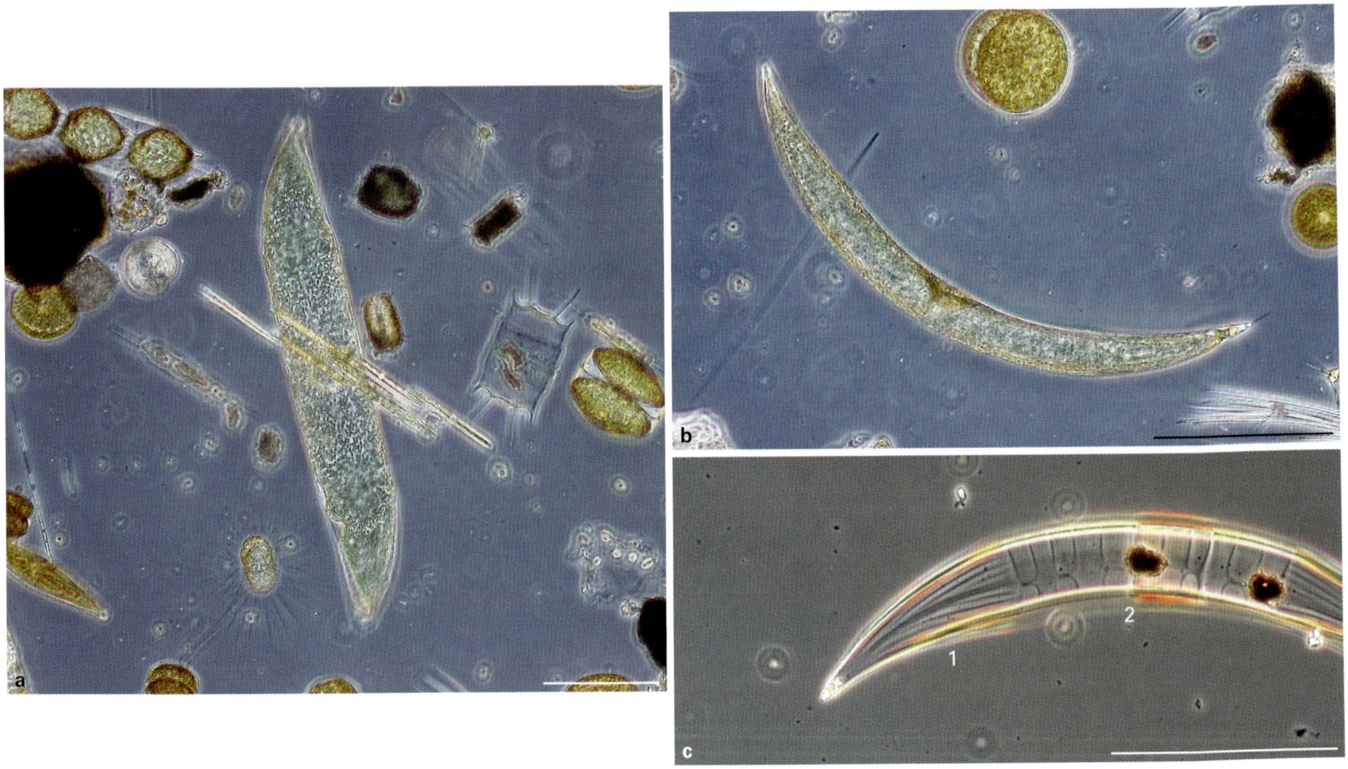

a, b, Phase contrast images of live cells of *Neocalyptrella robusta* in broad girdle view.
b, Live cell in narrow girdle view. Scale bars = 100 µm.
c, Phase contrast of empty frustule with longitudinal lines (1) and girdle band patterns (2) on valves. Scale bar = 200 µm.

Description

Cells usually solitary or occurring in pairs. Cells large and sickle- or s-shaped in lateral view with an elliptic cross-section. Valves curved conical and convex, with longitudinal lines and ending with a small process. Girdle segments arranged in latitudinal patterns. Chloroplasts numerous small and with grainy appearance often arranged in rows. Nucleus situated near the cell wall.

Size

Large diameter (apical axis): 50-400 µm
Small diameter (transapical axis): 25-220 µm
Length (pervalvar axis): up to 1200 µm

Distribution. This species is widespread in warm oceans. In European seas north of the English Channel the species can only be found occasionally to date.

Similar species: *Neocalyptrella robusta* is a very distinct species.

Synonyms:
Rhizosolenia robusta G. Norman ex Ralfs, 1861,
Calyptrella robusta (G. Norman ex Ralfs) D. U. Hernández-Becerril & M. E. Meave del Castillo, 1996.

Literature: Hernández-Becerril & Meave Del Castillo 1997, Sunesen & Sar 2007.

Leptocylindrus danicus Cleve, 1889

Family: Leptocylindraceae

Season				Trophic mode	Shape	Harmful	Bloom	Resting stage
W	S	S	A	autotrophic	cylinder	no	✿✿✿	yes

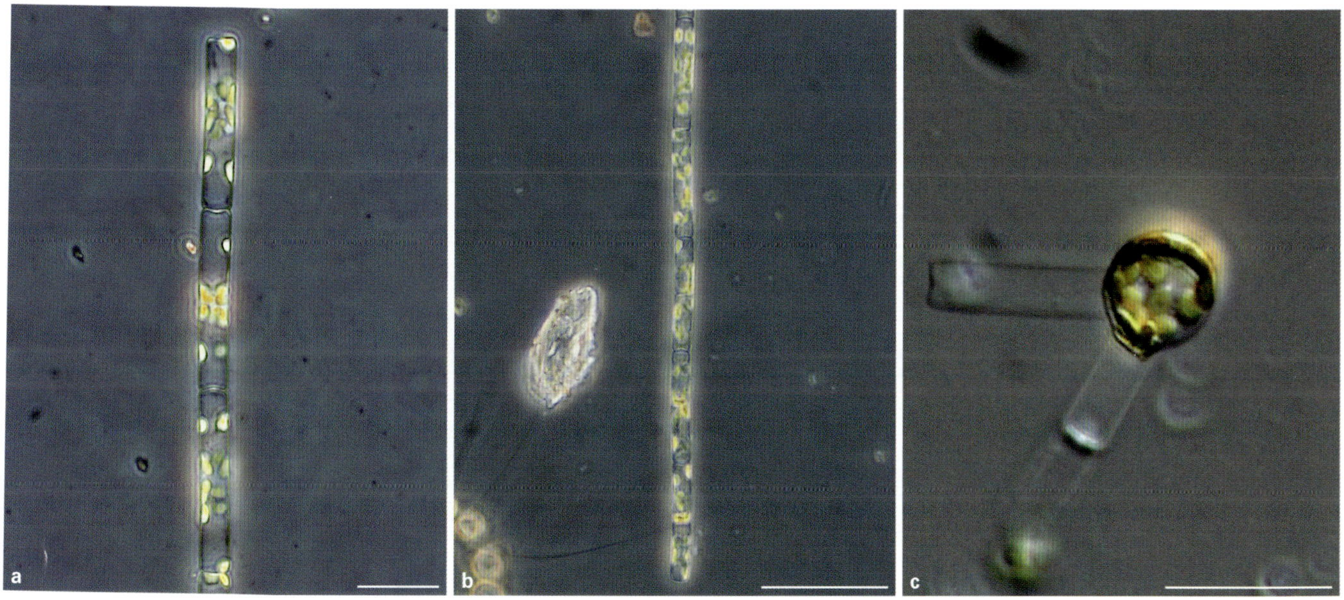

a, Part of a live chain of *Leptocylindrus danicus*. Scale bar = 20 µm.
b, Live chain at lower magnification. Scale bar = 50 µm.
Scale bar = 50 µm. Both images taken with phase contrast optics.
c, *L. danicus* auxospore. Scale bar = 20 µm.

Description

Cells long and cylindrical with a circular cross-section, connected into long close-set and straight chains. Valve face almost flat, slightly concave or convex. Weak patterns on the girdle segment not visible in the light microscope. Some pale elliptically lobed chloroplasts distributed throughout the cells.

Size

Diameter: 5-15 µm
Length (pervalvar axis): 20-50 µm

Distribution. *Leptocylindrus danicus* can be found in coastal waters between spring and autumn, often becoming very abundant in late spring.

Similar species: *Guinardia delicatula* has a similar shape and size. But its chloroplasts are larger and fewer in number. Cells have a small process on their valve margin which can normally be seen in the light microscope on the terminal cells in a chain.

Synonyms: none.

Literature: Davis et al. 1980; French III & Hargraves 1985, 1986.

Leptocylindrus minimus Gran, 1915 Family: Leptocylindraceae

Season				Trophic mode	Shape	Harmful	Bloom	Resting stage
W	S	S	A	autotrophic	cylinder	no	✿✿✿	yes

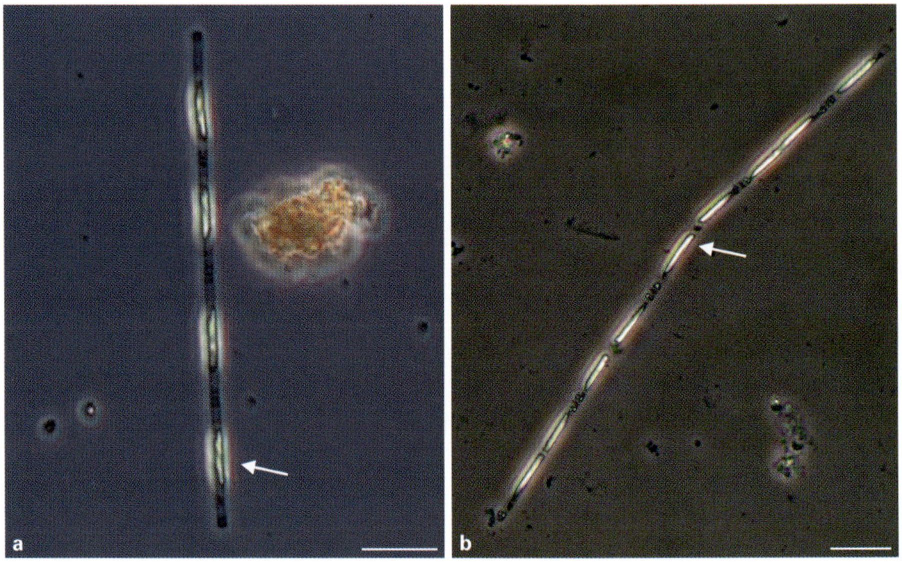

a, b, Live chain of *Leptocylindrus minimus*. Scale bars = 20 µm.

Description

Cells long and cylindrical with circular cross-section and united into long, straight or slightly undulating chains. Valve face flat. No spines or other processes are apparent in light microscopy. Cells with two characteristic, long and flattened chloroplasts on either side of the valve centre (along the pervalvar axis, see arrow).

Size

Diameter: 1.5-5 µm
Length (pervalvar axis): 15-50 µm

Distribution. The species can be found in coastal waters between spring and autumn but it is usually most abundant in late spring and early summer.

Similar species: Because of the small cell size and the two typical chloroplasts the species cannot be confused with others.

Synonym:
Leptocylindrus belgicus Meunier, 1915.

Literature: Hargraves 1990.

Brockmanniella brockmannii (Hustedt) Hasle, von Stosch, & Syvertsen, 1983

Family: Cymatosiraceae

Season				Trophic mode	Shape	Harmful	Bloom	Resting stage
W	S	S	A	autotrophic	elliptic cylinder	no	✿	no

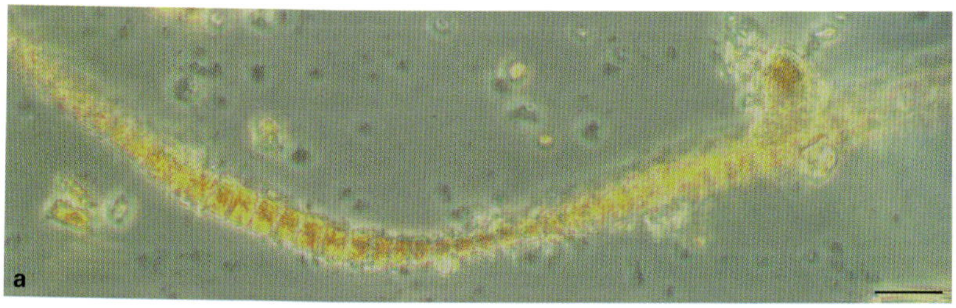

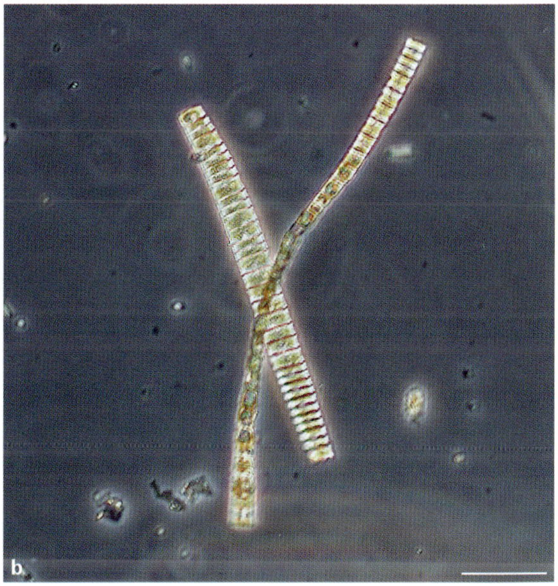

a, Lugol-fixed twisted chain of *Brockmanniella brockmannii*.
Scale bar = 20 µm.
b, Two live, twisted chains. Scale bar = 50 µm.
Both images taken with phase contrast optics.

Description

Cells narrow, rectangular in girdle view and elliptic in valve view forming long, partly twisted chains. Valve faces slightly undulating. Neighbouring cells abutting with their central part and with the valve poles, forming small apertures in between. Cells linked by small marginal spines. Marginal labiate process located on one of the two valves. An ocellus situated on the elevated poles. One chloroplast per cell located in the girdle area.

Size

Large diameter (apical axis): 4-36 µm
Small diameter (transapical axis): 3-10 µm
Height (pervalvar axis): 3-12 µm

Distribution. The species is typical of coastal sediments all over Northern Europe but is regularly found in the plankton. The chains are often covered by detritus and, therefore, can be difficult to see.

Similar species: *Cymatosira belgica* Grunow, 1881has a similar size and shape and can also form long twisted ribbons. But the neighbouring cells in this species are only connected by the central part of the valves and not with the poles.

Synonym:
Plagiogramma brockmannii Hustedt, 1939.

Literature: Hasle et al. 1983.

Plagiogrammopsis vanheurckii (Grunow) Hasle, von Stosch & Syvertsen, 1983

Family: Cymatosiraceae

Season				Trophic mode	Shape	Harmful	Bloom	Resting stage
W	S	S	A	autotrophic	rhomboid prism	no	no	no

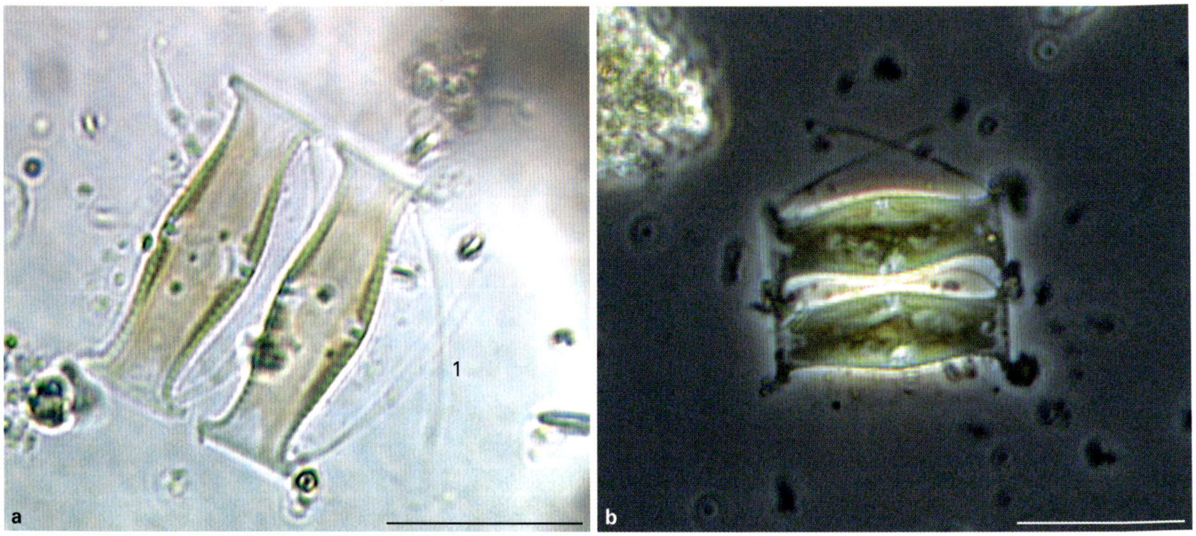

a, Bright field image of live cells of *Plagiogrammopsis vanheurckii* with the pili clearly discernable (1). Image courtesy of Hanne Halliger; **b,** DIC image of a live chain of *P. vanheurckii*. Scale bars = 20 µm.

Description

Cells strongly curved in girdle view, convex in the centre and constricted on both sides near the elevated poles causing the formation of large elliptic apertures between adjacent cells. Cells rhombic with elongated ends in valve view. Cells forming partially twisted chains, linked by marginal spines. A submarginal labiate process found on one of the two valves. An ocellus situated on the elevated poles. Long processes (termed pili) arising from the raised poles, pointing from their point of attachment towards the opposite valve pole (1) and reaching about two thirds along the long axis of the valve. One chloroplast situated in the girdle area.

Size

Large diameter (apical axis): 8-50 µm
Small diameter (transapical axis): 3-12 µm
Height (pervalvar axis): 3-12 µm

Distribution. *Plagiogrammopsis vanheurckii* is a littoral species but is also regularly found in the plankton.

Similar species: Because of the typical large elliptic apertures and the pilus the species is relatively easy to identify in plankton samples. But it is small and can be overlooked, particularly in sediment rich samples.

Synonym:
Plagiogramma vanheurckii Grunow, 1881.

Literature: Hasle et al. 1983, Nakata 1987.

Cerataulina pelagica (Cleve) Hendey, 1937

Family: Hemiaulaceae

Season				Trophic mode	Shape	Harmful	Bloom	Resting stage
W	S	S	A	autotrophic	cylinder	no	✿✿✿	no

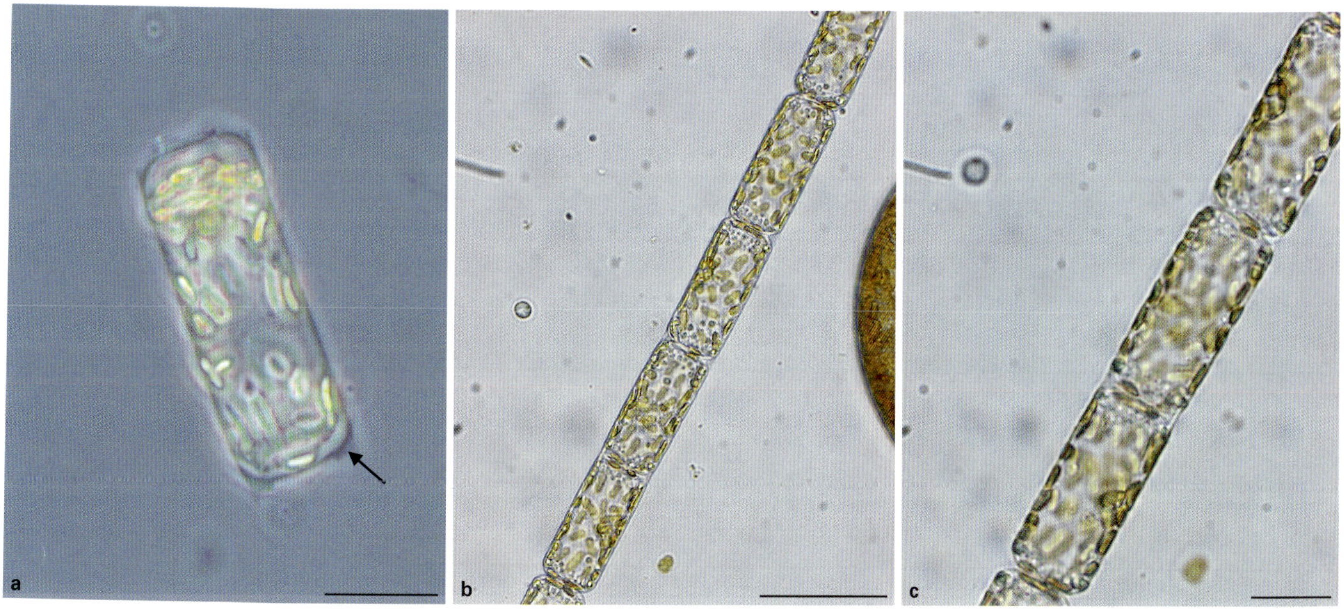

a, Phase contrast image of a single Lugol-fixed cell of *Cerataulina pelagica* showing the wing-like processes (arrow).
b, Bright field image of the same chain at higher magnification. a, b. Scale bars = 20 µm.
c, Bright field image of a live chain, in which the wing-like processes between adjacent cells are not seen. Scale bar = 50 µm.

Description

Cells long cylindrical with a circular cross-section and slightly twisted about the pervalvar axis, forming straight chains. Valve surface slightly convex with rounded edges and two short pervalvarly directed, winged processes on opposite valve margins (arrow) each with a fine spine. Wing-like structures easily seen using light microscopy. A ribbed and perforated plate (costate ocellus) located at the base of the process (not seen with light microscopy). Adjacent cells connected via these processes. Cells appearing heterovalvate due to the twisting about the pervalvar axis causing the processes to be in different positions on the two valves. Chloroplasts numerous and small. Nucleus situated close to the cell wall.

Size

Diameter: 10-60 µm
Length (pervalvar axis): 25-120 µm

Distribution. *Cerataulina pelagica* has a world-wide distribution in coastal areas with its distributional centre in temperate to warm waters. It can build blooms especially in spring and early summer.

Similar species: none.

Synonyms:
Cerataulus bergonii H. Peragallo, 1892,
Cerataulina bergonii (H. Peragallo) Schütt, 1896.

Literature: Hasle & Syvertsen 1980.

Eucampia zodiacus Ehrenberg, 1840

Family: Hemiaulaceae

Season				Trophic mode	Shape	Harmful	Bloom	Resting stage
W	S	S	A	autotrophic	elliptic cylinder	no		no

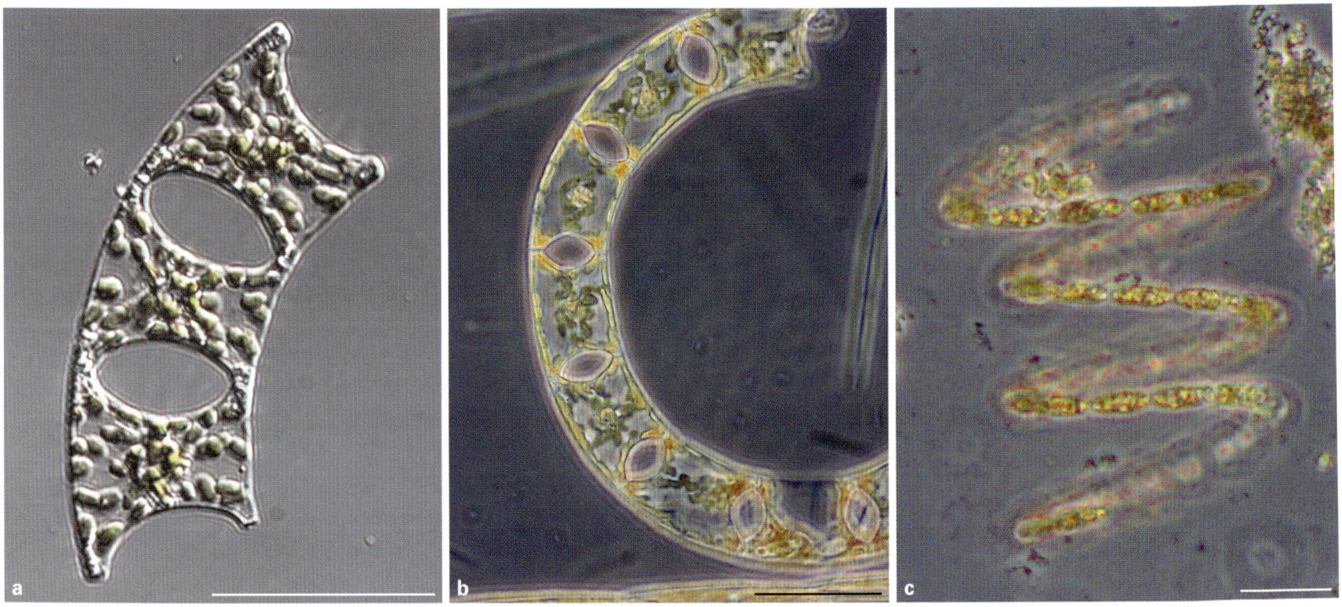

a, b, Phase contrast image of short live chains of *Eucampia zodiacus* in girdle view.
c, Phase contrast image of spirally twisted Lugol-fixed chain. Cells seen in narrow girdle view.
Scale bars = 50 µm.

Description

Trapezoid outline with a concave valve face in broad girdle view. Cells joined into helical chains by costate ocelli positioned at the tip of the horns arising from the valve margin. Aperture between adjacent cells very variable in width with an angular, elliptical to square shape. Single labiate process located in a depression in the valve centre. Chloroplasts numerous, small and elliptical.

Size

Large diameter (apical axis): 10-100 µm
Height (pervalvar axis): 8-60 µm

Distribution. *Eucampia zodiacus* has a world-wide distribution in coastal areas with its distributional centre in temperate to warm waters. It normally appears during spring and summer.

Similar species: The species can not be confused with species from other genera. Other *Eucampia* species from cold or warm water regions differ in cell shapes.

Synonyms:
Eucampia britannica W. Smith, 1853,
Eucampia nodosa A. Schmidt, 1888.

Literature: Syvertsen & Hasle 1983.

Ditylum brightwellii (T. West) Grunow in Van Heurck, 1885 Family: Lithodesmiaceae

Season				Trophic mode	Shape	Harmful	Bloom	Resting stage
W	S	S	A	autotrophic	triangular prism	no	✿✿	yes

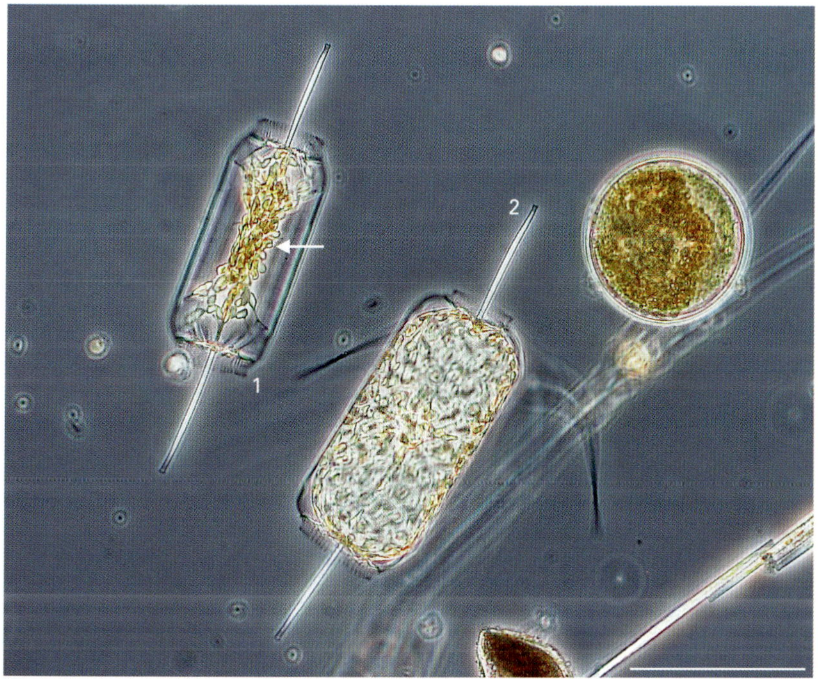

Phase contrast image of two live single cells of *Ditylum brightwellii*, the left one showing plasmolysis (arrow). Scale bar = 100 µm.

Description

A solitary species (sometimes joined into very short chains). Cells long rectangular in girdle view with a triangular cross-section. A marginal ridge consisting of small processes, called ansulae located at the valve margin (1). A thick central bilabiate process located in the centre of the flat or slightly pyramidal valves (2). Cells containing numerous small chloroplasts. Nucleus situated in the centre of the cell.
Cells very sensitive to plasmolysis (arrow). Resting spores formed at one end of the parent cell. Valves similar to those of the parent cell. Primary valve with long valve mantle and a central spine. Secondary valve without mantle.

Size

Diameter (edge length): 20-120 µm
Height (pervalvar axis): 40-300 µm

Distribution. This cosmopolitan species is usually most abundant in spring, but is sporadically found in coastal waters throughout the year.

Similar species: Because of the typical shape with the two thick spines in the centre of the valves this species is easily identified in samples from Northern Seas.

Synonyms:
Triceratium brightwellii T. West, 1860,
Ditylum trigonum L. W. Bailey ex L. W. Bailey, 1862,
Ditylum inaequale J. W. Bailey ex L. W. Bailey, 1862.

Literature: Hargraves 1982, Kaczmarska et al. 2006, Koester et al. 2007.

Helicotheca tamesis (Shrubsole) Ricard, 1987 — Family: Lithodesmiaceae

Season				Trophic mode	Shape	Harmful	Bloom	Resting stage
W	S	S	A	autotrophic	cuboid	no	no	no

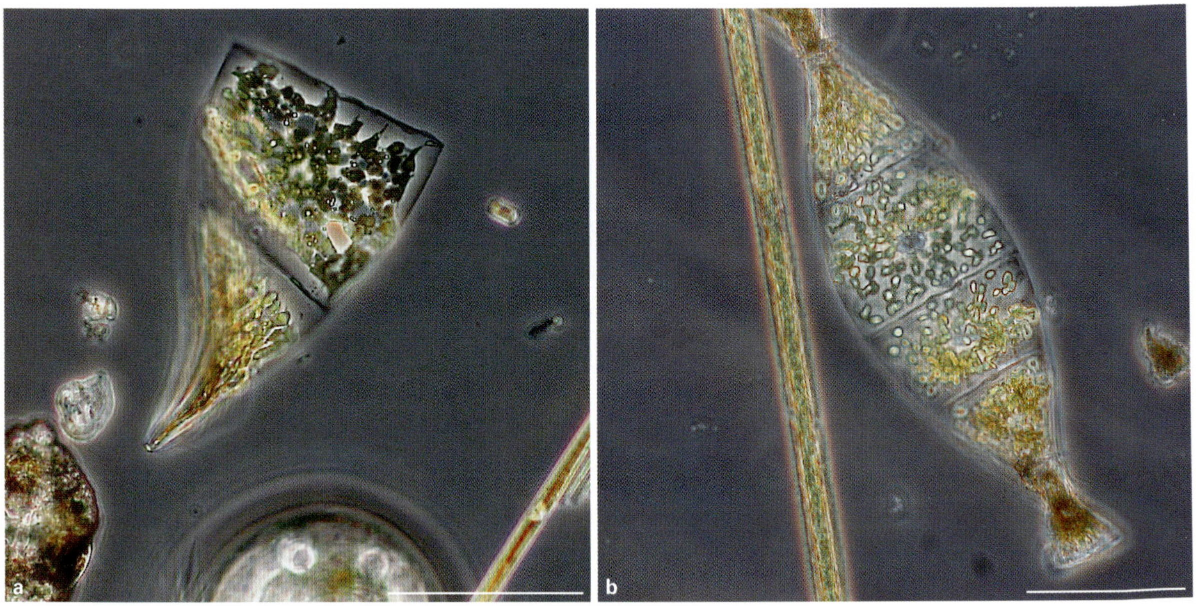

a, Two live cells of *Helicotheca tamesis*, one of them heavily twisted.
b, Live twisted chain. Both images taken with phase contrast optics.
Scale bars = 50 µm.

Description

Cells nearly rectangular in girdle view showing torsion around the pervalvar axis, resulting in spiral chains. In valve view cells small oblong with a nearly flat valve face so that chains are showing no gaps between the cells. Chloroplasts numerous, small, rounded and arranged radially from the central nucleus.

Size

Large diameter (apical axis): 20-160 µm
Small diameter (transapical axis): 5-12 µm
Height (pervalvar axis): 55-120 µm

Distribution. *Helicotheca tamesis* is distributed in coastal areas of temperate to warm waters. It is found sporadically throughout the year.

Similar species: The species can not be confused with other species.

Synonym:
Streptotheca tamesis Shrubsole, 1891.

Literature: Kaczmarska et al. 2006, von Stosch 1977.

Lithodesmium undulatum Ehrenberg, 1841 Family: Lithodesmiaceae

Season				Trophic mode	Shape	Harmful	Bloom	Resting stage
W	S	S	A	autotrophic	triangular prism	no	no	no

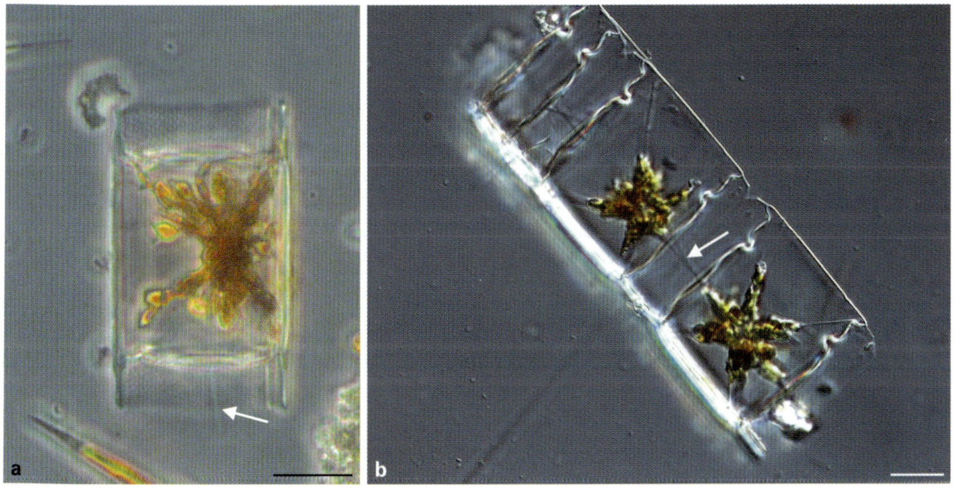

a, Phase contrast image of a Lugol-fixed single cell of *Lithodesmium undulatum* with the membranes at the valve margin and the central process with the external tube.
b, DIC image of a short live chain.
Scale bars = 20 µm.

Description

Cells short and rectangular in girdle view with a triangular cross-section. Pervalvarly directed marginal membranes (marginal ridge) on the valves connecting cells into straight chains with large gaps between the cells. Valves with a central bilabiate process with a long external tube. Chloroplasts numerous.

Size

Diameter (edge length): 40-90 µm
Height (pervalvar axis): 20-75 µm

Distribution. *Lithodesmium undulatum* is a neritic species in warm to temperate water regions, probably partly benthic.

Similar species: none.

Synonyms:
Ditylum intricatum (T. West) Grunow in Van Heurck, 1882-1885
Triceratium undulatum Brightwell, 1858,
Triceratium intricatum T. West, 1860,
Lithodesmium victoriae Karsten, 1906,
Ditylum undulatum Mann, 1907.

Literature: Kaczmarska et al. 2006, Manton & von Stosch 1966, von Stosch 1980.

Bellerochea malleus (Brightwell) Van Heurck, 1880 Family: Bellerocheaceae

Season				Trophic mode	Shape	Harmful	Bloom	Resting stage
W	S	S	A	autotrophic	triangular prism	no	no	no

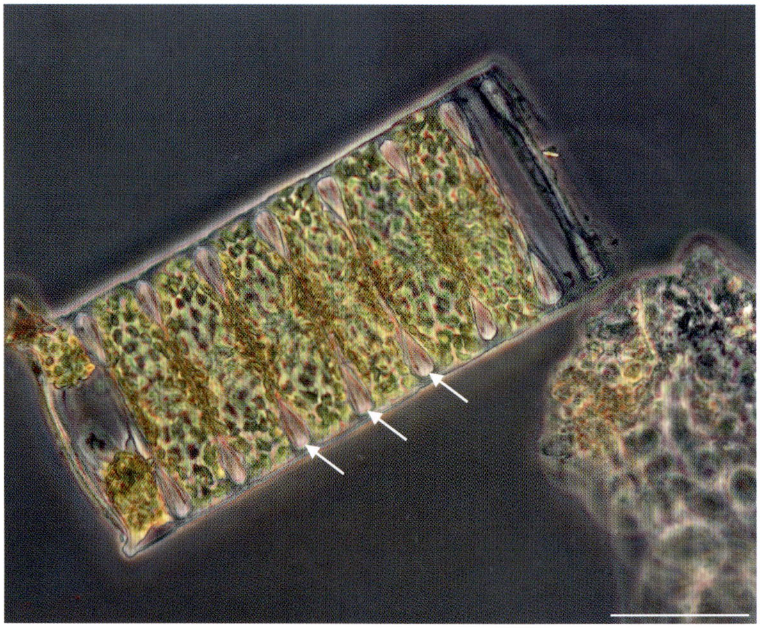

Live chain of *Bellerochea malleus* showing the teardrop shaped apertures between adjacent cells. Scale bar = 50 µm.

Description

Cells broadly rectangular in girdle view and triangular with concave sides in valve view, elliptical to rhomboid (f. *biangulata*) or quadrangular (f. *tetragona*) outlines also possible. Valves with elevated poles and central area connecting adjacent cells into straight chains, with small, more or less elliptical apertures on both sides when seen in girdle view. Chloroplasts small, numerous.

Size

Edge length: 50-180 µm
Height (pervalvar axis): 13-45 µm

Distribution. The species can be found in coastal areas of temperate waters.

Similar species: none.

Synonym:
Triceratium malleus Brightwell, 1858.

Literature: Kaczmarska et al. 2006, von Stosch 1977.

Mediopyxis helysia Kühn, Hargreaves & Halliger, 2006 incertae sedis

Season				Trophic mode	Shape	Harmful	Bloom	Resting stage
W	S	S	A	autotrophic	elliptic cylinder	no	✿✿	unknown

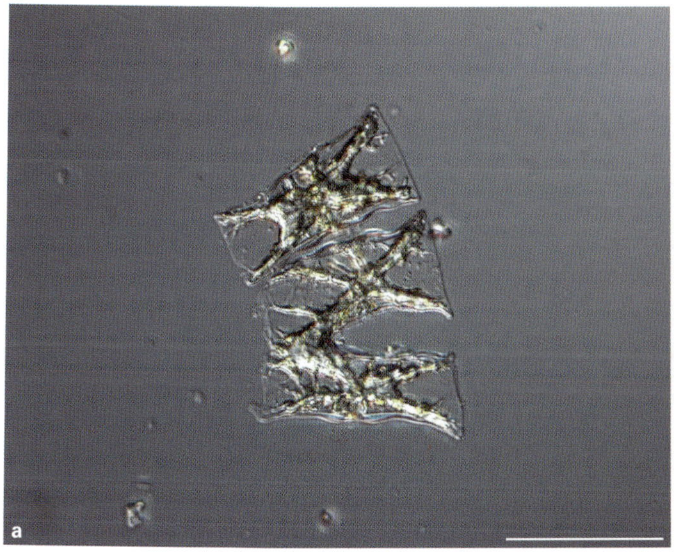

a, *Mediopyxis helysia*, live pair of cells. **b,** Short chain of 6 cells showing the differing valve ends in a heteropolar cells: 1, Short rounded end; 2, Elongated horn. Scale bars = 100 µm (a) and 50 µm (b).

Description

Cells solitary or in short chains of 2-6 cells. Valves weakly silicified and morphologically variable. Most cells heteropolar, with short round elevation at one end of the valve and a pointed elongated horn at the other, but isopolar cells with two rounded apices also found.

Broad girdle view: Valve centre raised, but with a central depression in the raised area. Adjacent cells abutting with valve centre and raised apices, producing two irregularly shaped apertures between adjacent cells. Numerous girdle bands present (only visible with electron microscopy).

Size

Large diameter (apical axis): 85-125 µm
Small diameter (transapical axis): 18-22 µm (cell centre), 2-4.5 µm (apices)
Height (pervalvar axis): 27-78 µm

Distribution. Few records are available, but *Mediopyxis helysia* appears to be a cold-temperate species. It has been found in the Gulf of Maine and near the islands of Sylt, Helgoland and Spiekeroog in the North Sea. The first record for this species in Helgoland was in 2003. In 2009 it was found in Helgoland from late spring and made its first appearance in 2010 in early February.

Mediopyxis helysia appears to become more abundant in several locations in the southern North Sea.

Similar species: *Helicotheca tamesis* (Shrubsole) Ricard, 1987 also forms ribbon-like chains with rectangular to square cells in broad girdle view. However, in *Helicotheca tamesis* there are no gaps between adjacent cells.

Synonyms: none.

Literature: Kühn et al. 2006.

Biddulphia alternans (J. W. Bailey) Van Heurck, 1885 — Family: Biddulphiaceae

Season				Trophic mode	Shape	Harmful	Bloom	Resting stage
W	S	S	A	autotrophic	triangular prism	no	no	no

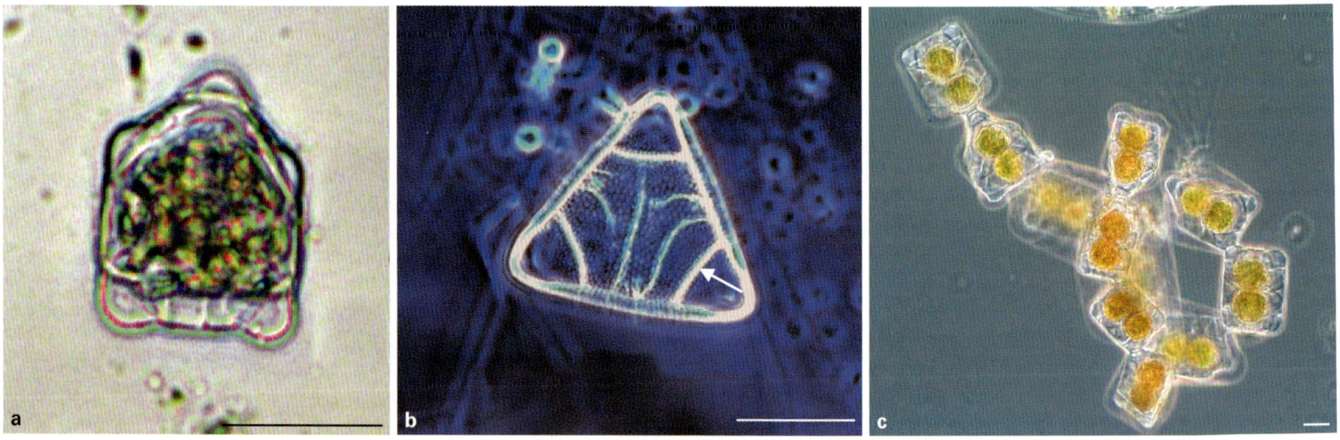

a, Bright field image of a Lugol-fixed single cell of *Biddulphia alternans* in girdle view with a partial view on the valve face. Ribs at the rounded corners are easily seen.
b, Valve face of empty valve from a permanent slide (arrow points to costae). The material was mounted in Naphrax.
c, Phase contrast image of a formalin-fixed zigzag chain (picture courtesy of Regina Hansen).
Cells are connected by their valve poles.
Scale bars = 20 µm.

Description

Cells solitary or in irregular zig-zag chains with adjacent cells connected by mucous secretions from the pseudocelli. Cells triangular or rectangular in valve view with a tripartite curved line occurring in the centre of the valve face. Three straight thickened lines (costae, see arrow in image c) separating the three valve corners with the pseudocelli from the central region. Three labiate processes located on the valve face each with a well developed internal stalked structure. Chloroplasts coccoid, several per cell.

Size

Edge length: 20-60 µm
Height (pervalvar axis): 20-60 µm.

Distribution. This littoral form is regularly found in plankton in coastal areas. It is tychopelagic.

Similar species: The typical form of the chains and the cell shape make this species distinctive in plankton samples.

Synonym:
Triceratium alternans J. W. Bailey, 1851.

Literature: Hoban 1983.

Odontella sinensis (Greville) Grunow, 1884 — Family: Triceratiaceae

Season				Trophic mode	Shape	Harmful	Bloom	Resting stage
W	S	S	A	autotrophic	elliptic cylinder	no	✿✿✿	no

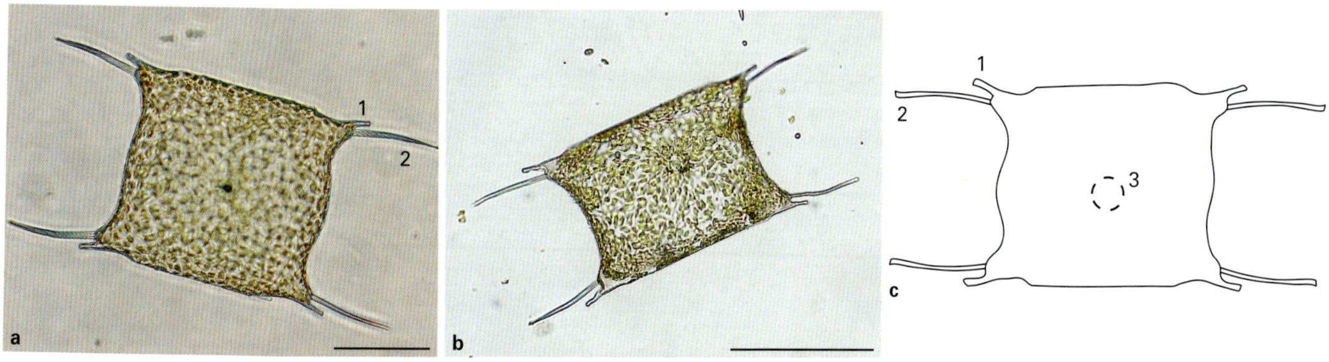

a, b, Two live single cells of *Odontella sinensis* of different cell proportions in bright field: almost square (**a**) vs elongate (**b**). Both are brightfield images. Scale bars 100 µm (a) and 200 µm (b). **c,** General schematic of an Odontella cell: 1, Horn; 2, Labiate process; 3, position of the nucleus.

Description

Cells solitary or connected by their processes into short chains. Cells cushion shaped in broad girdle view with a flat or concave valve face. Valve poles raised and drawn out to form straight to slightly curved horns (1). Two labiate processes per valve, their external tubes located between horn and valve centre (closer to the horn) exceeding the horns in length (2). Chloroplasts numerous, irregularly shaped.

Size

Large diameter (apical axis): 80-260 µm
Small diameter (transapical axis): 35-120 µm
Height (pervalvar axis): 80 400 µm

Distribution. Originally this species was native to the Pacific. At the beginning of the 20th century it was introduced probably into European waters by ship's ballast water. Today the species is cosmopolitan and can be found sporadically throughout the year but mainly during summer when it can form blooms.

Similar species: *Odontella sinensis* is easily to identify.

Synonym:
Biddulphia sinensis Greville, 1866.

Literature: Hendey 1964, Ross & Sims 1973, Round et al. 1990.

Odontella mobiliensis (J. W. Bailey) Grunow, 1884 Family: Triceratiaceae

Season				Trophic mode	Shape	Harmful	Bloom	Resting stage
W	S	S	A	autotrophic	elliptic cylinder	no		no

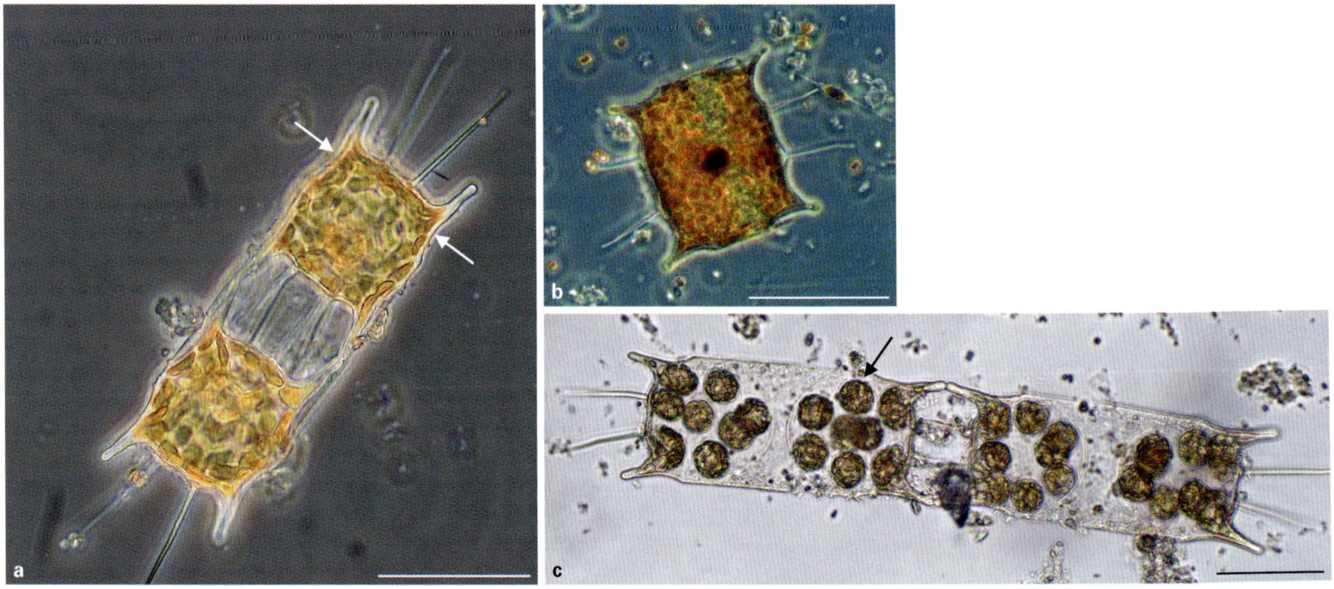

a, A pair of live cells of *Odontella mobiliensis*. **b,** A single Lugol-fixed cell. **c,** Two cells forming spermatogonia (arrow). a,c, Scale bars = 50 µm; b, scale bar = 100 µm. All images taken with phase contrast optics.

Description

Cells cushion shaped and roughly rectangular with a broad cingulum in girdle view. Cells elliptical to elliptical lanceolate in valve view. Valve poles drawn out into straight horns with blunt tips directed. Valve mantle distinctly constricted below these horns. Two external processes per valve arising from the labiate processes located closer to the centre of the valve face than in *Odontella sinensis* and *O. regia* (they divide the apical axis into three sections of approximately same width). The central valve face is slightly raised. Cells solitary or united into short chains via secretions from the ocelli at the distal ends of the horns. External tubes of labiate processes considerably longer than the horns. Chloroplasts numerous and small.

Size

Large diameter (apical axis): 40-200 µm
Small diameter (transapical axis): 18-90 µm
Height (pervalvar axis): 40-150 µm

Distribution. *Odontella mobiliensis* is a cosmopolitan species, which is found in plankton throughout the year with the highest abundance in late summer.

Similar species: *Odontella mobiliensis* closely resembles *O. regia*. In general *O. mobiliensis* is smaller, the outer processes are a bit longer and clearly directed outward and the inner spines are located more to the centre.

Synonyms:
Zygoceros mobiliensis J. W. Bailey, 1851,
Biddulphia baileyi W. Smith, 1856,
Biddulphia mobiliensis Grunow, 1882.

Literature: Hendey 1964, Ross & Sims 1973, Round et al. 1990.

Odontella regia (Schultze) Simonsen, 1974

Family: Triceratiaceae

Season				Trophic mode	Shape	Harmful	Bloom	Resting stage
W	S	S	A	autotrophic	elliptic cylinder	no		no

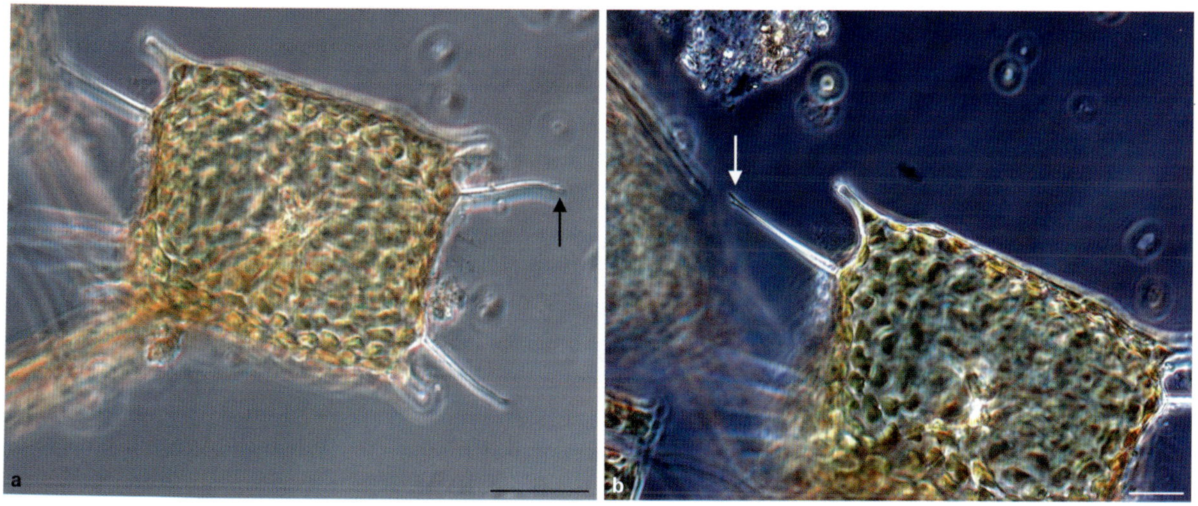

a, Bright field image of a single live cell of *Odontella regia*. Scale bar = 50 µm.
b, Phase contrast image showing the forked external tube of the labiate process. Scale bar = 20 µm.

Description

Valves rectangular in girdle view, with short horns at the valve poles. Ends of the horns rounded, terminating in an ocellus. Cells united into chains by mucus secretions from the ocelli. One labiate process situated just inside or slightly removed from each of the horns. External tubes of labiate processes initially diverging but distal ends curving inward again towards the centre of the valve. Tips of the processes split and sometimes hooked (arrow). Chloroplasts numerous and platelike. Large nucleus located in the centre of the cell.

Size

Large diameter (apical axis): 90-200 µm
Small diameter (transapical axis): 45-100 µm
Height (pervalvar axis): 100-250 µm

Distribution. *Odontella regia* is widely distributed in temperate seas.

Similar species: *Odontella regia* closely resembles *O. mobiliensis*. In general *O. regia* is larger, the horns are slightly shorter and more directed towards the pervalvar axis. The tubes of the labiate processes are located slightly nearer to the horns than in *O. mobiliensis*.

Synonyms:
Denticella regia Schultze, 1859,
Biddulphia regia (Schultze) Ostenfeld, 1908.

Literature: Hendey 1964, Ross & Sims 1973, Round et al. 1990.

Odontella rhombus (Ehrenberg) Kützing, 1849 — Family: Triceratiaceae

Season				Trophic mode	Shape	Harmful	Bloom	Resting stage
W	S	S	A	autotrophic	rhomboid prism	no	no	no

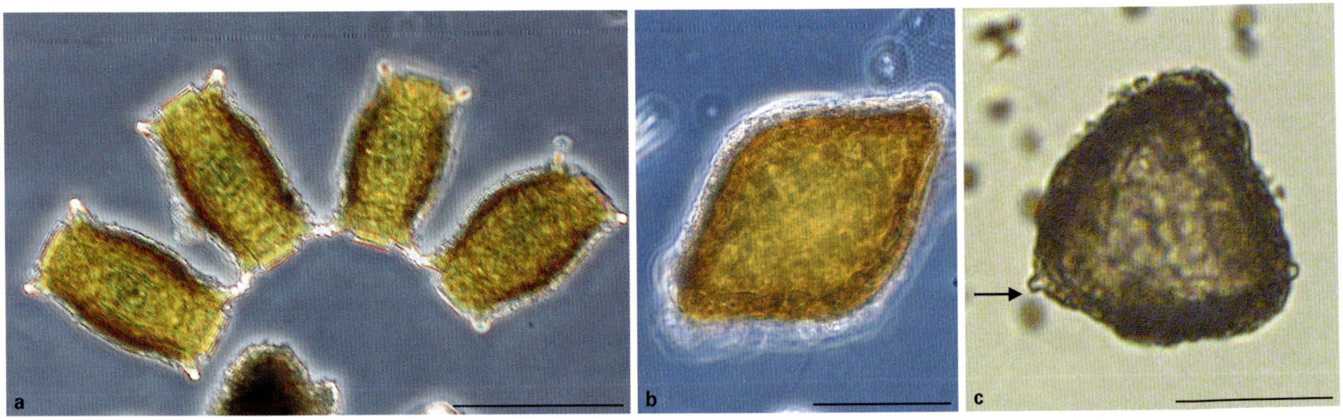

a, Phase contrast image of live chain of *Odontella rhombus* with cells in girdle view. Scale bar = 100 µm.
b, Phase contrast image of live cell in valve view.
c, Bright field image of a single, Lugol-fixed cell of *Odontella rhombus* f. *trigona* in valve view (arrow pointing to the labiate process terminating in an ocellus). b,c, Scale bars = 50 µm.

Description

Cells tin-shaped with an elliptical to rhomboid cross-section and a short pervalvar axis. Valves convex with two short pronounced, outwardly directed elevations at the poles. Valve mantle short and somewhat constricted below the horns. Whole valve covered with numerous small spikes and some larger ones at the transapical margins. Cells solitary or connected into short chains by one of the processes. Cells with numerous large rounded and densely packed chloroplasts giving the cells a brownish-yellow appearance.

Note

In addition to this typical form, the species can be found in an additional morphology. *Odontella rhombus* f. *trigona* (Cleve ex Van Heurck) R. Ross in Hartley, 1886 has a triangular cross-section.

Size

Large diameter (apical axis): 50-220 µm
Small diameter (transapical axis): 35-150 µm
Height (pervalvar axis): 50-150 µm

Distribution. *Odontella rhombus* is an obligate littoral form which is widely distributed. It can often be found in plankton after periods of water turbulence.

Similar species: none.

Synonyms:
Zygoceros rhombus Ehrenberg, 1839,
Biddulphia rhombus (Ehrenberg) W. Smith, 1856,
Biddulphia rhombus f. *trigona* Hustedt, 1927,
Biddulphia rhombus f. *tetragona* Hustedt, 1927.

Literature: Hendey 1964, Ross & Sims 1973, Round et al. 1990, Sar et al. 2007.

Odontella aurita (Lyngbye) C. A. Agardh, 1832 — Family: Triceratiaceae

Season				Trophic mode	Shape	Harmful	Bloom	Resting stage
W	S	S	A	autotrophic	elliptic cylinder	no	✿✿	yes

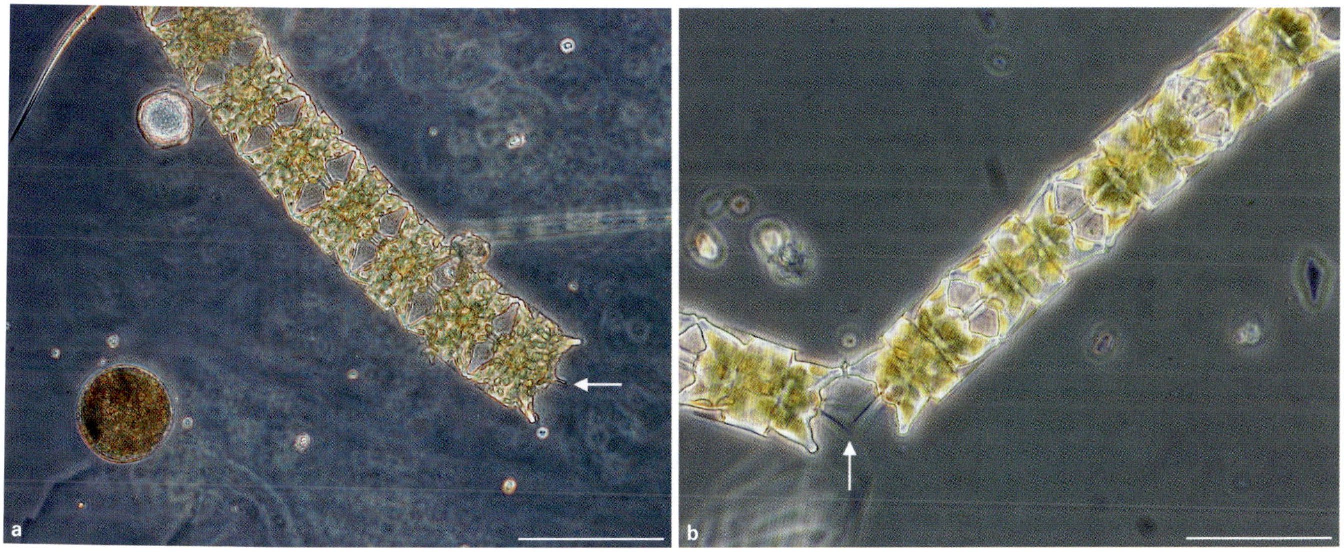

a, b, Phase contrast images of live chains of *Odontella aurita* with cells of different cell proportions. The arrow is pointing to the central labiates process. Scale bars = 100 µm (a) and 50 µm (b).

Description

Cell proportions very variable: from a very short pervalvar axis (apical axis can be twice that of the pervalvar axis) to an extremely elongate one. Valves more or less elliptical with two elevations (very short horns) at the poles and a distinct convex part in the valve centre. Cells usually united into chains by their valve elevations. External, slightly diverging tubes of labiate processes arising from the valve centre. A distinct constriction situated at the base of the valve mantle. Cells with numerous kidney shaped or elliptical chloroplasts.

Size

Large diameter (apical axis): 10-100 µm
Small diameter (transapical axis): 5-45 µm
Height (pervalvar axis): 15-80 µm

Distribution. *Odontella aurita* is a common temperate species, occurring, e.g., in the North Sea, Baltic and English Channel. It is a neritic and littoral species, with chains attached to the substratum. They are found more often in the water column after stormy periods and can become abundant in late winter and early spring.

Similar species: none.

Synonyms:
Diatoma auritum Lyngbye, 1819,
Biddulphia aurita (Lyngbye) Brébisson, 1838.

Literature: Hendey 1964, Ross & Sims 1973, Round et al. 1990, Takano 1984.

Triceratium favus Ehrenberg, 1840

Family: Triceratiaceae

Season				Trophic mode	Shape	Harmful	Bloom	Resting stage
W	S	S	A	autotrophic	triangular prism	no	no	no

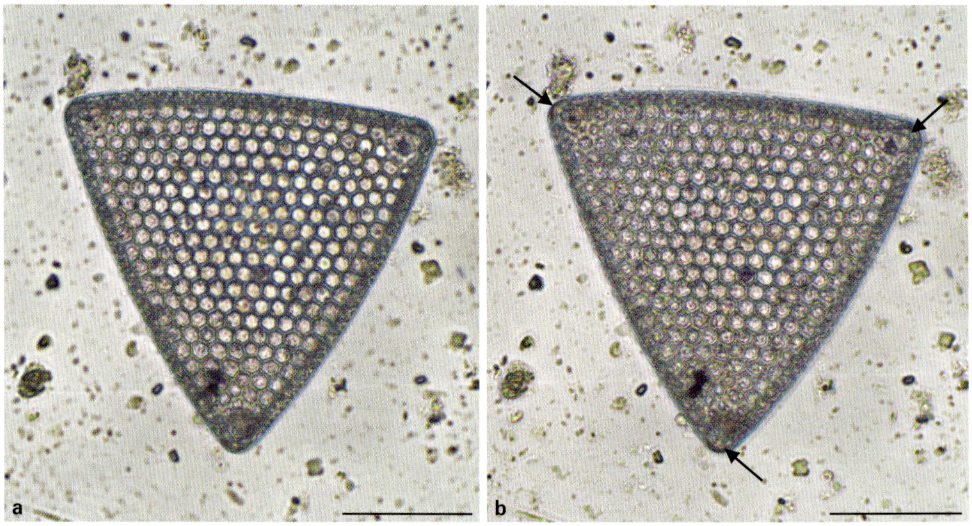

a, Bright field image of an empty frustule of *Triceratium favus* focussed on the valve margin.
b, The same cell focussed on the valve centre showing the large hexagonal areolae and location of the ocelli (arrows). Both are bright field images. Scale bars = 50 µm.

Description

Cells solitary, box-shaped in girdle view with a triangular cross-section and normally short pervalvar axis. Valve edges straight or slightly convex with rounded corners. A process (ocellus) situated in each corner. Valve area between the processes flat or slightly arched showing large regularly arranged hexagonal areolae. Chloroplasts numerous small and densely packed granules giving live cells a dark colour.

Size

Edge length: 40-350 µm
Height (pervalvar axis): 20-150 µm

Distribution. *Triceratium favus* is a wide spread littoral form which, in the North Sea occurs regularly in plankton after periods of water turbulence.

Similar species: Because of the coarse structure of its hexagonal areolae this species is easily identified.

Synonyms:
Triceratium comptum Ehrenberg, 1844,
Triceratium muricatum Brightwell, 1853,
Triceratium fimbriatum Wallich, 1858,
Biddulphia favus (Ehrenberg) Grunow, 1883,
Triceratium ferox Castracane, 1886,
Triceratium sarcophagus Castracane, 1886.

Literature: Miller & Collier 1978

Pennate Diatoms

Bacillaria paxillifer (O. F. Müller) Hendey, 1951 — Family: Bacillariaceae

Season				Trophic mode	Shape	Harmful	Bloom	Resting stage
W	S	S	A	autotrophic	elliptic cylinder	no	no	no

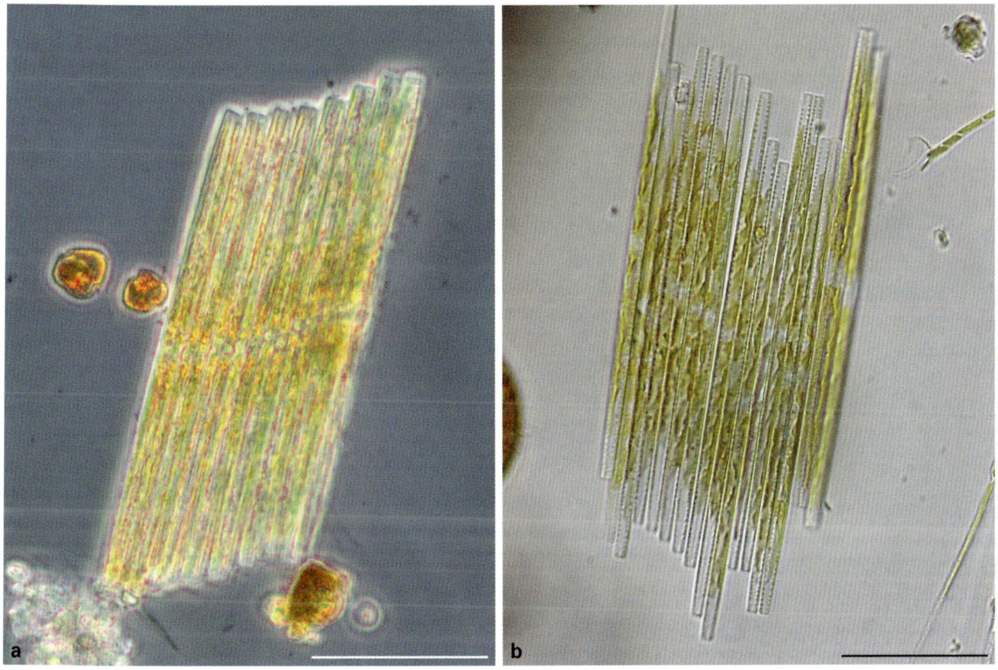

a, Phase contrast image of a Lugol-fixed colony of *Bacillaria paxillifer*.
b, Bright field image of a live colony with an irregular cell pattern. Scale bars = 50 µm.

Description

Cells long, lanceolate with slightly produced ends in valve view and rectangular in girdle view. Raphe system running from pole to pole. Striation strong. Cells united into colonies and motile, sliding along each other and causing a variable colony shape. In fixed samples colonies appearing as stacks of cells or as long bands with little overlap between cells. Two large plate-like chloroplasts per cell. Nucleus positioned centrally.

Size

Length (apical axis): 70-200 µm
Width (transapical axis): 5-8 µm
Height (pervalvar axis): 5-10 µm

Distribution. *Bacillaria paxillifer* is probably a cosmopolitan species which can be found in plankton throughout the year with a main emphasis in winter and spring.

Similar species: *Bacillaria paxillifer* has a very characteristic colony form. However, in fixed samples it often occurs as individual cells. It then looks superficially similar to other pennate diatoms.

Synonyms:
Vibrio paxillifer O. F. Müller, 1783,
Bacillaria paradoxa J. F. Gmelin, 1788,
Oscillaria paxillifera (O. F. Müller) Schrank, 1823,
Nitzschia paxillifer (O. F. Müller) Heiberg, 1863,
Nitzschia paradoxa (J. F. Gmelin) Grunow, 1880,
Oscillatoria paxillifera (O. F. Müller) Schrank ex Gomont, 1892,
Homoeocladia paxillifer (O. F. Müller) Elmore, 1921.

Literature: Jahn & Schmid 2007, Schmid 2007.

Cylindrotheca closterium (Ehrenberg) Reimann & Lewin, 1964 Family: Bacillariaceae

Season				Trophic mode	Shape	Harmful	Bloom	Resting stage
W	S	S	A	autotrophic	elliptic cylinder	slime	✿	no

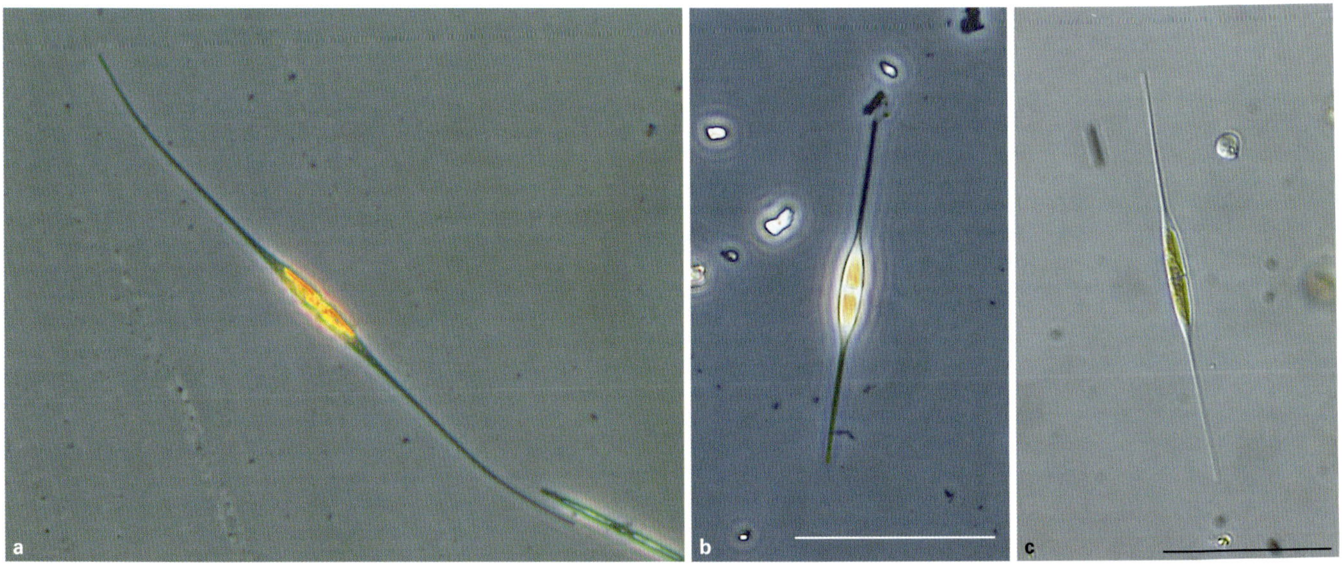

a, Lugol-fixed cell of *Cylindrotheca closterium*. **b, c,** Two morphologically different live cells. Scale bars = 50 µm. All images taken in phase contrast optics.

Description

Cells with long extended needle-shaped edges which can be curved or slightly sigmoid, central part of the cell spindle-shaped. Two chloroplasts present. Slime production during *Cylindrotheca* blooms is thought to be harmful for fisheries

Size

Total length (apical axis): 30-400 µm
Length of cell body: 6-50 µm
Width (transapical axis): 2.5-8 µm
Height (pervalvar axis): 2.5-8 µm

Distribution. *Cylindrotheca closterium* is cosmopolitan. It can be found in plankton and also on substrates like seaweeds.

Similar species: There are other species with a very similar shape. *Lennoxia faveolata* H. A. Thomsen & K. R. Buck, 1993 is smaller (apical axis 10-22 µm; transapical axis 1.5-2 µm) and has only one chloroplast. A special fusiform cell type of *Phaeodactylum tricornutum* Bohlin, 1897 can also be confused with *C. closterium*. *P. tricornutum* also has only one chloroplast.

Synonyms:
Ceratoneis closterium Ehrenberg, 1841,
Nitzschia closterium W. Smith, 1853,
Nitzschiella tenuirostris Mereschkowsky, 1901.

Literature: Round et al. 1990.

Pseudo-nitzschia pungens (Grunow ex Cleve) Hasle, 1965   Family: Bacillariaceae

Season				Trophic mode	Shape	Harmful	Bloom	Resting stage
W	S	S	A	autotrophic	rhomboid prism	ASP toxin	✿✿	no

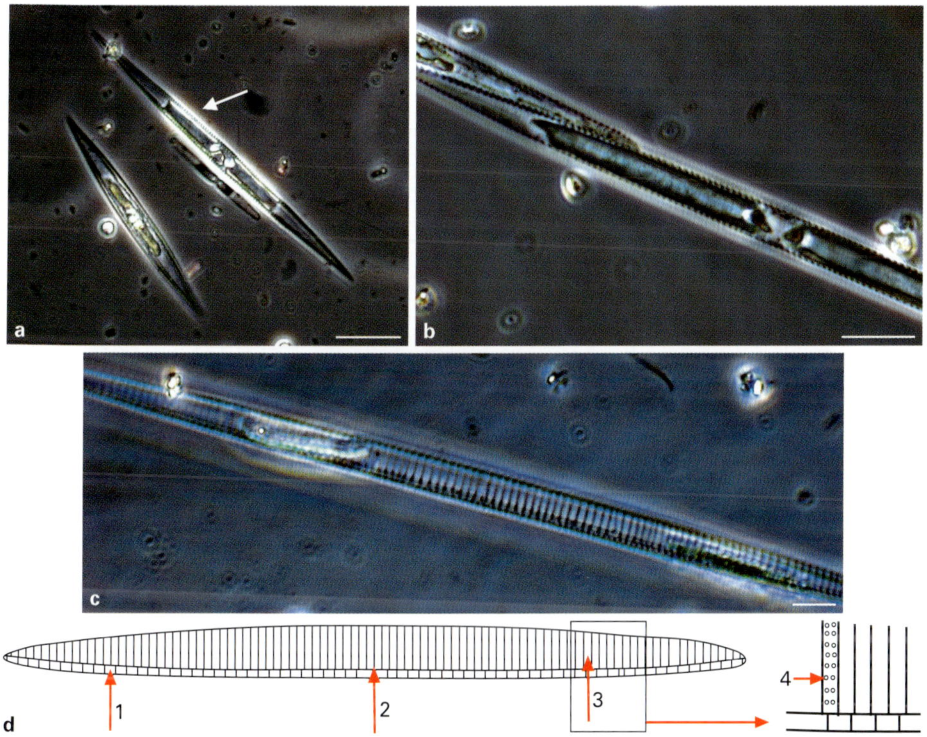

a, Bright field image of part of a single cell of *Pseudo-nitzschia pungens* in girdle view. **b,** Phase contrast image of the area of overlap between two cells in girdle view. a, b, Scale bars = 10 µm. **c,** Oil immersion image (with phase contrast) of the central area of the cell in valve view. The striae and fibulae are visible. Scale bar = 5 µm. **d,** Schematic drawing of a *Pseudo-nitzschia* valve face: 1, Fibula; 2, Central interspace; 3, Interstriae; 4, Poroids. Schematic adapted from Hasle & Syvertsen 1997.

Description

Cells spindle-shaped in girdle view and linear to spindle-shaped in valve view. Transapical axis smaller than in *P. seriata* and usually the pervalvar axis longer than the transapical axis. The ends of the cells distinctly pointed in valve view. Cells heavily silicified with interstriae and poroids just visible with light microscopy. Central interspace absent. Cells connected into stepped chains with an overlapping of the cells of about one third or more. In valve view both margins symmetrical in central part of valve. Cells containing two plate-like chloroplasts lying in the girdle area, one on each side of the transapical plane.

Size

Length (apical axis): 75-140 µm
Width (transapical axis): 3-4.5 µm
Height (pervalvar axis): 5-8 µm

Distribution. *Pseudo-nitzschia pungens* is cosmopolitan and can be found from spring to autumn with a main emphasis in summer.

Similar species: *Pseudo-nitzschia seriata* and *Pseudo-nitzschia multiseries* (Hasle) Hasle, 1995 are very similar species. *P. pungens* is much more heavily silicified and can, therefore, be identified by light microscopy using high magnification oil immersion lenses. Unless these are available, cells should be identified as belonging to the "*Pseudo-nitzschia seriata* complex".

Synonym:
Nitzschia pungens Grunow ex Cleve, 1897.

Literature: Chepurnov et al. 2005; Hasle 1994, 1995; Hasle et al. 1996.

Pseudo-nitzschia seriata (Cleve) H. Peragallo, 1908   Family: Bacillariaceae

Season				Trophic mode	Shape	Harmful	Bloom	Resting stage
W	S	S	A	autotrophic	rhomboid prism	ASP toxin		no

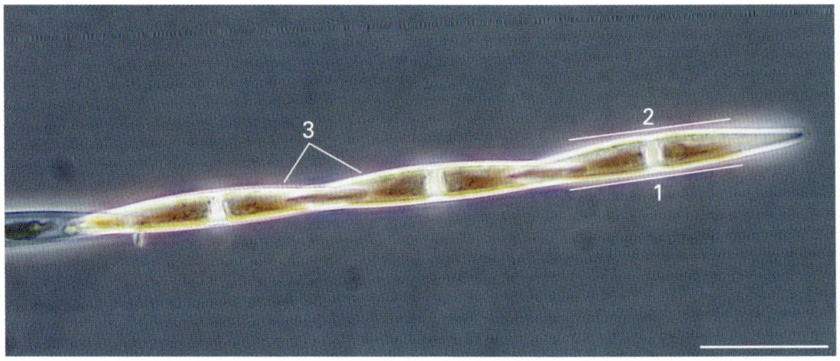

Phase contrast image of a short live chain of *Pseudo-nitzschia seriata* in valve view. One valve edge is nearly straight (1) in the central part of the valve, the other is convex (2). The area of overlap is also discernable (3). Scale bar = 50 µm.

Description

Cells fusiform in girdle view with a long apical axis and a short transapical and pervalvar axis. Valve ends distinctly pointed. In valve view, one side nearly straight in the centre (1), the other side convex (2), effect most pronounced in central part of the cell. Transapical axis normally larger than the pervalvar axis. Cells forming stepped chains in which valves are overlapping by one third to one fourth of their length. Poles more or less rounded. Two plate-like chloroplasts in the girdle area, one on each side of the transapical plane. Valve structures (interstriae) visible in water mounts.

Note

The presence of the neurotoxin ASP has been demonstrated in cultures and field samples of a number of *Pseudo-nitzschia* species including *P. australis*, *P. pungens*, *P. multiseries* (Orlova et al. 2008) and *P. seriata* (Fehling et al 2004).

Size

Length (apical axis): 90-160 µm
Width (transapical axis): 5-8 µm
Height (pervalvar axis): 3-6 µm

Distribution. *Pseudo-nitzschia seriata* as a neritic species is distributed in northern cold and temperate waters. It can be found from late spring to autumn mainly during summer.

Similar species: *Pseudo-nitzschia seriata* is very similar to *P. pungens* and other species from the genus. It is very difficult to distinguish these species during routine analysis with a light microscope. They should be identified to the genus level or as "*Pseudo-nitzschia seriata* complex" (> 3 µm).

Synonym:
Nitzschia seriata Cleve, 1883.

Literature: Hasle 1994, Hasle et al. 1996.

Pseudo-nitzschia delicatissima (Cleve) Heiden, 1928

Family: Bacillariaceae

Season				Trophic mode	Shape	Harmful	Bloom	Resting stage
W	S	S	A	autotrophic	rhomboid prism	ASP toxin	✿✿✿	no

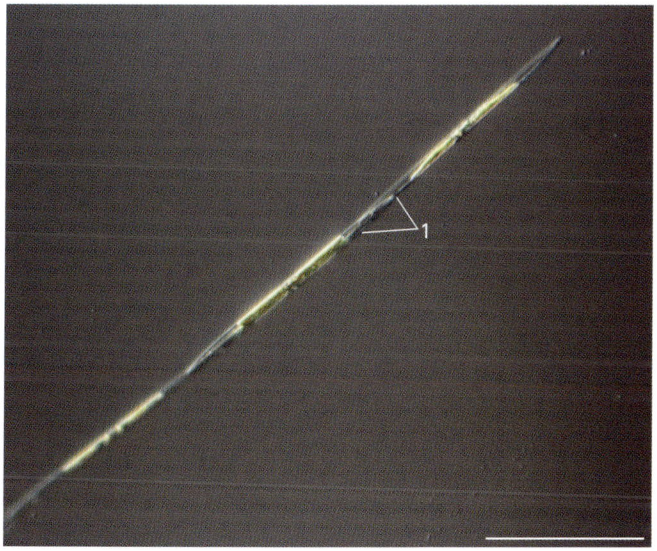

DIC image of part of a Lugol-fixed chain of *Pseudo-nitzschia delicatissima* in valve view, showing the area of overlap between two cells (1). Scale bar = 50 µm. Image courtesy of Lars Edler.

Description

Cells narrow, linear with slightly sigmoid valve ends in girdle view. Very short overlap between adjacent cells in the stepped chains (less than 1/8). Cells narrowly elliptical to linear with rounded valve ends in valve view. Valve structure hardly visible with a light microscope. Only interspace and fibulae just discernable in acid cleaned valves. Two plate-like chloroplasts per valve.

Size

Length (apical axis): 40-75 µm
Width (transapical axis): ~2 µm
Height (pervalvar axis): 1.5-2 µm

Distribution. *Pseudo-nitzschia delicatissima* is probably cosmopolitan and is found in oceanic and coastal areas mainly during summer when it can be a dominant species building higher biomasses.

Similar species: There are several small *Pseudo-nitzschia* species which cannot be distinguished in girdle view and only with difficulty in valve view with the light microscope. These small species are summarized as the "*Pseudo-nitzschia delicatissima* complex" (< 3 µm) and should be identified to that level.

Synonyms:
Nitzschia delicatissima Cleve, 1897,
Nitzschia actydrophila Hasle, 1965.

Literature: Amato et al. 2005, Hasle 1994, Hasle et al. 1996, Lundholm et al. 2003.

Meuniera membranacea (Cleve) P. C. Silva in Hasle & Syvertsen, 1996 Family: Bacillariaceae

Season				Trophic mode	Shape	Harmful	Bloom	Resting stage
W	S	S	A	autotrophic	elliptic cylinder	no	no	no

a, Bright field image of a live chain of *Meuniera membranacea* with two cells in division. Scale bar = 100 µm.
b, Partial chain at high magnification. Arrow points to the ribbon shaped chloroplast. Scale bar = 20 µm.

Description

Cells broadly rectangular in girdle view and narrowly elliptic in valve view forming short close-set chains. Valve face flat or slightly concave with pointed ends. Each cell with four ribbon-like chloroplasts. Nucleus located centrally.

Size

Large diameter (apical axis): 40-90 µm
Height (pervalvar axis): 30-50 µm

Distribution. *Meuniera membranacea* occurs in temperate waters.

Similar species: none.

Synonyms:
Stauropsis membranacea (Cleve), Meunier, 1910,
Navicula membranacea Cleve, 1897,
Stauroneis membranacea (Cleve) Hustedt, 1959.

Literature: Hasle & Syvertsen 1997, Cox & Reid 2004.

Pleurosigma / Gyrosigma

Family: Pleurosigmataceae

Season				Trophic mode	Shape	Harmful	Bloom	Resting stage
W	S	S	A	autotrophic	rhomboid prism/elliptic cylinder	no	no	no

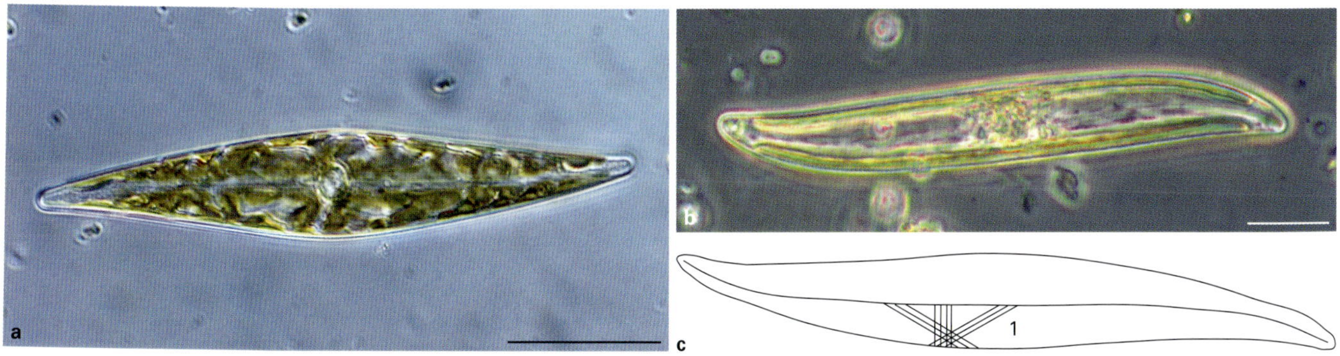

a, Live cell of a *Pleurosigma* species in valve view. Scale bar = 50 µm.
b, Lugol-fixed cell of a *Gyrosigma* species in valve view. Scale bar = 20 µm. Both images taken with phase contrast optics.
c, Schematic drawing of a typical *Pleurosigma* cell showing the typical striation pattern in relation to the raphe system (1). Redrawn from Hasle & Syvertsen 1997.

Description

Species from both genera usually littoral forms, some occasionally found in coastal or brackish water plankton. Valves of both genera more or less sigmoid to nearly straight with rounded valve poles. The two genera can be distinguished with certainty by their valve striation. *Pleurosigma* has transverse and two oblique striae, *Gyrosigma* a transverse and a longitudinal one. *Pleurosigma* cells usually rhomboid to elliptical in valve view; *Gyrosigma* often more elliptical with nearly parallel margins in the centre of the valve area.

Size

Depending on the species the length of the apical axis can be 600 µm or more.

Distribution. Most of the species from both genera are benthic (some are planktonic) but are regularly found in the plankton throughout the year.

Literature: Boalch & Harbour 1977, Cox & Reid 2004.

Asterionellopsis glacialis (F. Castracane) F. E. Round, 1990 — Family: Fragilariaceae

Season				Trophic mode	Shape	Harmful	Bloom	Resting stage
W	S	S	A	autotrophic	triangular prism	no		no

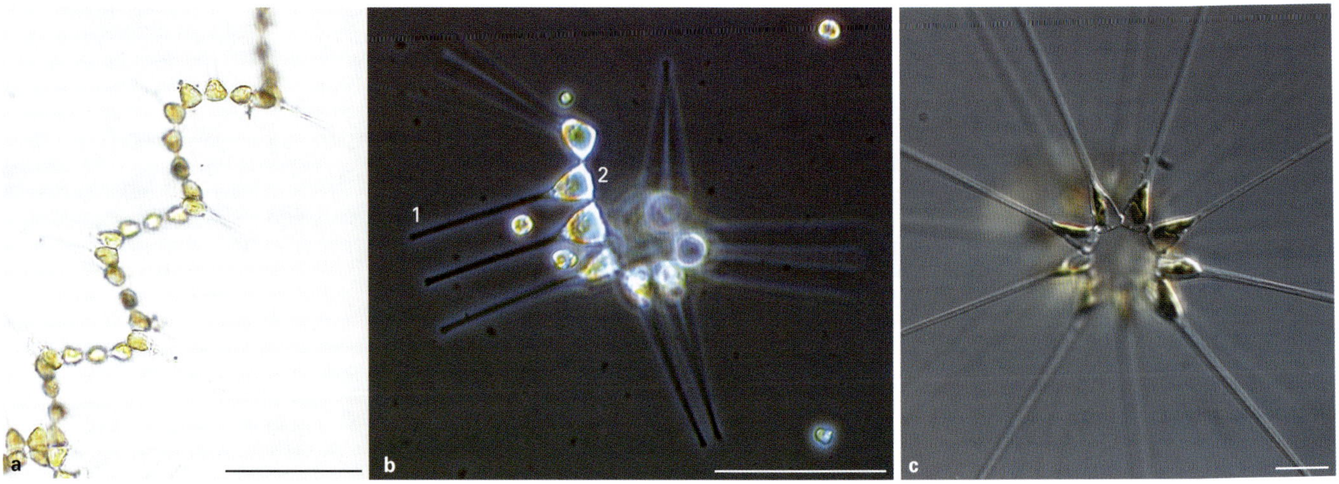

a, Bright field image of a live spiral colony of *Asterionellopsis glacialis*. Scale bar = 50 µm.
b, High magnification phase contrast image of Lugol-fixed *A. glacialis*. Scale bar = 25 µm.
c, DIC image of a live colony. Scale bar = 20 µm.

Description

Cells narrow linear in girdle view with a large expanded triangular foot pole. The opposite end of the cell is needle like and termed the head pole (1). Foot pole (2) rounded to elliptic in valve view containing one or two chloroplasts. Valve face of the expanded foot pole connecting cells into long spiralling chains. Therefore, cells usually seen in girdle view.

Size

Total length (apical axis): 30-150 µm
Length of expanded section: 10-25 µm
Width of basal part (transapical axis): 7-18 µm
Height of basal part (pervalvar axis): 5-15 µm

Distribution. *Asterionellopsis glacialis* is a cosmopolitan species appearing in the plankton throughout the year mainly in coastal waters of temperate and cold regions where it can be very abundant especially in late spring.

Similar species: *Asterionellopsis glacialis* cannot be confused with others, both as chain and as single cell. But in fixed samples single cells are easily overlooked.

Synonyms:
Asterionella glacialis F. Castracane, 1886,
Asterionella japonica Cleve, 1908.

Literature: Hasle & Syvertsen, 1997.

Asteroplanus karianus (Grunow) C. Gardner & R. M. Crawford, 1990 Family: Fragilariaceae

Season				Trophic mode	Shape	Harmful	Bloom	Resting stage
W	S	S	A	autotrophic	rotation ellipsoid	no	no	no

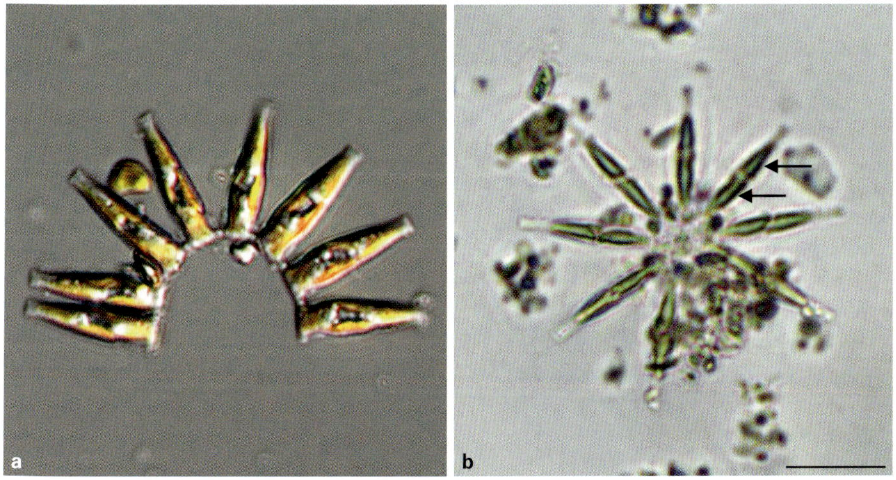

a, Live colony of *Asteroplanus karianus*. Scale bar = 20 µm.
b, Lugol-fixed star-shaped colony of *A. karianus* with two chloroplasts per cell (arrows).

Description

Like *Asterionellopsis glacialis*, a heteropolar species, i. e., with two morphologically dissimilar valve poles. Cells anvil shaped at the foot pole. Adjacent cells forming curved colonies with cells attached to each other via mucus secretions from an ocellus, on the foot pole. A single labiate process located at the head pole. In settled samples, cells usually seen in girdle view.

Size

Length (apical axis): 10-70 µm
Width (transapical axis): 3-8 µm
Height (pervalvar axis): 3-10 µm

Distribution. *Asteroplanus karianus* is sporadically found in northern temperate to cold coastal waters usually in spring.

Similar species: none

Synonyms:
Asterionella kariana Grunow ,1880,
Asterionellopsis kariana (Grunow) F. E. Round, 1990.

Literature: Crawford & Gardner 1997.

Striatella unipunctata (Lyngbye) C. Agardh, 1830 — Family: Striatellaceae

Season				Trophic mode	Shape	Harmful	Bloom	Resting stage
W	S	S	A	autotrophic	elliptic cylinder	no		no

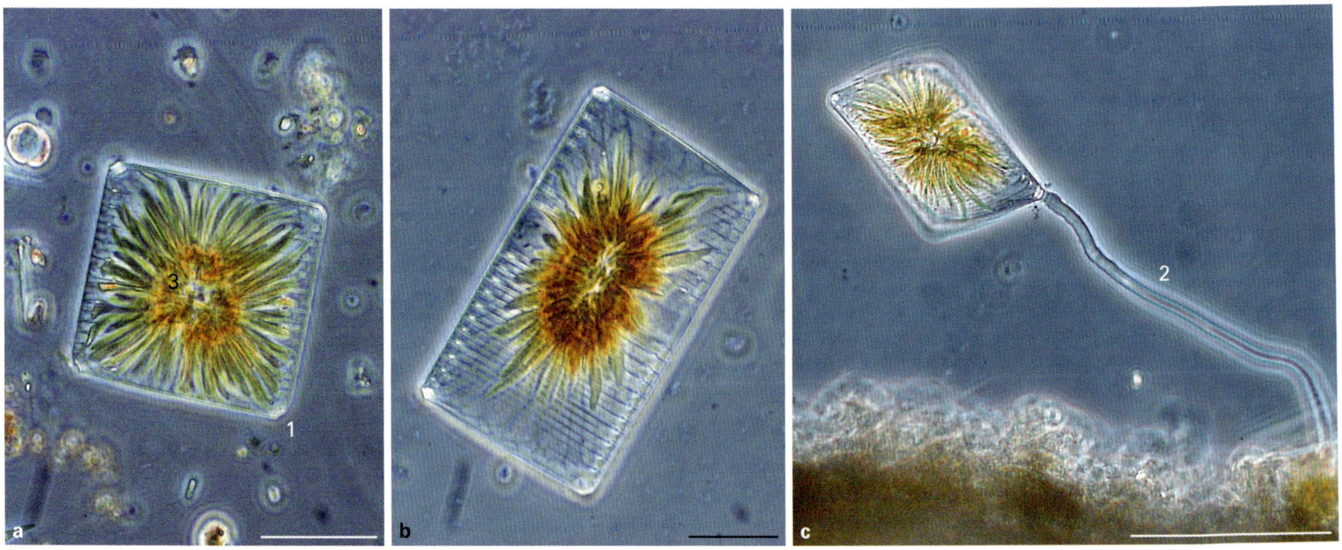

a, b, Two single live cells of *Striatella unipunctata* in girdle view with the distinct girdle bands and the radially arranged chloroplasts (3). Scale bars = 20 µm.
c, Single live cell attached to a substratum with a mucilage stalk (2). Scale bar = 100 µm.

Description

Cells almost square in girdle view and lanceolate in valve view. Cells araphid but with a distinct sternum and striae arranged at right angles to this sternum. The four corners (1) appearing diagonally cut off in girdle view, due to the presence of slightly sunken apical pore fields (ocelli) at the valve poles. Several girdle bands (6-10 in 10 µm) with narrow septa clearly recognizable. Numerous oblong chloroplasts arranged radially from the centre of the cell. Cells solitary or united into straight or zigzag ribbons attached to the substratum with a mucilage stalk (2).

Size

Large diameter (apical axis): 35-125 µm
Small diameter (transapical axis): 6-20 µm
Height (pervalvar axis): 35-150 µm

Distribution. *Striatella unipunctata* is a littoral species in temperate waters and is attached to the substratum with a mucilage stalk (2). As this connection is relatively sensitive it can often be found in the plankton during periods of water turbulence.

Similar species: Because of the typical girdle band structure and arrangement of chloroplasts this species is very distinct.

Synonym:
Fragilaria unipunctata Lyngbye, 1819.

Literature: Roth & de Francisco 1977.

Delphineis surirella (Ehrenberg) G. W. Andrews, 1981 — Family: Raphoneidaceae

Season				Trophic mode	Shape	Harmful	Bloom	Resting stage
W	S	S	A	autotrophic	elliptic cylinder	no	no	no

a, Phase contrast image of two live cells of *Delphineis surirella* in girdle view attached to a sand grain. **b,** Bright field image of several short chains attached to a sand grain. Scale bars = 50 µm.

Description

Cells flat and rectangular in girdle view and usually joined into ribbons of varying length. Valve outline narrowly to broadly elliptical, sometimes with very slightly elongated valve ends. Small individuals nearly circular in cross-section. Striation of areolae parallel or slightly radial leaving a median sternum, which is enlarged towards the valve ends. Cells with several small chloroplasts.

Size

Length (apical axis): 15-55 µm
Width (transapical axis): 8-25 µm
Height (pervalvar axis): 5-15 µm

Distribution. *Delphineis surirella* is a sublittoral form in temperate waters. Ribbons are often attached to sand grains (1). After turbulence the species can regularly be found in plankton, especially in spring.

Similar species: An attached single cell in girdle view can be confused with *Rhaphoneis amphiceros*. A correct identification is only possible in valve view.

Synonyms:
Zygoceros surirella Ehrenberg, 1840,
Rhaphoneis surirella (Ehrenberg) Grunow, 1881.

Literature: Andrews 1977, 1981.

Rhaphoneis amphiceros (Ehrenberg) Ehrenberg, 1844 Family: Raphoneidaceae

Season				Trophic mode	Shape	Harmful	Bloom	Resting stage
W	S	S	A	autotrophic	elliptic cylinder	no	no	no

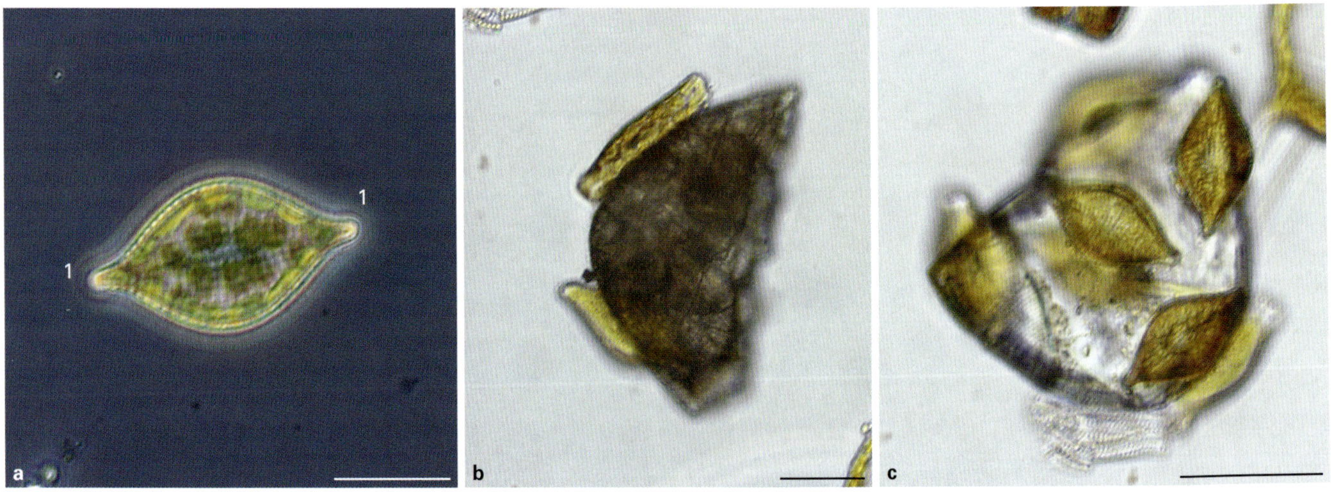

a, Phase contrast image of a single live cell of *Rhaphoneis amphiceros* in valve view.
b, Bright field image of two live cells (one in girdle view) attached to a particle. a,b, Scale bars = 20 µm.
c, Bright field image of several live cells in valve view attached to a particle. Scale bar = 50 µm.

Description

Cells rectangular in girdle view, with a short pervalvar axis. Elliptical to nearly circular in valve view with distinctly drawn out apices (1) letting cells appear lemon shaped. Areolation coarse, striae parallel or radial. Distinct sternum in a median position. Numerous small chloroplasts present.

Size

Length (apical axis): 20-100 µm
Width (transapical axis): 18-25 µm
Height (pervalvar axis): 5-20 µm

Distribution. This cosmopolitan species is benthic and usually found attached to sediment (see figures above) or empty larger diatom frustules. Occasionally this species occurs unattached in the plankton.

Similar species: In girdle view *Rhaphoneis amphiceros* closely resembles a single cell of *Delphineis surirella* from which it can be distinguished for certain only in valve view.

Synonyms:
Cocconeis amphiceros Ehrenberg, 1840,
Rhaphoneis rhombus Ehrenberg, 1844,
Doryphora amphiceros (Ehrenberg) Kützing, 1844,
Rhaphoneis lanceolata Ehrenberg, 1845,
Rhaphoneis amphiceros var. *rhombica* Grunow ex Van Heurck, 1881.

Literature: Andrews 1975, Hendey 1974.

Thalassionema nitzschioides (Grunow) Mereschkowsky, 1902

Family: Thalassionemataceae

Season				Trophic mode	Shape	Harmful	Bloom	Resting stage
W	S	S	A	autotrophic	elliptic cylinder	no	✿✿	no

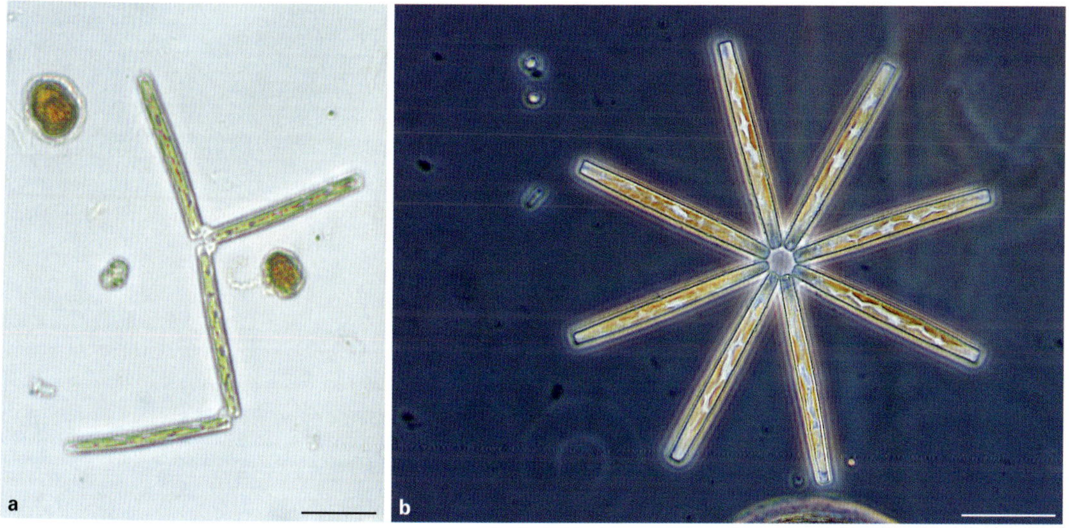

a, Bright field image of a Lugol-fixed zigzag colony of *Thalassionema nitzschioides* with the mucous pads between connected cells.
b, Phase contrast image of a live star-shaped colony. Scale bars = 20 µm.

Description

Cells arranged in typical star-shaped or zigzag colonies. Cells rectangular in girdle view. and narrowly elliptical with one row of marginal areolae in valve view. Sternum wide. One labiate process on each valve end. Sometimes a small apical spine also present. Cells connected by mucous pads (1). Chloroplasts small, numerous scattered throughout cell.

Size

Length (apical axis): 10-110 µm
Width (transapical axis): 2-4 µm
Height (pervalvar axis): 3-8 µm

Distribution. *Thalassionema nitzschioides* is a cosmopolitan species with the exception of polar regions. It can be found throughout the year with a main emphasis in spring when it can form blooms.

Similar species: In contrast to other *Thalassionema* species cells of *T. nitzschioides* are equally lanceolate with two similar valve ends in shape and width.

Synonyms:
Synedra nitzschioides Grunow 1862,
Thalassiothrix nitzschioides (Grunow) Grunow, 1881,
Thalassiothrix nitzschioides var. *javanica* Grunow, 1881,
Synedra nitzschioides var. *minor* Cleve, 1883,
Thalassiothrix curvata Castracane, 1886,
Thalassiothrix fraunfeldii var. *nitzschioides* (Grunow) Jörgensen, 1900.

Literature: Hasle 2001.

Other diatom species occasionally occurring in the German Bight

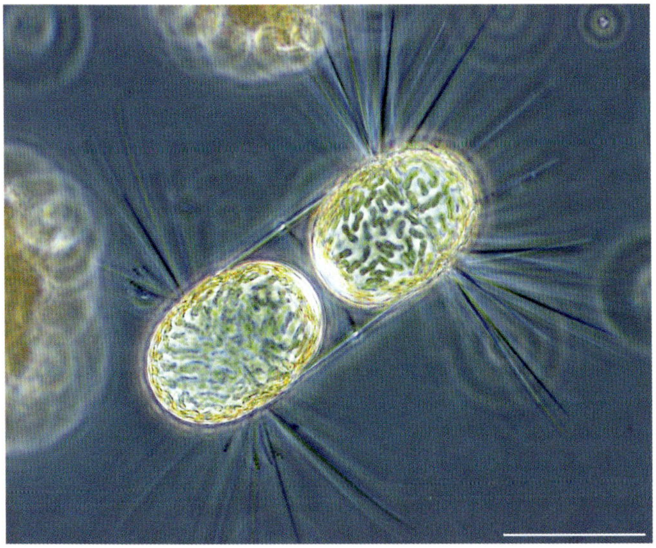

Corethron criophilum Castracane, 1886,
phase contrast image of a live cell.
Scale bar = 50 µm.

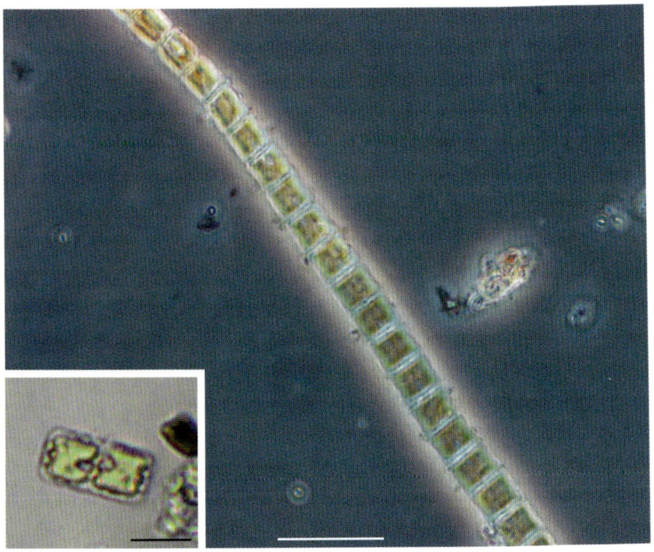

Eunotogramma dubium Hustedt;
Phase contrast image of live chain. Scale bar = 20 µm.
The insert shows a single cell. Scale bar = 10 µm.

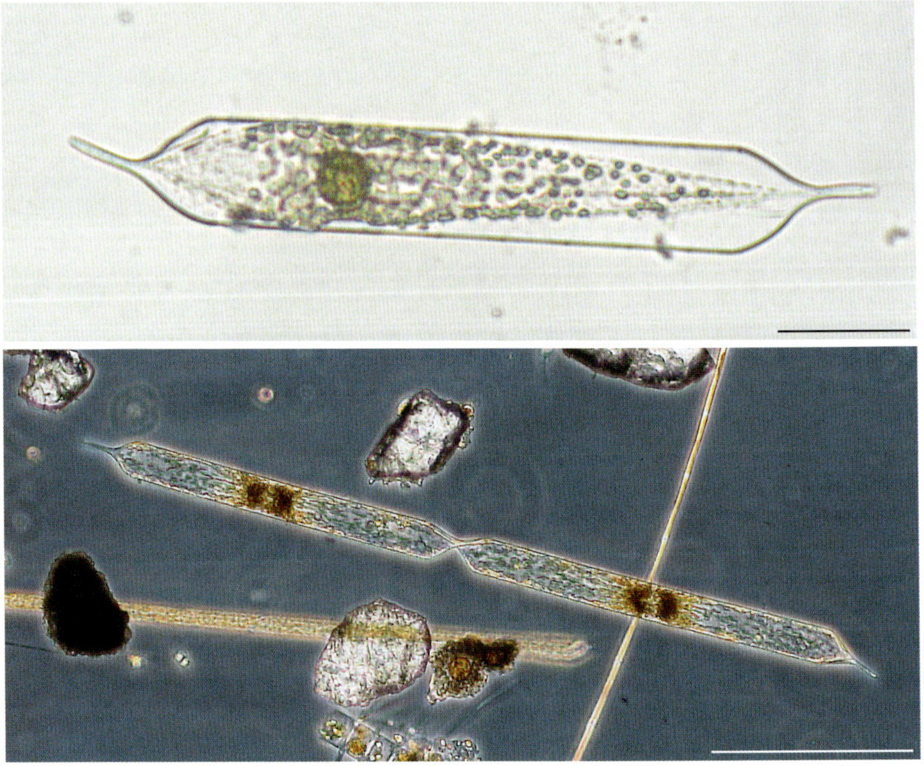

Proboscia truncata (G. Karsten) Nöthig & Ligowski, 1991,
Lugol-fixed cell showing plasmolysis. Scale bar = 50 µm.

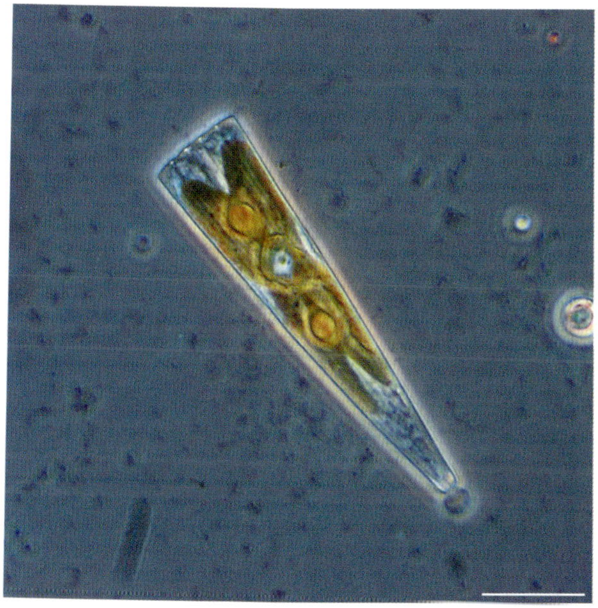

Licmophora sp. C. Agardh,
phase contrast image of a single live cell in girdle view.
Scale bar = 20 µm.

Licmophora sp., phase contrast image of a live colony.
Scale bar = 50 µm.

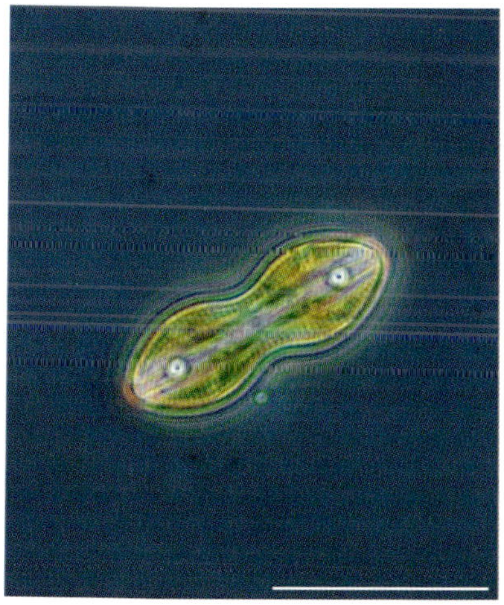

Diploneis sp. Ehrenberg ex Cleve, 1894, phase contrast image of a live cell. Scale bar = 50 µm.

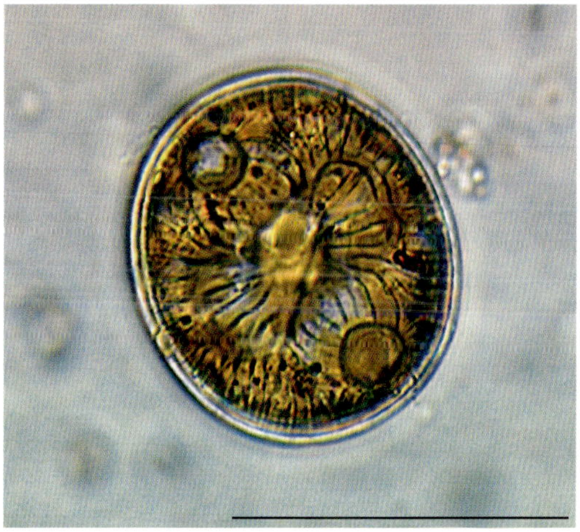

Auliscus sculptus (W. Smith) Ralfs ex Pritchard, 1861, bright field image. Scale bar = 50 µm.

Akashiwo sanguinea (Hirasaka) G. Hansen & Moestrup, 2000 — Family: Gymnodiniaceae

Season				Trophic mode	Shape	Harmful	Bloom	Resting stage
W	S	S	A	autotrophic	ellipsoid	fishkills	✿✿✿	no

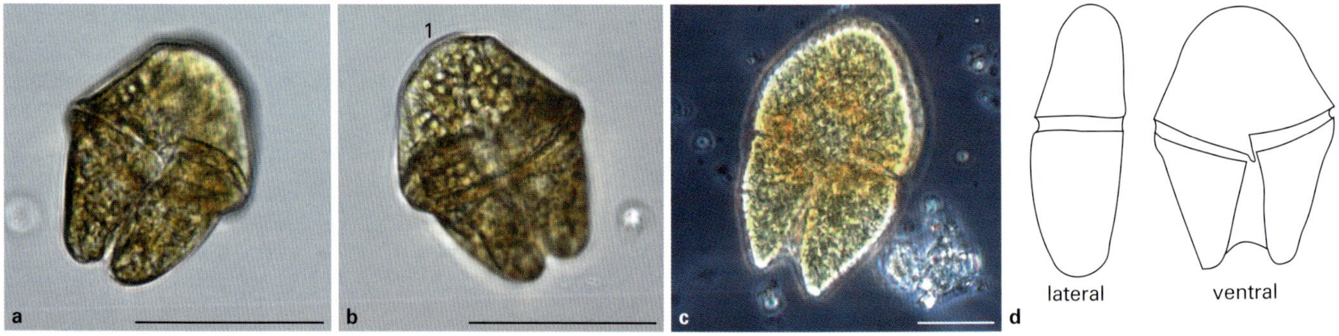

Akashiwo sanguinea: **a,** Ventral side; **b,** Dorsal side; **c,** Lugol-fixed cell. Scale bars = 50 µm.
d, Schematic drawing of *A. sanguinea* in lateral and ventral view respectively, showing the dorso-ventral compression.

Description

A large naked dinoflagellate with variable morphology, particularly in preserved samples. But cells usually dorso-ventrally flattened. Sides of the epicone straight or very slightly concave in ventral view. Apex rounded. Girdle in a median position with only a slight displacement (descending). Sulcus deeply indented posteriorly, producing a bilobed hypocone. Sulcus not extending into the epicone.
An autotrophic species with many small, round chloroplasts clearly visible in water mounts (1).

Size

Length: 40-80 µm

Distribution. *Akashiwo sanguinea* occurs in temperate and tropical regions usually in coastal and estuarine areas. This species can form dense blooms, for instance around Helgoland in autumn 2007.

Similar species: Despite some morphological variability this is a distinctive species that can be identified in live as well as Lugol- or formalin-fixed samples.

Synonym:
Gymnodinium sanguineum K. Hirasaka, 1922.

Literature: Daugbjerg et al. 2000, Takano & Horiguchi 2006.

Amphidinium carterae Hulburt, 1957 Family: Gymnodiniaceae

Season				Trophic mode	Shape	Harmful	Bloom	Resting stage
W	S	S	A	autotrophic	flattened ellipsoid	ichthyotoxic	✿	no

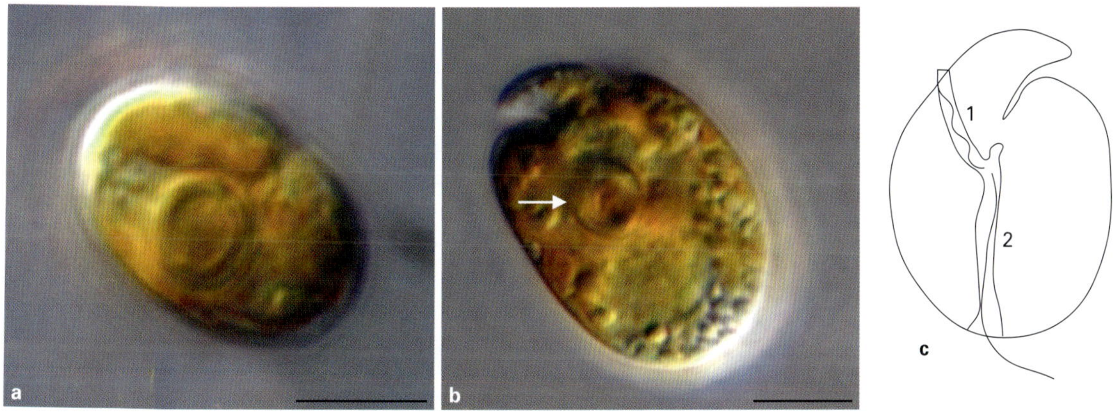

Amphidinium carterae. **a,** Schematic drawing showing the location of cingulum (1) and sulcus (2); Flagella not drawn. **b,** Live cell in dorsal view. **c,** Live cell in antapical view. Scale bars = 5 µm. Photos courtesy of Daniel Vaulot.

Description

Cells oval and dorso-ventrally flattened. Cingulum lying close to the anterior part of the cell, reducing the epicone to a fingerlike projection (1). Antapex rounded. Sulcus narrow, almost straight and located on the right side of the hypocone (2). *Amphidinium carterae* contains one large chloroplast with many branches and a centrally located pyrenoid (arrow). Nucleus situated in the posterior half of the cell.

Size

Diameter: 12-17 µm
Length: 7 10 µm

Distribution. *Amphidinium carterae* is often described as a tropical species but is also found in temperate regions for instance the North Sea and English Channel. It is predominantly benthic but can be very abundant in rock pools or sandy beaches. However, it is not often found in the water column.

Similar species: none.

Synonyms:
Amphidinium microcephalum R. E. Norris, 1961,
Amphidinium klebsii Carter, 1937.

Literature: Hulburt 1957, Jørgensen et al. 2004, Murray et al. 2004.

Lepidodinium chlorophorum (Elbrächter & Schnepf) Hansen, Botes & de Salas, 2007

Family: Gymnodiniaceae

Season				Trophic mode	Shape	Harmful	Bloom	Resting stage
W	S	S	A	autotrophic	triaxial ellipsoid	discolouration	❀❀❀	no

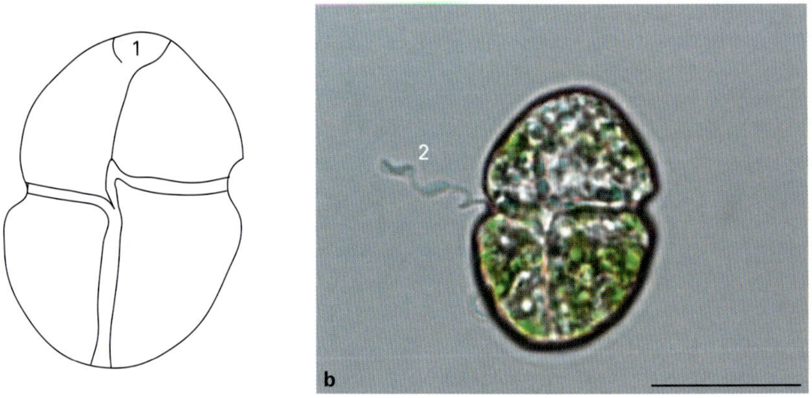

a, Schematic drawing of *Lepidodinium chlorophorum* showing the shape of the apical groove (1).
b, DIC image of a single live cell in ventral view. The transverse flagellum is visible (2). Scale bar = 20 µm.

Description

Cells small, athecate. Epicone somewhat shorter than the hypocone. Cells slightly dorso-ventrally flattened. Cingulum displaced by 1.5-2 times its width and descending. Sulcus extending into the epicone as a narrow thread running towards and around the apex, forming the apical groove (1). Above features sometimes visible with light microscopy when using high magnification oil immersion lenses, but seen in more detail with the electron microscope.
Lepidodinium chlorophorum has a green colour due to the presence of chlorophyll b as one of its major pigments.

Size

Diameter: 13-21 µm
Length: 16-24 µm

Distribution. The species was originally described from the North Sea around Helgoland, but has since also been described from other locations, e.g., the Adriatic. *Lepidodinium chlorophorum* can form extensive blooms that produce green scums on the water surface. However, adverse effects have so far not been reported.

Similar species: Several naked dinoflagellates, e.g., *Karenia mikimotoi*, look superficially similar, but *Lepidodinium chlorophorum* is distinguished from these by its green colour. Note: In Lugol-fixed samples one cannot differenciate between species on the basis of pigmentation. Moreover, *L. chlorophorum* rapidly disintegrates in fixed samples.

Synonym:
Gymnodinium chlorophorum Elbrächter & Schnepf, 1996.

Literature: Elbrächter & Schnepf 1996, Honsell & Talarico 2004.

Gyrodinium spirale (Bergh) Kofoid & Swezy, 1921 — Family: Gymnodiniaceae

Season				Trophic mode	Shape	Harmful	Bloom	Resting stage
W	S	S	A	heterotrophic	2 cones	no	✿✿	no

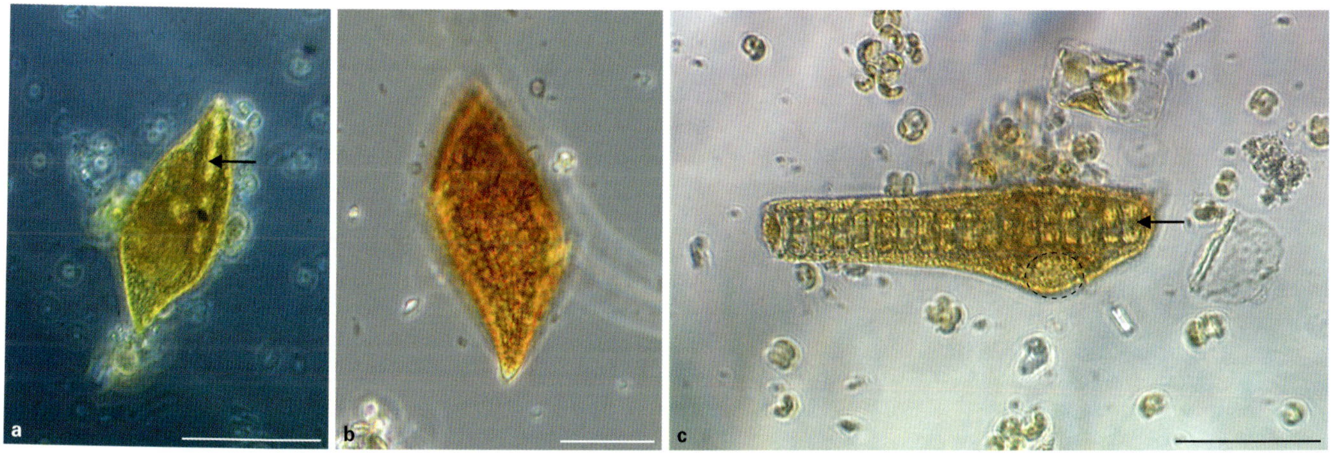

a, Lugol-fixed image of *Gyrodinium spirale* showing an ingested cell of the diatom genus *Navicula*. **b,** Lugol-fixed cell without ingested food. **c,** A Lugol-fixed cell with ingested chain of the diatom *Paralia sulcata.* Arrows point to ingested food particles. Circle indicates the position of the nucleus. Scale bars = 50 µm.

Description

Cells medium-sized. Epicone longer and narrower than the hypocone. Hypocone with parallel sides in the area closest to the cingulum and convex sides posteriorly with an antapical sulcal notch. Cingulum narrow and displaced by up to five times its width, clearly visible in light microscopy. Sulcus slightly curved and less conspicuous. Cell surface with longitudinal striations visible under the light microscope. Shape of the nucleus round or ellipsoid, usually located on the left side of the cell when seen in ventral view, position variable in cells with food vacuoles.

Gyrodinium spirale is not pigmented and ingested prey is frequently visible within the cells. Ingested prey can cause considerable cell deformation.

Size

Length: 70-100 µm
Diameter: 30-38 µm

Distribution. *Gyrodinium spirale* has been recorded from the North Sea and North Atlantic, Adriatic, Irish Sea Indian Ocean and Pacific.

Similar species: It resembles other *Gyrodinium* species but can be distinguished by the sulcal notch.

Synonyms:
Gymnodinium spirale Bergh, 1881,
Spirodinium spirale Entz, 1884.

Literature: Hansen 1992, Hansen & Daugbjerg 2004, Hulburt 1957, Kofoid & Swezy 1921.

Gyrodinium undulans Hulburt, 1957

Family: Gymnodiniaceae

Season				Trophic mode	Shape	Harmful	Bloom	Resting stage
W	S	S	A	heterotrophic	two cones	no	no	no

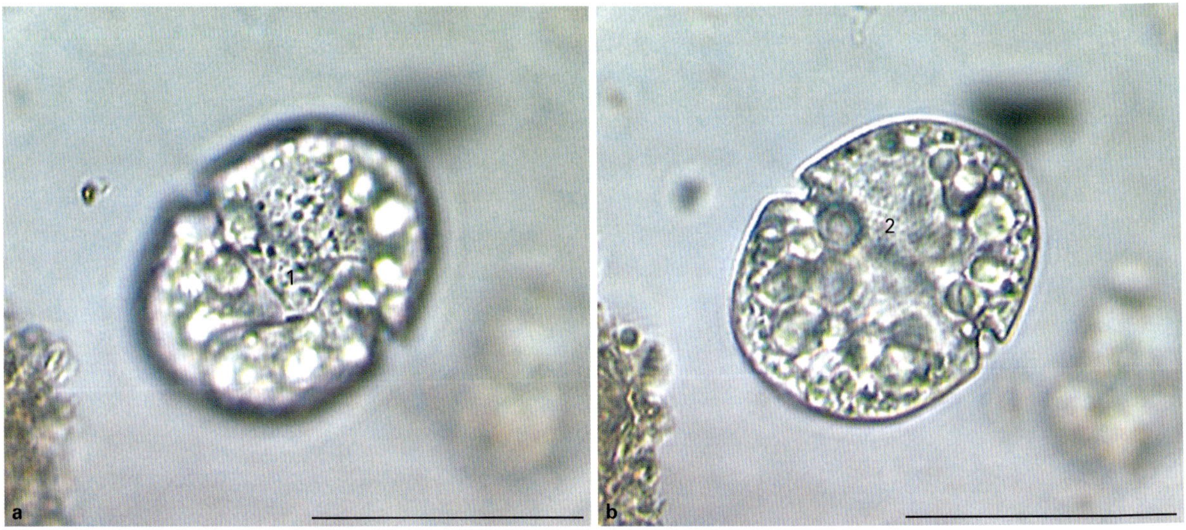

a, Live cell of *Gyrodinium undulans* showing the shape of the sulcus (1).
b, The same cell with focus on the nucleus (2).
Scale bars = 40 µm. Both images courtesy of Hanne Halliger.

Description

Cells ovoid and slightly dorso-ventrally flattened. Apex usually rounded and antapex obliquely flattened. Cingulum wide, deep, displaced by twice its width and descending. Sulcus undulating (1) extending considerably into the epicone. Sulcus narrow in the epicone but widening to where it meets the cingulum. Posterior part of the sulcus curving right to form a characteristic overlapping lobe before making a leftward turn towards the antapex. Nucleus situated in the epicone (2). Species not pigmented.

Size

Length: 27-38 µm
Diameter: 21-31 µm
Sizes as stated in the original reference.

Distribution. *Gyrodinium undulans* is probably a coastal species which has been found in the Atlantic near Woods Hole and in the North Sea, e.g., around the island of Sylt, but records are infrequent. The distribution is probably linked to that of its preferred prey species *Odontella aurita*.

Similar species: The bi-sigmoidally curved sulcus makes this a distinct species.

Synonyms: none.

Literature: Hulburt 1957.

Sclerodinium calyptoglyphe (Lebour) Dodge, 1981

Family: Gymnodiniaceae

Season				Trophic mode	Shape	Harmful	Bloom	Resting stage
W	S	S	A	heterotrophic	2 cones triaxial ellipsoid	no	✿✿	no

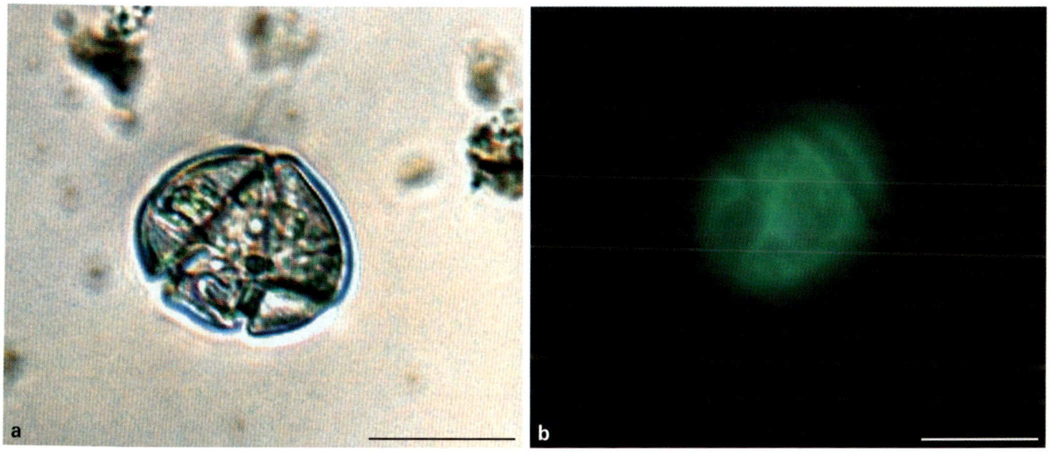

a, Bright field image of *Sclerodinium calyptoglyphe*, showing the displaced cingulum and irregular sulcus.
b, Epifluorescence image (UV-light) showing the green autofluorescence.
Both images are taken from formalin-fixed field samples. Scale bars = 20 µm.

Description

Small unpigmented cell with green autofluorescence under UV excitation. Cell ovoid in ventral view with the cingulum displaced by approximately twice its width. Cingulum making 1 1/4 turns around the cell. Sulcus running along the entire length of the cell, partially covered by a lobe from the right side of the epicone where the sulcus meets the cingulum. Sulcus further obscured by two overhangs near the posterior end of the cell. Nucleus situated in the hypocone.

Size

Diameter: 28-30 µm

Distribution. *Sclerodinium calyptoglyphe* has been reported from the Northern North Sea and also from the waters around Helgoland.

Similar species: none.

Synonyms:
Gyrodinium calyptoglyphe Lebour, 1925,
Gyrodinium calyptoglyphe Schiller, 1933,
Gyrodinium calyptoglyphe Drebes, 1974.

Literature: Dodge 1981, 1982.

Karenia mikimotoi (Miyake & Kominami ex Oda) G. Hansen & Ø. Moestrup, 2000

Family: Gymnodiniaceae

Season				Trophic mode	Shape	Harmful	Bloom	Resting stage
W	S	S	A	autotrophic	triaxial ellipsoid	haemolytic	✿✿✿	no

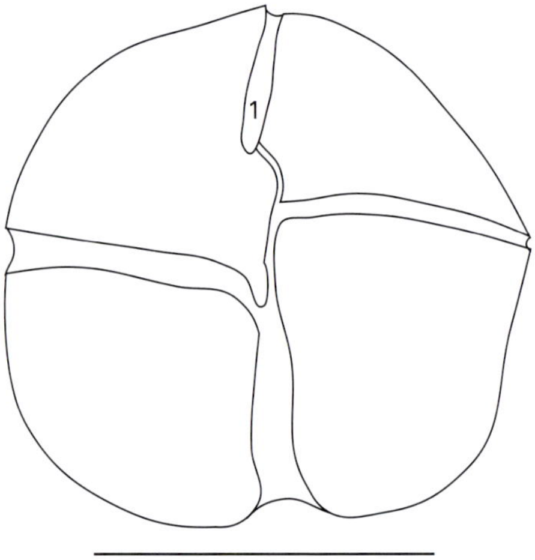

Schematic drawing of *Karenia mikimotoi* in ventral view showing the arrangement of cingulum, sulcus and the apical groove (1). Scale bar = 20 µm.

Description

Cells oval-round in ventral view and dorso-ventrally compressed. Cells straight, longitudinal apical groove present (1). Groove crossing the cell apex from the ventral to the dorsal side of the cell forms a small indentation. The cingulum is displaced by approximately 1/5 of the cell length. Sulcus straight, invading the epicone. Sulcus widening slightly towards the antapex.

Size

Diameter: 14-35 µm
Length: 18-37 µm

Distribution. *Karenia mikimotoi* is widely distributed in Northern Europe, having formed massive blooms, e.g., in Norway and the west coast of Ireland. Large scale blooms have not yet been reported from the vicinity of Helgoland.

Similar species: There is some taxonomic confusion surrounding this species. This species was often referred to as *Gyrodinium aureolum* in European waters. However, it is morphologically distinct from the species originally described as *Gyrodinium aureolum* from Woods Hole. The latter has now been renamed *Gymnodinium aureolum* (E. M. Hulburt) G. Hansen, 2000.

Synonyms:
Gymnodinium mikimotoi Miyake & Kominami ex Oda, 1935,
Gymnodinium nagasakiense Takayama & Adachi 1985.

Literature: Daugbjerg et al. 2000, Hansen et al. 2000.

Katodinium glaucum (Lebour) Loeblich III, 1965 — Family: Gymnodiniaceae

Season				Trophic mode	Shape	Harmful	Bloom	Resting stage
W	S	S	A	heterotrophic	2 cones	no		no

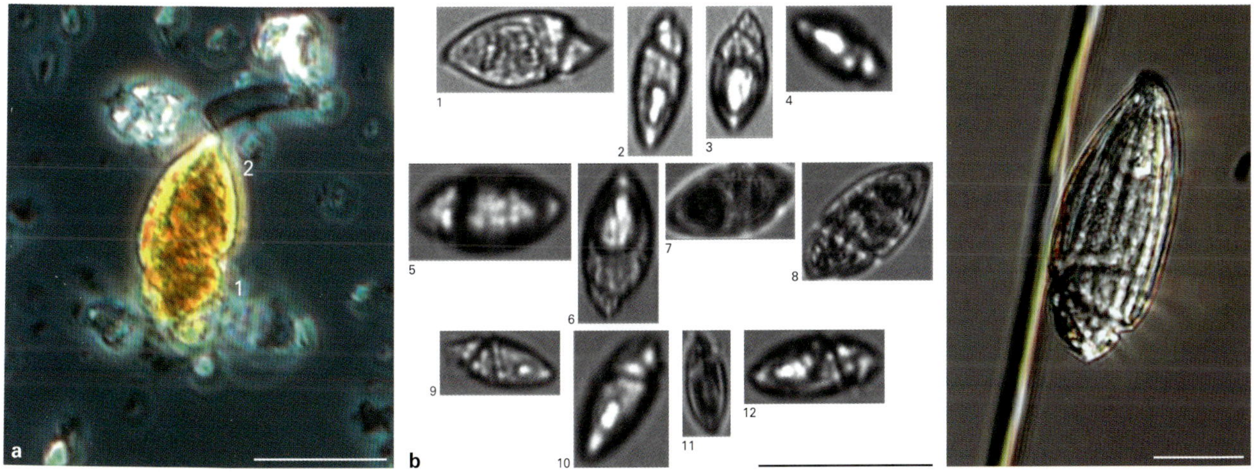

a, Phase contrast image of a Lugol-fixed cell of *Katodinium glaucum* showing the short hypocone (1) and long epicone (2). Scale bar = 20 µm. **b,** Composite flowCAM image of *K. glaucum*. Scale bar = 40 µm. **c,** Live image cell of *K. glaucum*. Scale bar = 20 µm

Description

Cell spindle-shaped with a very short hypocone and long epicone (making up 4/5 of the entire cell length). Epicone with many longitudinal striations, two to three striations on the hypocone. Cingulum displaced by 3-4 times its width. Sulcus runs from just above the anterior end of the girdle to the antapex. *Katodinium glaucum* is heterotrophic, often with conspicuous food vacuoles near the cell apex.

Size

Diameter: ~16 µm
Length: 36-62 µm

Distribution. *Katodinium glaucum* probably has a world wide distribution.

Similar species: none.

Synonyms:
Spirodinium glaucum Lebour, 1917,
Gyrodinium glaucum (Lebour) Kofoid & Swezy, 1921,
Massartia glauca (Lebour) Schiller, 1937.

Literature: Dodge 1982, Loeblich III 1965.

Polykrikos schwartzii Bütschli, 1873

Family: Polykrikaceae

Season				Trophic mode	Shape	Harmful	Bloom	Resting stage
W	S	S	A	heterotrophic	triaxial ellipsoid	no	no	yes

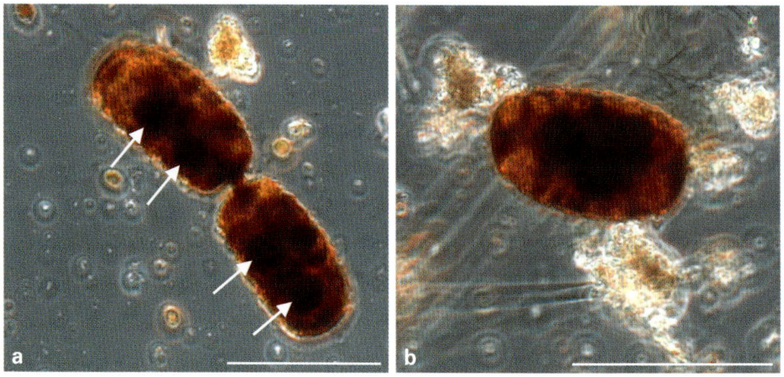

a, Paired cells of *Polykrikos schwartzii*. **b,** single cell. Both are phase contrast images of Lugol-fixed cells (arrows pointing to nuclei).
Scale bar = 100 µm. Images courtesy of Martin Loeder.

Description

Large, pseudocolonial species, with each colony consisting of several zooids. Usually one nucleus per two zooids. Number of zooids varying: 2, 4, 8, or 16. Cell outline in ventral view roughly cylindrical. Apex and antapex flattened. Each zooid with cingulum (no displacement) in a median position. Sulcus continuous between the zooids and located in a central position along the apical-antapical axis. Nuclei located in the area of the sulcus. Nematocysts and food vacuoles can be present.

Polykrikos schwartzii produces a very large oval cyst with a distinct network of protrusions, forming a mesh on the cell surface.

Size

Diameter: 100-160 µm (colony with 8 zooids)
Length: 60-100 µm

Distribution. This is probably a cosmopolitan species but occurs most frequently in coastal waters.

Similar species: In *Polykrikos kofoidii* Chatton, 1914 the hypocone is striated. In *P. schwartzii* it is smooth. The sulcus is also straight rather than oblique in *P. schwartzii*. In *P. kofoidii*, the sulcus is displaced towards the left side of the cell.

Synonyms: none.

Literature: Dodge 1982, Throndsen et al. 2007.

Nematodinium armatum (Dogiel) Kofoid & Swezy, 1921 — Family: Warnowiaceae

Season				Trophic mode	Shape	Harmful	Bloom	Resting stage
W	S	S	A	heterotrophic	spheroid	no	✿	no

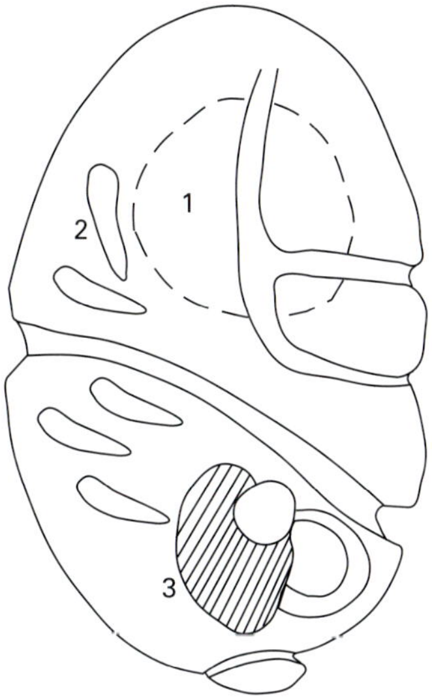

Schematic drawing of *Nematodinium armatum:* 1, Nucleus; 2, Nematocysts; 3, Eyespot.
Drawing adapted from Kofoid & Swezy, 1921.

Description

Cell outline ovoid in ventral view. Hypocone and epicone of similar size with rounded apex and antapex. Cingulum displaced by approximately half the cell length. Eyespot located in the posterior part of the cell. Eyespot consisting of a pigment mass associated with a lense-like structure. 5-15 elongated nematocysts present, visible with light microscopy. Cells only weakly pigmented with few yellowich chromatophores according to Francis (1967) and Dodge (1982) but described as unpigmented by Horner (2002). Cells occasionally enveloped by a wall or membrane.

Size

Diameter: 30-60 µm
Length: 40-90 µm

Distribution. *Nematodinium armatum* is widely distributed and has been found in the Eastern and Western Atlantic, the North Sea and Mediterranean.

Similar species: *Warnowia parva*: In contrast to *Nematodinium armatum*, this species has no nematocysts.

Synonym:
Pouchetia armata Dogiel, 1906.

Literature: Kofoid & Swezy 1921, Francis 1967.

Torodinium robustum Kofoid & Swezy, 1921 Family: Gymnodiniaceae

Season				Trophic mode	Shape	Harmful	Bloom	Resting stage
W	S	S	A	heterotrophic	flattened ellipsoid	no	✿	no

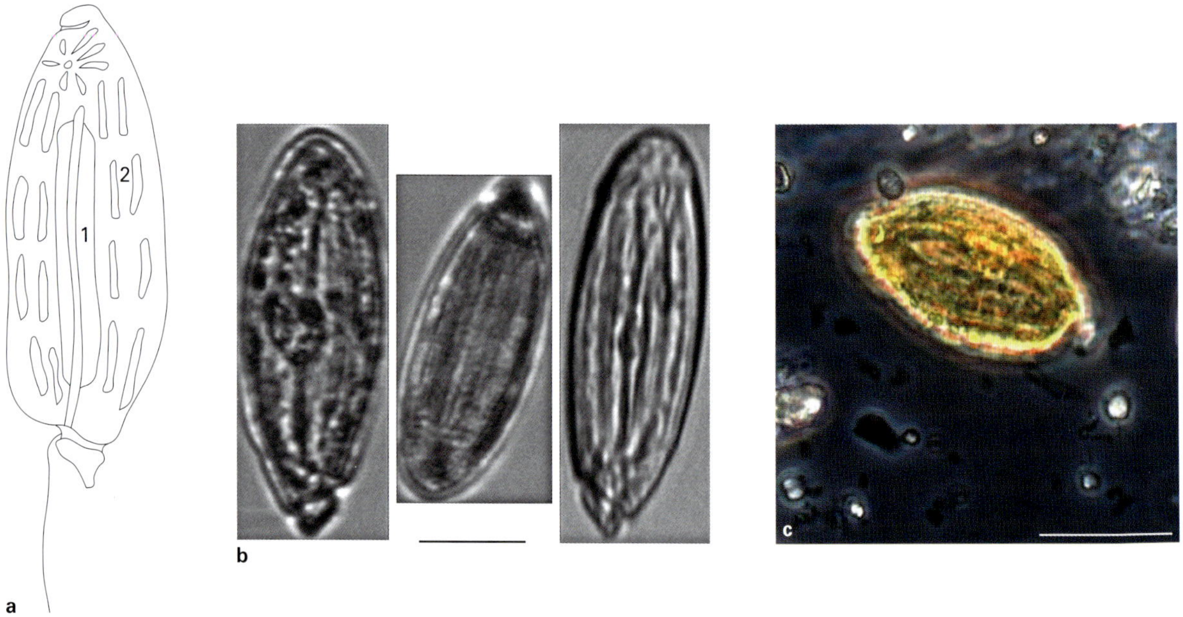

a, Schematic drawing of *Torodinium robustum* showing the position of nucleus (1) and chloroplasts (2), redrawn from Lebour (1925).
b, Images from a FlowCAM showing the arrangement of the chloroplasts. Image courtesy of Florian Hantzsche. **c,** Lugol-fixed cell. Scale bars = 20 µm.

Description

Cells about 3 times longer than wide. Epicone very long in relation to hypocone (approximately 4/5 of the body length). Cingulum only weakly displaced. Sulcus twisted by one turn apically running almost the entire length of the epicone, but not the hypocone. Nucleus elongate and located centrally in an apical-antapical direction. Many greenish to brown chloroplasts, arranged into longitudinal rows.

Size

Diameter: 20-25 µm
Length: 60-90 µm

Distribution. *Torodinium robustum* is widely distributed and has been reported from several locations in the North Sea, English Channel and Irish Sea and the Pacific.

Similar species: *Torodinium teredo* (Pouchet) Kofoid & Swezy, 1921 is more slender than *Torodinium robustum*. It occurs in temperate and tropical regions.

Synonyms: none.

Literature: Dodge 1982, Hansen & Larsen 1992, Steidinger & Tangen 1997.

Noctiluca scintillans (Macartney) Kofoid & Swezy, 1921 Family: Noctilucaceae

Season				Trophic mode	Shape	Harmful	Bloom	Resting stage
W	S	S	A	heterotrophic	sphere	no	✿✿✿	no

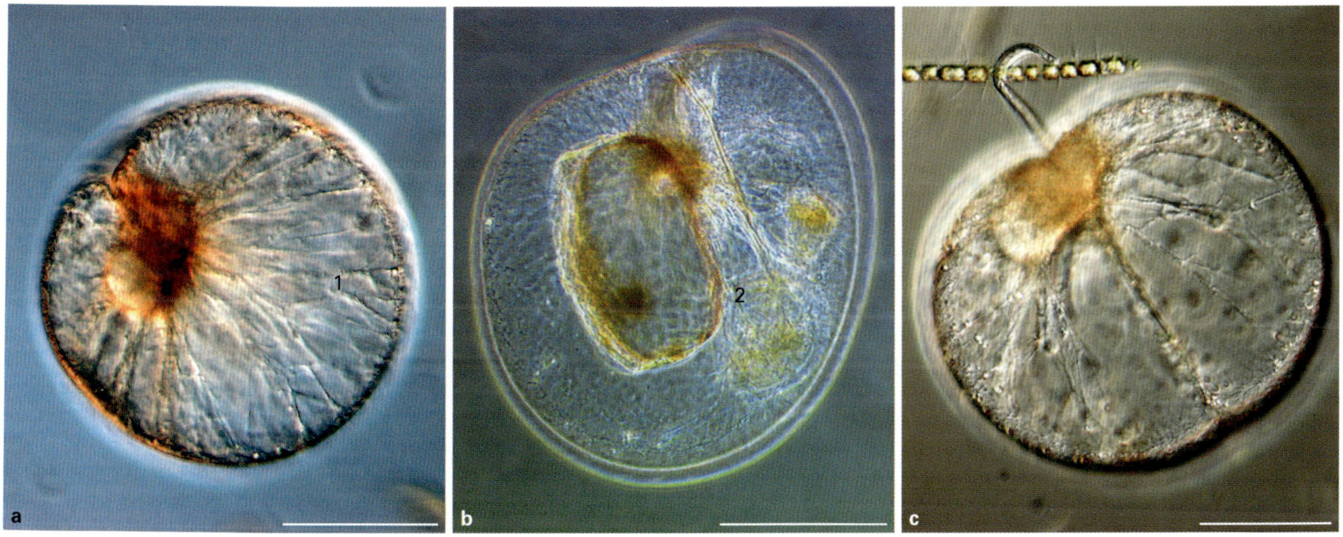

a, Single cell of *Noctiluca scintillans*, showing the cytoplasmic strands (1). Scale bar = 200 µm.
b, Single cell with ingested food particles. Scale bar = 200 µm.
c, Live cell with extended tentacle in contact with a diatom chain. Scale bar = 100 µm.

Description

Distinctly shaped athecate dinoflagellate species with cell not divided into epitheca and hypotheca. Cells very large, inflated (balloon-like) and subspherical. Ventral groove deep and wide, and housing a flagellum, tooth (specialized extension of the cell wall) and striated tentacle. Only one flagellum present, equivalent to the transverse flagellum in other dinoflagellates. Nucleus located close to ventral groove with cytoplasmic strands radiating from nucleus towards the cell periphery. Reproduction asexually by binary fission and also sexually via formation of isogametes.

In contrast to other dinoflagellate species vegetative cells of *Noctiluca scintillans* are diploid. The gametes are gymnodinioid with dinokaryotic nuclei.

Size

Diameter: 200-2000 µm

Distribution. *Noctiluca scintillans* is widely distributed throughout the world, although in tropical regions it often appears green due to the presence of internal symbionts.

Similar species: none.

Synonyms:
Medusa marina Slabber, 1771,
Medusa scintillans Macartney, 1810,
Noctiluca millaris Suriray, 1816,
Mammaria scintillans Ehrenberg, 1834,
Noctiluca marina Ehrenberg, 1834.

Literature: Buskey 1995, Kiørboe & Titelmann 1998.

Dinophysis acuta Ehrenberg, 1839 Family: Dinophysiaceae

Season				Trophic mode	Shape	Harmful	Bloom	Resting stage
W	S	S	A	mixotrophic	triaxial ellipsoid	DSP toxins	✿✿	no

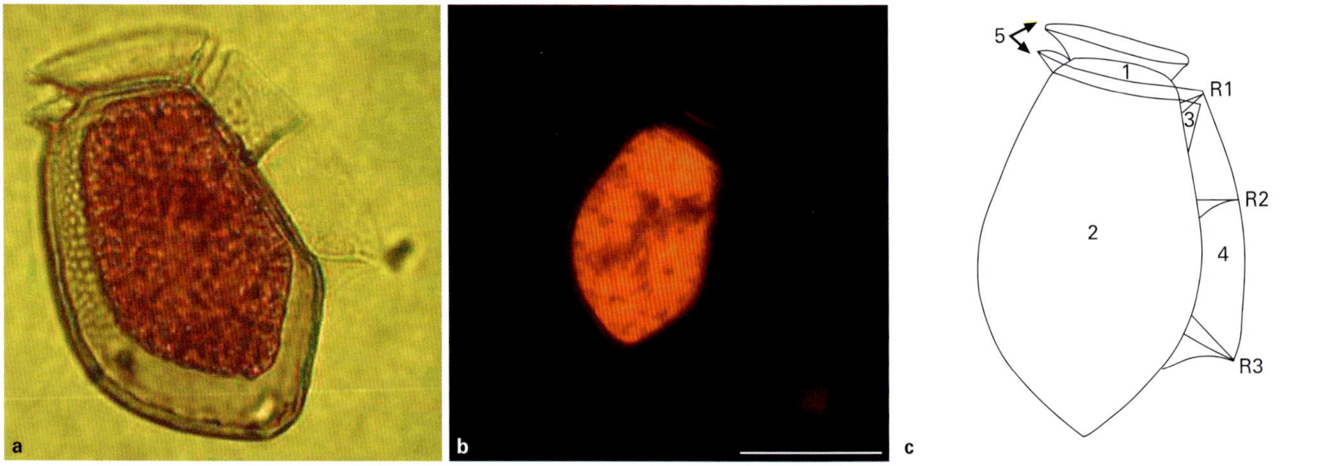

a, Bright field image of a Lugol-fixed cell in lateral view. **b,** Epifluorescence image (UV light) of a live cell in lateral view. **c,** Schematic of *Dinophysis acuta* showing the typical features of the genus *Dinophysis*: 1, Epitheca; 2, Hypotheca; 3, Right sulcal list; 4, Left sulcal list, 5, Cingular lists; R1-3, Ribs of sulcal list (R1, anterior rib; R3, posterior rib, pointing towards the posterior end of the cell). The orientation of these ribs in an anterior-posterior direction is a diagnostic feature. Scale bars = 50 µm.

Description

Cells strongly laterally flattened. Cells in lateral view broadest below the midpoint between anterior and posterior margin. Sulcal lists well developed, forming a short right wing and longer left wing. Left sulcal list broad and long, ending near the broadest part of the cell with the third rib (R3) pointing towards the antapex. Posterior end pointed or blunt with its dorsal and ventral portions forming a right angle. *Dinophysis acuta* is a phototrophic cell with reddish/orange chloroplasts when placed under UV light (this is characteristic of cryptophyte pigment). The thecal plates are strongly areolated. Reproduction asexually by binary fission.

Note

Hansen (1993) has suggested that sexual reproduction with the occurrence of dimorphic cells also occurs. The species *Dinophysis dens* has been considered a gamete in the life cycle of *D. acuta*.

Size

Diameter: 43-60 µm
Length: 54-94 µm

Distribution. *Dinophysis acuta* has been reported widely in Europe, e.g., from the North Sea, Skagerrak, Baltic, Irish Sea, Channel and the Mediterranean Sea.

Similar species: *Dinophysis norvegica* (see below).

Synonyms: none.

Literature: Berland et al. 1995, Hansen 1993, Reguera et al. 1995.

Dinophysis norvegica Claparede & Lachmann, 1859 — Family: Dinophysiaceae

Season				Trophic mode	Shape	Harmful	Bloom	Resting stage
W	S	S	A	mixotrophic	triaxial ellipsoid	DSP toxins	✿	no

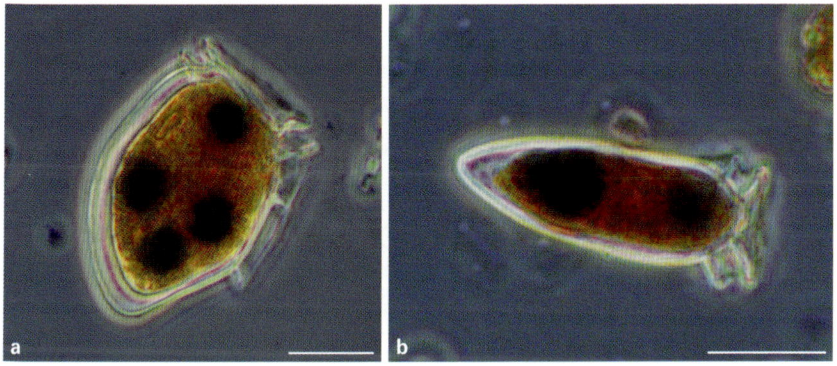

Single cell of *Dinophysis norvegica* in lateral (**a**) and ventral (**b**) view.
Both are phase contrast images of Lugol-fixed cells. Scale bars = 20 µm.

Description

Cells laterally flattened and deepest at the midpoint of the hypotheca. Epitheca small and not extending beyond the cingular lists. Cells pointed at the posterior end. Cingular lists well developed. Left sulcal list medium sized with its first and second rib pointing anteriorly and the third rib posteriorly. Thecal plates strongly areolated.

Note

This is a morphologically variable species. Dorsal and posterior margins sometimes have thick thecal extensions. Under UV light cells have an orange fluorescence characteristic of cryptophyte pigments.

Size

Diameter: 39-70 µm (dorso-ventral width)
Length: 48-80 µm

Distribution. This species is widely distributed. Blooms of this species have been reported, e.g., from the US, Scandinavia and the British Isles.

Similar species: *Dinophysis acuta*: In *Dinophysis norvegica* the widest part of the cell (when seen in lateral view) is around halfway along the apical axis, whereas it is below the halfway point in *D. acuta.*

Synonyms:
Dinophysis norvegica var. *debilior* Paulsen, 1907,
Dinophysis debilior Paulsen, 1949.

Literature: Carpenter et al. 1995, Carvalho et al. 2008, Dodge 1982, Hansen 1993, Jacobsen & Anderson 1994.

Dinophysis acuminata Claparède & Lachmann, 1859 — Family: Dinophysiaceae

Season				Trophic mode	Shape	Harmful	Bloom	Resting stage
W	S	S	A	mixotrophic	triaxial ellipsoid	DSP toxins	✿✿	no

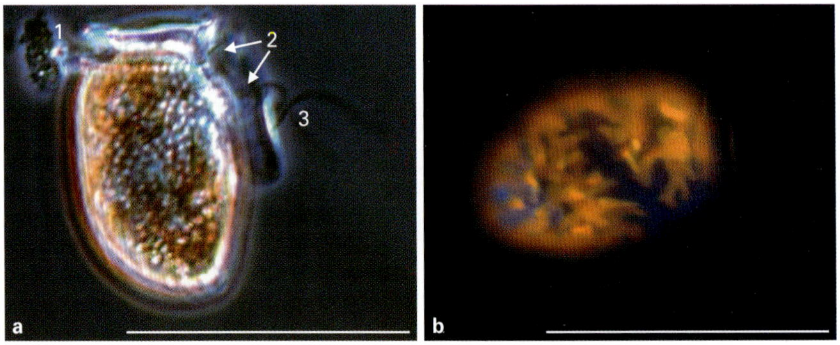

a, Phase contrast image of *Dinophysis acuminata* in lateral view showing the cingular (1) and sulcal list (2) and flagella (3).
b, Epifluorescence image (in UV light) of the same cell showing the orange autofluorescence typical for this species.
Scale bars = 50 µm.

Description

Variable cell outline in lateral view but often oval or elliptical. Posterior portion of the cell rounded sometimes with small protrusions. Cells, as in all *Dinophysis* species, laterally compressed. Theca covered with pronounced areolae, each containing a pore. Cingulum located very close to the cell apex, also bearing pronounced lists. Epitheca not extending beyond the cingular lists. Left sulcal list well developed, relatively narrow, variously sculptured and extending beyond the midpoint of the cell. Third sulcal rib usually the longest of the three ribs and pointing posteriorly.
Cells are clearly pigmented. A posteriorly positioned pyrenoid is associated with the large chloroplast. The nucleus is located centrally.

Size

Diameter: 38-58 µm
Length: 30-40 µm

Distribution. *Dinophysis acuminata* is regarded as neritic and has been reported widely including the Irish Sea, North Sea and the Baltic (but see below). Bloom formation by *D. acuminata* has also been widely reported. It can become very abundant in sheltered, enclosed areas such as fjords.

Similar species: The taxonomic position of this species is still uncertain with many, very similar species existing which could be sub-species of *Dinophysis acuminata*. Examples are *Dinophysis sacculus* Stein, 1883, *Dinophysis punctata* Jörgensen, 1923 and *Dinophysis ovum* Schütt, 1895.
D. aucminata is most commonly confused with *D. sacculus*. In *D. sacculus* the cell outline in lateral view is more elongate and the surface areolation is less conspicuous.

Synonyms:
Dinophysis ellipsoides Kofoid, 1907,
Dinophysis lachmannii Paulsen, 1949,
Dinophysis skagii Paulsen, 1949,
Dinophysis borealis Paulsen, 1949,
Dinophysis boehmi Paulsen, 1949,
Dinophysis lachmanii Solum, 1962.

Literature: Hallegraeff & Lucas 1988, Jacobsen & Anderson 1994, Zingone et al. 1998.

Phalacroma rotundatum (Claparede & Lachmann) Kofoid & Michener, 1911

Family: Dinophysiaceae

Season				Trophic mode	Shape	Harmful	Bloom	Resting stage
W	S	S	A	heterotrophic	ellipsoid	DSP toxin	no	no

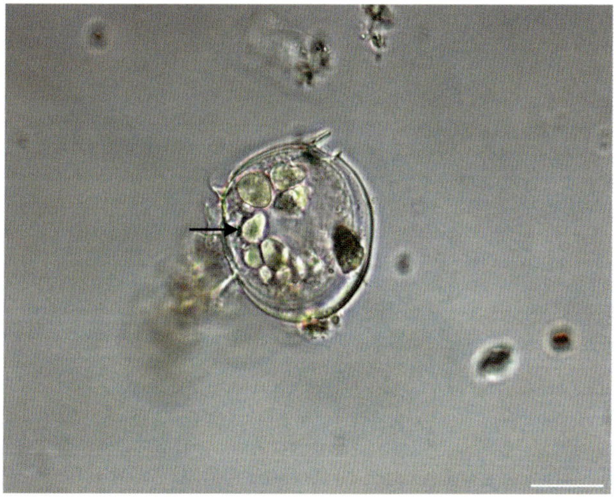

Phase contrast image of a live cell of *Phalacroma rotundatum* in lateral view showing the food vacuoles inside the cell (arrow). Scale bar = 20 µm.

Description

Cells spherical to ovoid in lateral view. Cells strongly laterally compressed, with convex sides. In contrast to *Dinophysis* species epitheca clearly protruding beyond the cingular lists. Large main thecal plates covered by small pores. Left sulcal list of moderate width, widening slightly posteriorly. The first two ribs (R1 and R2) positioned more closely together than R2 and R3. Right sulcal list very long. Cells not pigmented, food vacuoles often present. Green autofluorescence in UV light.

Size

Diameter: 40-45 µm
Length: 45-50 µm

Distribution. *Phalacroma rotundatum* is sometimes regarded as an oceanic species, but clearly also occurs in more coastal areas. It was reported from many areas in the North Sea including the area around the islands of Helgoland and Sylt. It has a worldwide distribution with records also from Mediterranean and the Pacific Ocean

Similar species: *Phalacroma* species are still regarded by many as members of the genus *Dinophysis* (but see Edvardsen et al. 2003). Both genera are in need of revision.

Synonyms:
Dinophysis rotundata Claparède & Lachmann, 1859,
Prodinophysis rotundatum (Claparéde & Lachmann) Balech, 1944,
Dinophysis whittingae Balech, 1971.

Literature: Edvardsen et al. 2003, Hallegraeff & Lucas 1988, Larsen & Moestrup 1992.

Mesoporos perforatus (Gran) Lillick, 1937 Family: Prorocentraceae

Season				Trophic mode	Shape	Harmful	Bloom	Resting stage
W	S	S	A	not known	sphere-10 %	no	✿✿	no

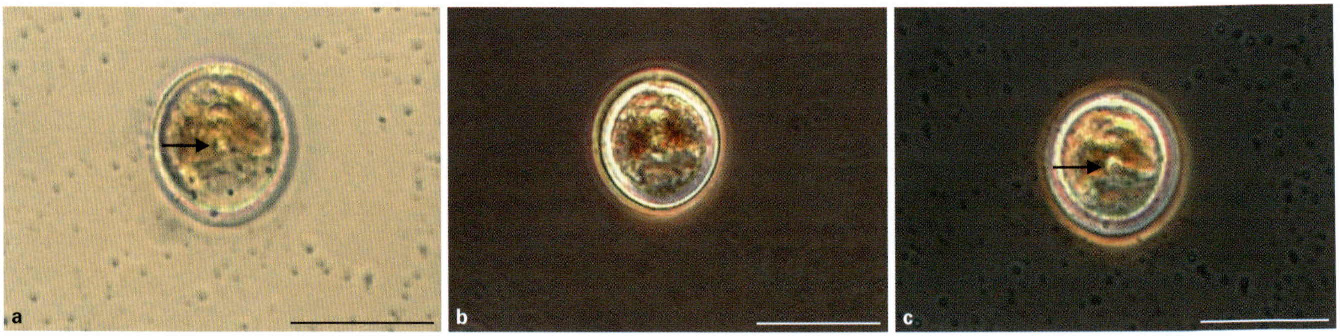

a, Bright field image of *Mesoporos perforatus* (arrow pointing to the central depression).
b, c, Phase contrast images of the same cell. Scale bars = 20 µm.

Description

Oval to round cell in broad lateral view with a central depression on both valves (not a pore proper). Depth of depressions variable. Small pores distributed around cell periphery. The entire valve, except of the region around the central depression, covered by tightly packed small papillae. Two inconspicuous spines border the flagellar pores (best seen using oil immersion on empty valves). Two reddish brown chloroplasts per cell. Nucleus located posteriorly.

Size

Diameter: 18-21 µm
Length: 14-27 µm

Distribution. *Mesoporos perforatus* probably has a worldwide distribution including Polar Regions. It has been reported, e.g., from the North Atlantic, North Sea, Baltic, Adriatic and from the Antarctic and Arctic.

Similar species: *Prorocentrum* species such as *P. balticum* are similar in general cell outline but lack the central depression on the valves.

Note. In the literature the genus *Mesoporos* is commonly also referred to as *Mesoporus*.

Synonyms:
Dinoporella perforata (Gran) Halim, 1960,
Exuviaella perforata Gran, 1915,
Porella perforata Schiller, 1928,
Porella adriatica Schiller, 1928,
Porella bisimpressa Schiller, 1928,
Porella asymmetrica Schiller, 1933,
Porotheca perforata (Gran) Silva, 1960.

Literature: Dodge 1982.

Prorocentrum micans Ehrenberg, 1833 — Family: Prorocentraceae

Season				Trophic mode	Shape	Harmful	Bloom	Resting stage
W	S	S	A	mixotrophic	flattened ellipsoid	no		no

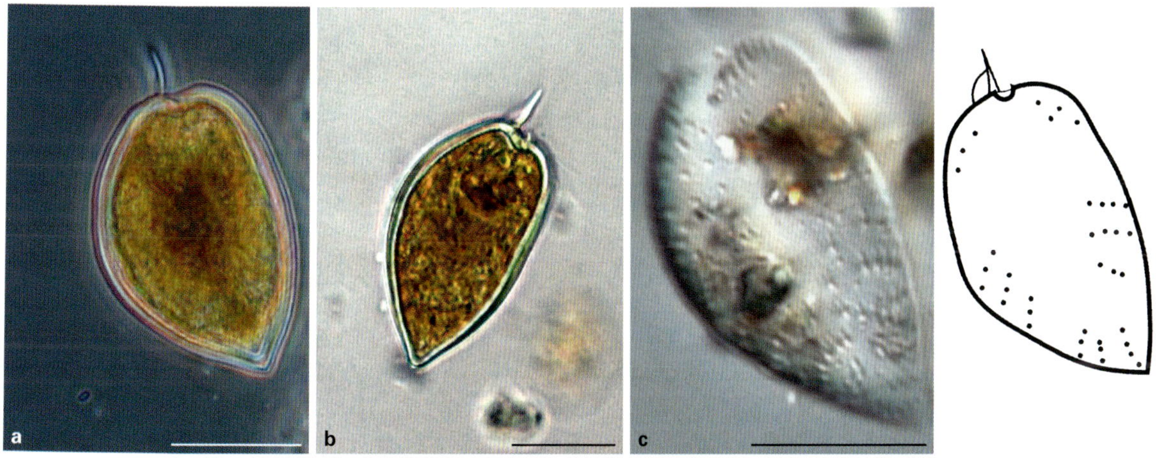

a, Light micrograph (phase contrast) of a *Prorocentrum micans* cell in lateral view. **b,** Formalin-fixed cell. **c,** Light micrograph (DIC) and schematic drawing of a single valve of *P. micans* showing the pattern of small diagonal rows of pores. Scale bars = 20 µm.

Description

Cells strongly laterally flattened with a pyriform to heart-shape outline in lateral view. Both flagella emerge from a pore field near the anterior end of the cell. A prominent spine arising from the anterior end of the cell. Both valves covered by a series of pores and depressions. Small pores in distinct rows running from cell margin towards the cell centre. Larger pores present at posterior end. Depressions or closed pores visible at valve margin. Cells clearly pigmented with a large v-shaped nucleus located in the posterior half of the cell.

Size

Diameter: 20-50 µm
Length: 35-65 µm

Distribution. *Prorocentrum micans* is probably a cosmopolitan species and has been reported from the North Sea and Atlantic as well as the Mediterranean and Pacific.

Similar species: *Prorocentrum gracile* Schütt, 1895: The latter species usually has a narrower diameter in lateral view, although according to Cohen-Fernandez et al 2006 this is not actually a diagnostic feature. A clearer characteristic is the pore pattern and morphology, as only in *P. micans* are small and large pores found.

Synonyms:
Prorocentrum schillerii Böhrn, 1933,
Prorocentrum levantinoides Bursa, 1959,
Prorocentrum pacificum Wood, 1963.

Literature: Bursa 1959, Cohen-Fernandez et al 2006.

Prorocentrum minimum (Pavillard) J. Schiller, 1933 — Family: Prorocentraceae

Season				Trophic mode	Shape	Harmful	Bloom	Resting stage
W	S	S	A	mixotrophic	triaxial ellipsoid	discolouration		no?

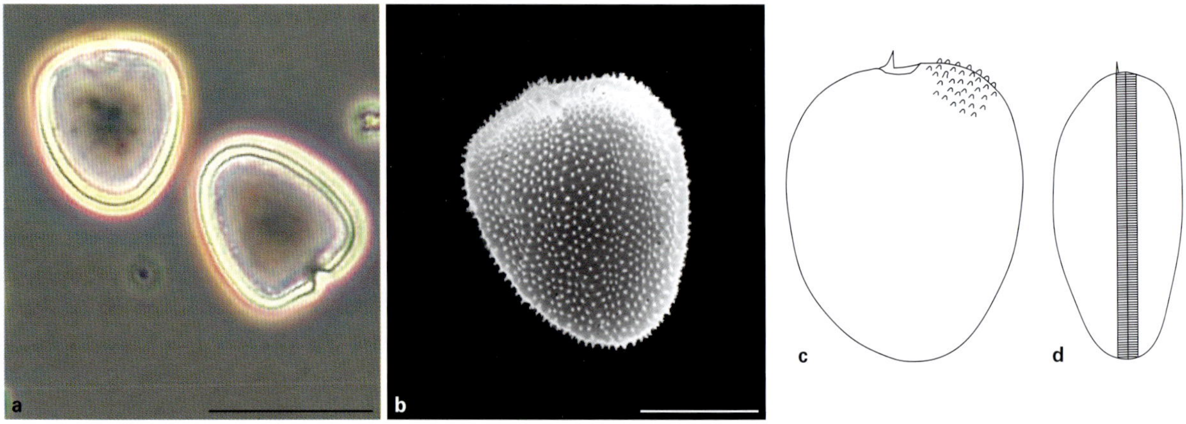

a, Two empty *Prorocentrum minimum* valves. **b,** SEM of a whole cell showing the papillae on the cell surface. **c,** Schematic drawing of *P. minimum* in lateral view, redrawn from Steidinger & Tangen, 1997. **d,** The megacytic growth zone.

Description

Cell outline varying from triangular to oval to heart shaped. An intercalary band with transverse striations located between the two valves. Valves covered by regularly arranged spines or papillae. A short apical spine present. Two types of pores visible on the valves: Small pores scattered all over the valve, larger pores situated at the base of the spines. Two yellowish brown chloroplasts per valve, located in the cell periphery.

Size

Diameter: 14-22 µm
Length: 10-15 µm

Distribution. *Prorocentrum minimum* is widely distributed in temperate waters of the Northern hemisphere. It was first recorded in the North Sea in 1976. Blooms have been recorded in a wide range of environmental conditions but appear to be most intense in brackish waters.

Similar species: *Prorocentrum balticum*: *P. minimum* is larger in size than *P. balticum*. It is also more angular in outline.

Synonyms:
Exuviaella minima Pavillard, 1916,
Prorocentrum triangulatum Martin, 1929,
Exuviaella minima Schiller, 1933,
Exuviaella marie-lebouriae Parke & Ballantine, 1957,
Prorocentrum cordiformis Bursa, 1959,
Prorocentrum marie-lebouriae (Parke & Ballantine) Loeblich III, 1970.

Literature: Pertola et al. 2003, Stoecker et al. 1997.

Prorocentrum triestinum J. Schiller, 1918 — Family: Prorocentraceae

Season				Trophic mode	Shape	Harmful	Bloom	Resting stage
W	S	S	A	autotrophic	triaxial ellipsoid	no		no

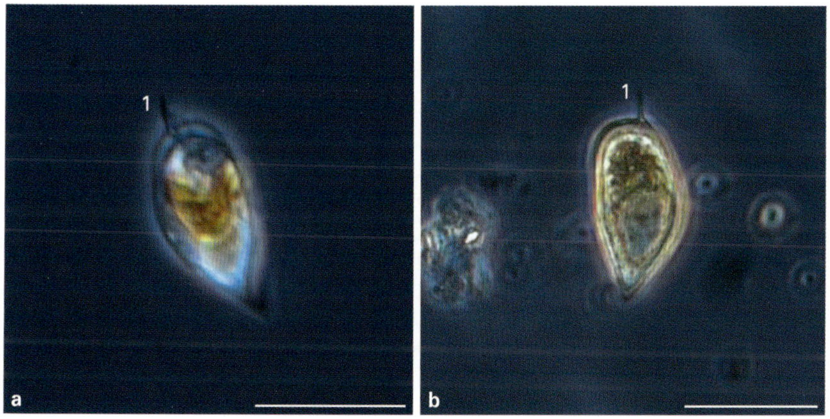

a, Combined phase contrast and UV light image of *Prorocentrum triestinum* in lateral view. **b,** Lugol-fixed cell. Scale bars = 20 µm.

Description

Small *Prorocentrum* species, with a cell outline resembling *P. micans* in lateral view. Valves only weakly ornamented, depressions on the valve surface usually positioned near the valve margin. A thin spine, without wings, emerges from the cell apex (1). Each cell with a single chloroplast.

Size

Diameter: 6-11 µm
Length: 18-22 µm

Distribution. *Prorocentrum triestinum* is widely distributed. In Europe it has been reported, e.g., from the North Sea (e.g., Helgoland, Sylt, Norderney), from the Irish Sea and the Mediterranean.

Similar species: *Prorocentrum redfieldii* has previously been reported as a similar species. However, it is now regarded as synonym of *P. triestinum*.

Synonyms:
Prorocentrum redfieldii Bursa, 1959,
Prorocentrum pyrenoideum Bursa, 1959.

Literature: Bursa 1959, Faust et al. 1999, Schiller 1918.

Ceratium furca (Ehrenberg) Claparède & Lachmann, 1859 — Family: Ceratiaceae

Season				Trophic mode	Shape	Harmful	Bloom	Resting stage
W	S	S	A	mixotrophic	cone	no		no

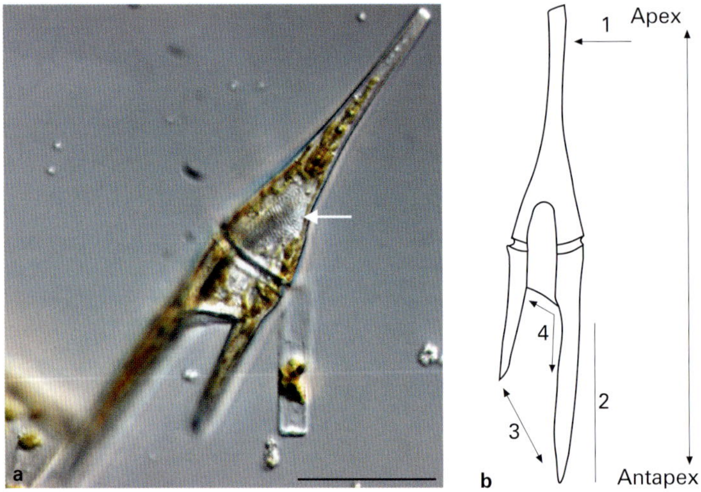

a, DIC image of live cell in ventral view showing the large condensed nucleus (arrow). Scale bar = 20 µm.
b, General schematic of *Ceratium furca*, indicating features important for species identification in the genus: 1, Direction of apical horn (straight or curved); 2, Direction of the antapical horns (straight or curved away from the apical antapical axis); 3, Difference in length of right and left antapical horn; 4, The angle formed between the antapical margin of the main cell body and the antapical horns.

Description

Cell slightly dorso-ventrally flattened and widest in the region of the cingulum. Epitheca tapering gradually into the anterior horn. The hypotheca sub-trapezoid, extending into a long left and a short right antapical horn. Antapical horns usually straight in line with the cell and sometimes toothed along the sides. Left antapical horn usually twice as long as the right one. Thecal plates ornamented with a reticulum of ridges and pores. Plates in the central area of the ventral side very delicate. Nucleus situated in the epitheca. Cells containing numerous yellow-brown chloroplasts.
Plate formula: Po, cp?, 4', 6'', 5-6, 2+, 6''', 2''''

Size

Diameter: 30-35 µm
Length: 150-230 µm

Distribution. *Ceratium furca* has a worldwide distribution. It is found in Helgoland roughly between June and October.

Similar species: *Ceratium hircus* Schröder, 1909 and *C. incisum* (Karsten) E. G. Jørgensen, 1911: In the latter the posterior margin of the cell is u-shaped rather than v-shaped as in *C. furca* and the left antapical horn in *C. incisum* is curved towards the ventral side of the cell. These species have not yet been recorded from the North Sea.

Synonym:
Peridinium furca Ehrenberg, 1836.

Literature: McDermott & Raine 2006.

Ceratium fusus (Ehrenberg) Dujardin, 1841 — Family: Ceratiaceae

Season				Trophic mode	Shape	Harmful	Bloom	Resting stage
W	S	S	A	mixotrophic	2 cones	fish kills	✿ ✿ ✿	no

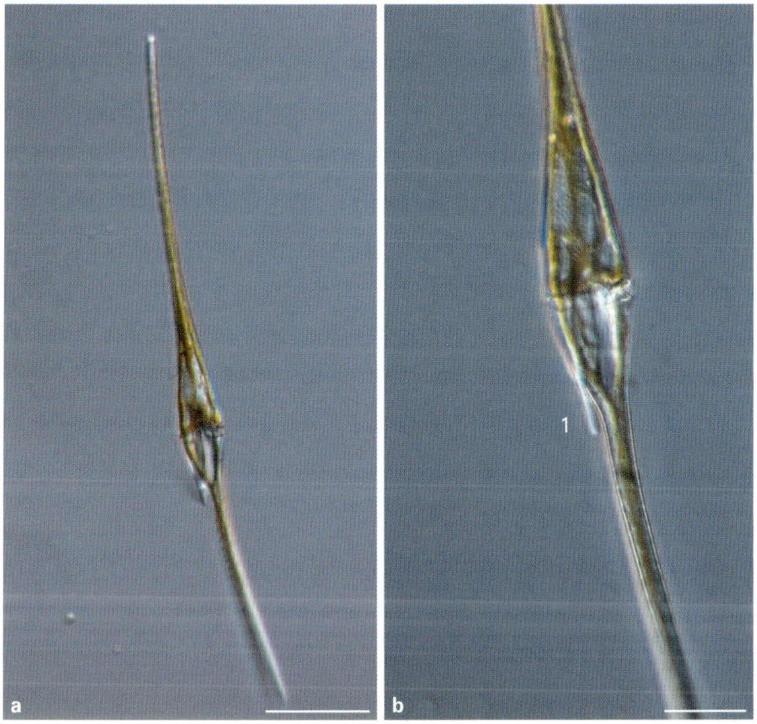

a, DIC image of a live cell of *Ceratium fusus* in ventral view. **b,** The same cell at higher magnification showing the rudimentary right antapical horn (1). Scale bars = 50 (a) and 20 µm (b).

Description

Cells needle-shaped, elongated with curved outline, widest in the area of the cingulum. Epitheca gently tapering into a long anterior horn. Hypotheca extending into a very long left and a rudimentary tooth-like (1) right antapical horn. Thecal plates ornamented. Plates in the central area on the ventral side very delicate. Nucleus situated in the epitheca. Cells with numerous yellow-brown chloroplasts and sometimes food vacuoles. Apical and antapical horns occasionlly bear spines.

Note

Ceratium fusus has been shown to negatively affect invertebrate larvae, although the mechanisms for this have not yet been elucidated. There have also been reports of fish kills during mass occurrences of the dinoflagellate.

Size

Diameter: 15-30 µm
Length: 150-230 µm

Distribution. *Ceratium fusus* is probably a cosmopolitan species. It commonly occurs in the North Sea (including around Helgoland) with blooms occurring during the months of June and July and again in the autumn.

Similar species: *Ceratium extensum* (Gourret) Cleve, 1900 and *C. inflatum* (Kofoid) E. G. Jørgensen, 1911: Both are longer (> 1 mm) than *C. fusus,* and *C. inflatum* has an enlarged epitheca.

Synonym:
Peridinium fusus Ehrenberg, 1834.

Literature: Hallegraeff, Anderson & Cembella (eds.) 2003, Dujardin 1841, McDermott & Raine 2006.

Ceratium horridum (Cleve) Gran, 1902

Family: Ceratiaceae

Season				Trophic mode	Shape	Harmful	Bloom	Resting stage
W	S	S	A	autotrophic?	cone	no	no	no

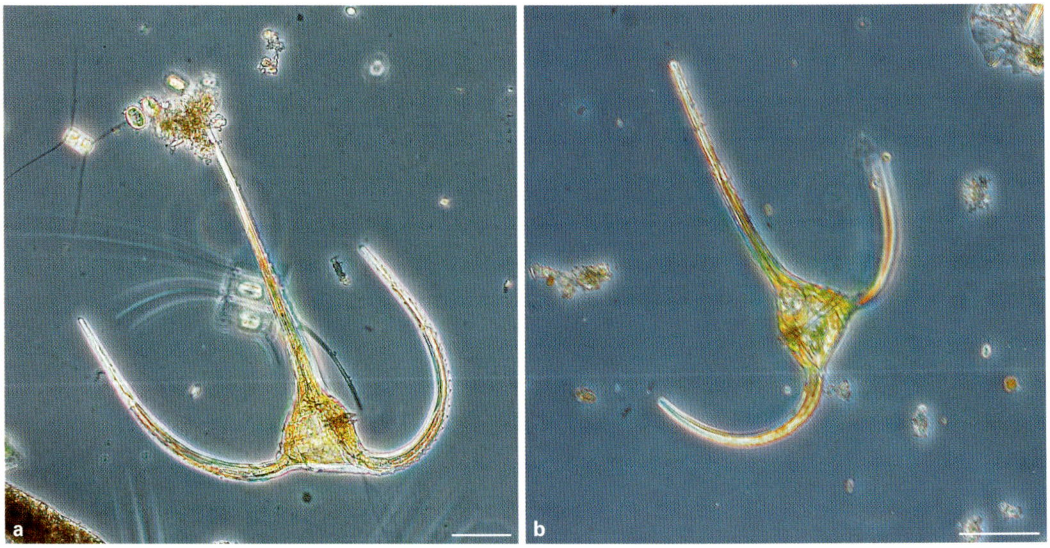

a, b, Phase contrast images of *Ceratium horridum*. Scale bars = 50 µm.

Description

Cells dorso-ventrally flattened. Epitheca asymmetric, with rounded left and steep right side; forming an almost straight anterior open-ended horn, directed slightly to the right. Hypotheca extending into two open-ended antapical horns, directed anteriorly beyond the main cell body. The right horn almost parallel with the apical horn. Thecal plates ornamented with a reticulum of ridges and pores. Plates in the central area of the ventral side very delicate. Nucleus situated in the epitheca. Cells with numerous yellow-brown chloroplasts.

Size

Length: 40-50 µm (not including horns)

Distribution. *Ceratium horridum* has a worldwide distribution in cold to warm temperate coastal and oceanic waters.

Similar species: *Ceratium horridum* is a very variable species, difficult to distinguish from *Ceratium longipes* (Bailey) Gran, 1902. In *C. longipes* the apical horn is more curved to the right.

Synonyms:
Ceratium tripos var. *horridum* Cleve, 1897,
Ceratium intermedium (Jorgensen) Jorgensen, 1905,
Ceratium batavum Paulsen, 1908,
Ceratium tenue (Ostenfeld & Schmidt) Jorgensen, 1911.

Literature: Sournia 1967.

Ceratium lineatum (Ehrenberg) Cleve, 1899 — Family: Ceratiaceae

Season				Trophic mode	Shape	Harmful	Bloom	Resting stage
W	S	S	A	autotrophic	cone	no	✿	no

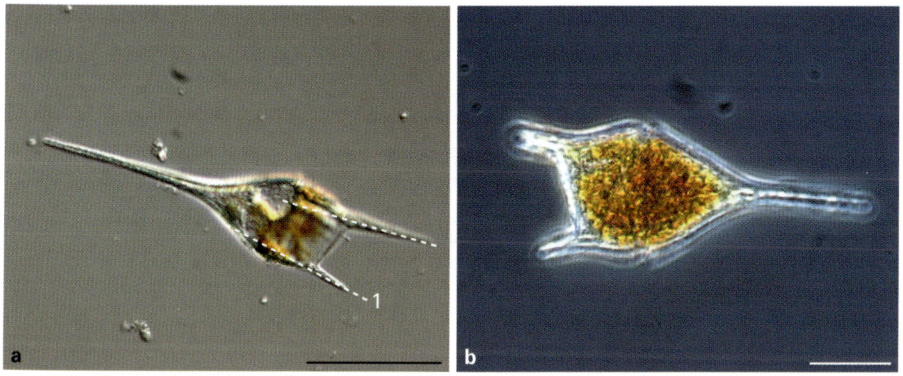

a, DIC image of live cell of *Ceratium lineatum* indicating position of antapical horns in relation to the main cell body. Scale bars = 50 µm.
b, Lugol-fixed cell. Scale bar = 20 µm.

Description

Cells pentagonal in ventral outline, widest at the cingulum and flattened dorso-ventrally. Epitheca triangular, forming a long anterior horn. Hypotheca sub-trapezoid, extending into a long left and a short right closed antapical horn. Both are straight, nearly in line with the main cell body (1) or slightly diverging. Thecal plates ornamented with delicate ridges and pores. Nucleus situated in the epitheca or centrally. Cells containing numerous yellow-brown chloroplasts.

Size

Diameter: 22-41 µm
Length: 105-151 µm (including horns)

Distribution. *Ceratium lineatum* is a widely distributed species (neritic to oceanic) in cold temperate to tropical waters

Similar species: Several warm water species, e.g., *Ceratium pentagonum* Gourret, 1883 have a similar morphology but often have more delicate and less structured thecae. *C. lineatum* species also resembles *C. furca* but is smaller and more delicate.

Synonym:
Peridinium lineatum Ehrenberg, 1854.

Literature: Dodge 1982, McDermott & Raine 2006.

Ceratium macroceros (Ehrenberg) Vanhöffen, 1897 — Family: Ceratiaceae

Season				Trophic mode	Shape	Harmful	Bloom	Resting stage
W	S	S	A	autotrophic	cone	no	no	no

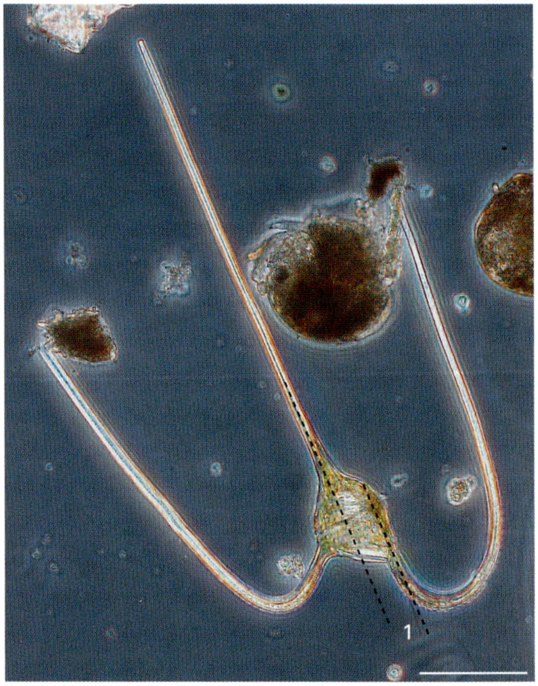

Phase contrast image of *Ceratium macroceros* with lines indicating parallel orientation of apical and left antapical horns. Scale bar = 50 µm.

Description

Cells are dorso-ventrally flattened with open horns. Epitheca rounded with a sharp transition to the long, straight, slender anterior horn. Hypotheca extending into two antapical horns at its posterior edges. Antapical horns emerging in a posterior direction, before curving anteriorly, with the left antapical horn becoming nearly parallel with the apical horn (1). Thecal plates ornamented. Plates in the central area on the ventral side are very delicate. Nucleus situated in the epitheca. Cells containing numerous yellow-brown chloroplasts

Size

Diameter: 50-60 µm (main cell body)
Length: 300-500 µm

Distribution. *Ceratium macroceros* is a cosmopolitan species and has also been recorded from the North Sea.

Similar species: There are several similar species, which are regarded as warm water species, but their presence in Atlantic/North Sea waters is insufficiently investigated. These species include *Ceratium massiliense* (Gourret) E. G. Jørgensen, 1911 and *Ceratium trichoceros* (Ehrenberg) Kofoid, 1908. They are distinguished from *C. macroceros* mainly by the orientation of their antapical (particularly in relation to the apical horn).

Synonym:
Peridinium macroceros Ehrenberg, 1840.

Literature: Dodge 1982, McDermott & Raine 2006, Sournia 1967.

Ceratium tripos (O. F. Müller) Nitzsch, 1817 — Family: Ceratiaceae

Season				Trophic mode	Shape	Harmful	Bloom	Resting stage
W	S	S	A	mixotrophic?	cone	no	✿✿	no

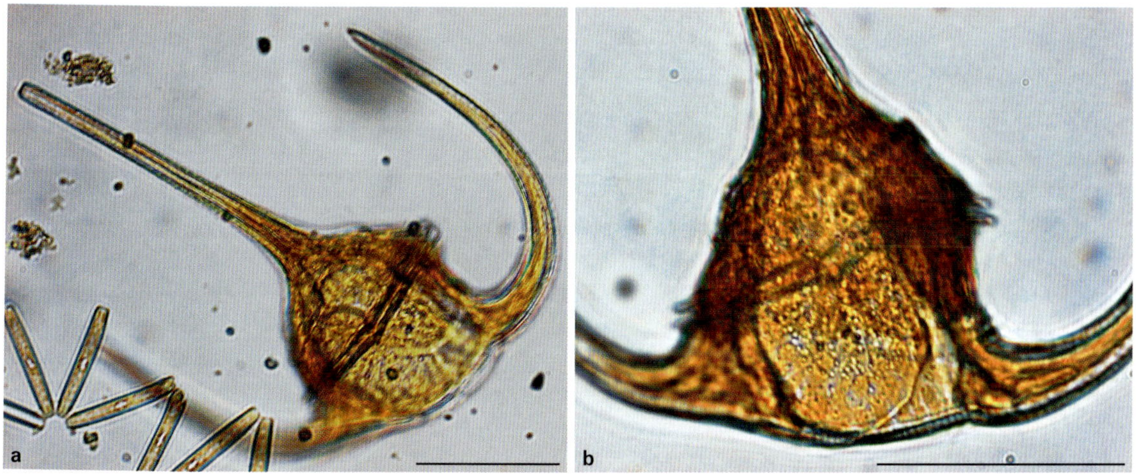

a, Bright field image of whole cell of *Ceratium tripos* in ventral view.
b, The same cell at higher magnification, showing details of the ventral area. Scale bars = 50 µm.

Description

Cells somewhat asymmetrical with the left ventral side slightly shorter than the right side. Apical horn straight. Short antapical horns arising in a straight line from the antapex of the cell before curving upwards until they are almost parallel to the apical horn. Distal end of the apical horns closed. Cells pigmented.

Size

Diameter: 50-80 µm (main cell body)
Length: up to 300 µm

Distribution. *Ceratium tripos* is a common species in temperate coastal waters. In the North Sea around Helgoland it can occur sporadically all year round, but particularly in the autumn.

Similar species: *Ceratium pulchellum* Schröder, 1906 resembles *C. tripos* very closely. However, the antapical horns are often short and stouter than in *C. tripos*, with the right antapical horn shorter than the left. But *C. tripos* is a very variable species with many varieties described and misidentifications are probably not uncommon.

Synonym:
Cercaria tripos O. F. Müller, 1781.

Literature: Dodge 1982, McDermott & Raine 2006, Sournia 1967.

Peridiniella danica (Paulsen) Y. B. Okolodkov & J. D. Dodge, 1995

Family: Cladopyxiaceae

Season				Trophic mode	Shape	Harmful	Bloom	Resting stage
W	S	S	A	heterotrophic	triaxial ellipsoid	no	✿	no

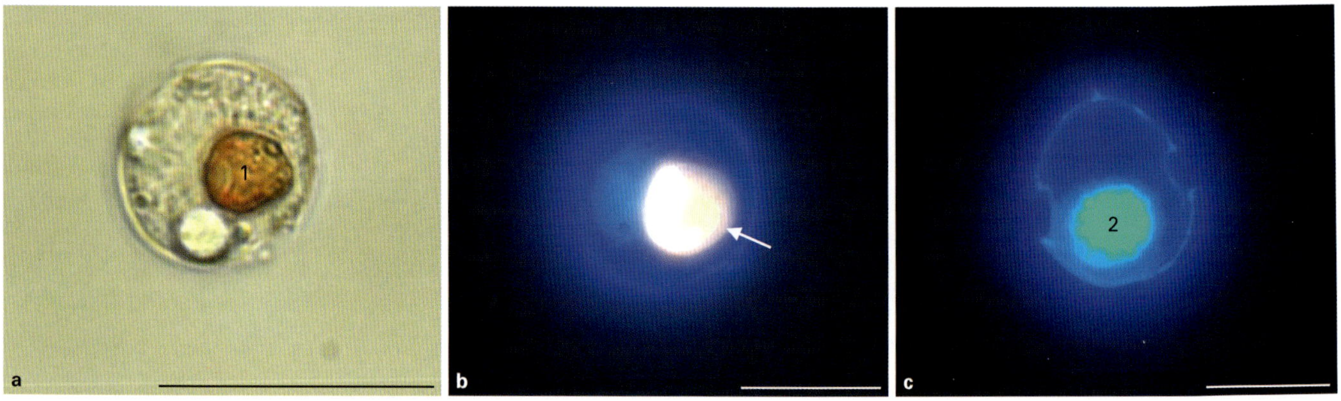

a, Bright field image of formalin-fixed cell of *Peridiniella danica* showing ingested prey (1). Scale bar = 20 µm. **b,** The same cell under UV light showing the pale orange fluorescence of the ingested prey (arrow). **c,** Formalin-fixed cell stained with DAPI to reveal shape and position of the nucleus (2). b,c. Scale bars = 10 µm.

Description

Cell shape somewhat variable but usually round in outline, when seen in ventral view and slightly dorso-ventrally compressed. Cingulum displaced by about one girdle width, located in a median position, deeply excavated and bordered by small lists. Sulcus bordered by lists on both sides, more prominent on the left than the right hand side of the sulcus. Apical pore complex conspicuous with a central pore within a horseshoe shaped plate. Pore plate also with several small pores. Collar formed by the anterior margins of the apical plates. Collar sometimes visible in Lugol-fixed cells as two small notches on either side of the cell apex.
The plate formula is: P0, X, 4', 3a, 7'', 6c, 4s, 6''', 2'''', according to the re-description of Okolodkov & Dodge, 1995, who placed the taxon in the family Cladopyxiaceae.

Note

As this species also shows similarities to the genus *Gonyaulax* it is placed in the family Gonyaulacaceae by some authors.

Size

Diameter: 18-20 µm
Length: 18-25 µm

Distribution. The presence of *Peridiniella danica* has been confirmed from many locations in the North Sea and North East Atlantic and has also been reported from the Baltic (http://planktonnet.awi.de/index.php?contenttype=image_details&itemid=16110). It was recorded at Helgoland in 2005.

Similar species: *Peridiniella danica* is a very small species with a delicate theca that can even be mistaken for an athecate species in Lugol-fixed samples. It also superficially resembles the outline of *Alexandrium* cells but these are usually larger.

Synonym:
Glenodinium danicum Paulsen, 1907.

Literature: MacKenzie 1991, Okolodkov & Dodge 1995.

Gonyaulax spinifera (Claparède & Lachmann) Diesing, 1866 Family: Gonyaulacaceae

Season				Trophic mode	Shape	Harmful	Bloom	Resting stage
W	S	S	A	autotrophic	cone + half sphere	Yessotoxin		yes

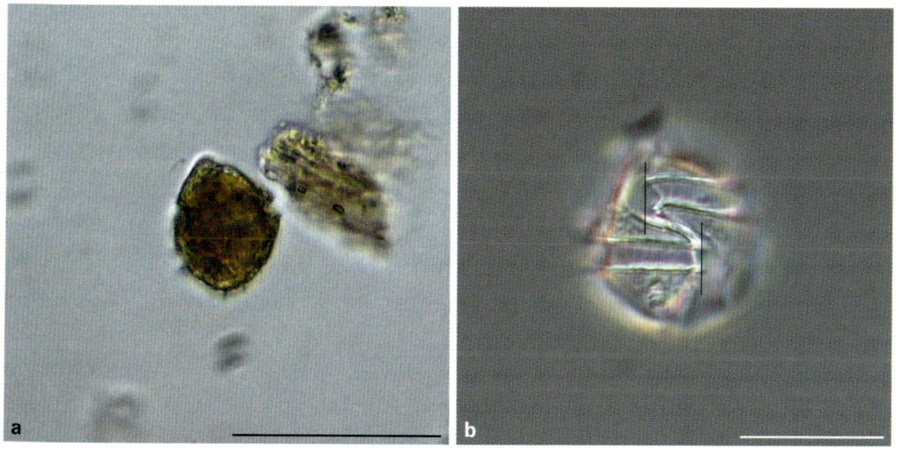

a, Formalin-fixed cell of *Gonyaulax spinifera*. **b,** Empty theca showing the considerably displaced cingulum and the overhang (vertical lines). Scale bars = 50 µm (a) and 20 µm (b).

Description

Cells slightly longer than broad with a small apical horn and a varying number of antapical spines. Cingulum in a median position and descending, displaced by at least twice its width. Cingulum with a pronounced overhang (see vertical lines in image b). Sulcus broad posteriorly but narrowing anteriorly with the narrowest point between the two ends of the girdle. Thecal plates relatively heavily ornamented with reticulations around the many trichocyst pores.

Note

Gonyaulax spinifera is a cyst producer with several cyst types (genus name of cysts: *Spiniferites*) attributed to this species.

Size

Diameter: 30-40 µm
Length: 24-50 µm

Distribution. *Gonyaulax spinifera* probably has a world wide distribution. It is regularly found in the North and Irish Sea and is occasionally reported from the Baltic.

Similar species: *Gonyaulax spinifera* is highly variable in its morphology and probably represents a species complex. It can be confused with *Gonyaulax digitale* (Pouchet) Kofoid, 1911, which has a more pronounced apical horn.

Synonyms:
Spiniferites mirabilis (M. Rossignol) Sarjeant, 1970,
Nematosphaeropsis labyrinthea (Ostenfeld) Reid, 1974,
Tectatodinium pellitum Wall, 1967,
Spiniferites ramosus (Ehrenberg) Mantell, 1854,
Peridinium spiniferum Claparède & Lachmann, 1859.

Literature: Dodge 1989, Ellegaard et al. 2003, Rhodes et al. 2006.

Alexandrium tamarense (Lebour) E. Balech, 1995

Family: Gonyaulacaceae

Season				Trophic mode	Shape	Harmful	Bloom	Resting stage
W	S	S	A	autotrophic	spheroid	PSP toxins	✿✿	yes

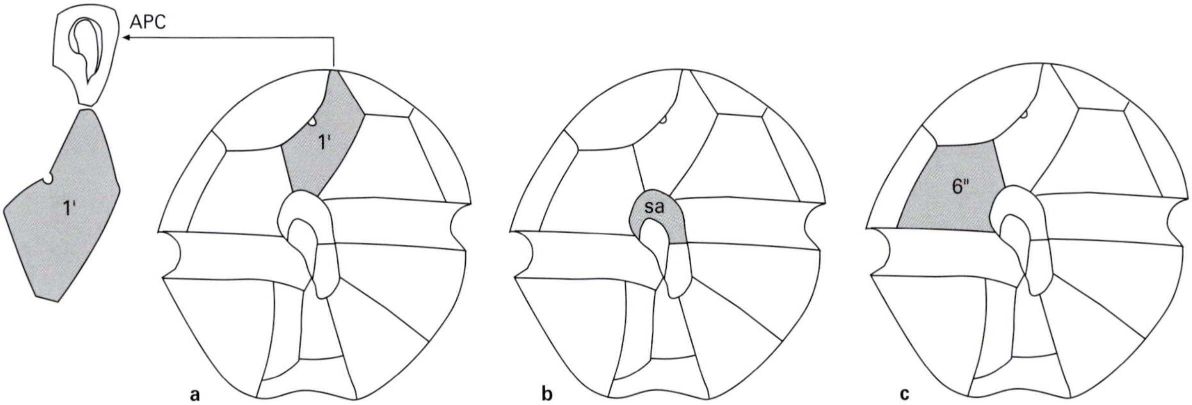

Schematic drawings of the ventral side of *Alexandrium tamarense*.
Plates important for species identification are marked in the three images:
a, First apical plate (1') and apical pore complex. **b,** Anterior sulcal (sa). **c,** 6th precingular plate.
Schematics redrawn from Balech 1995.

Description

Cells almost spherical, slightly angular in ventral view. First apical plate, with a small pore approximately halfway along the right hand suture of the plate with the neighbouring plate. Sixth precingular plate of similar height and width. Cingulum displaced (descending) slightly more than one girdle width. Sulcus widening posteriorly and bordered on both sides by conspicuous lists. Cells pigmented and appearing very dark in Lugol-fixed samples.
Alexandrium tamarense produces ovoid resting cysts with a smooth wall. The resting cysts usually contain reddish lipid bodies and are covered by mucus.

Note

Although this species is regarded as toxic, both toxic and non-toxic strains have been found. In the southern North Sea only the non-toxic strain has been detected so far.

Size

Diameter: 22-51 µm
Length: 17-44 µm

Distribution. *Alexandrium tamarense* is widely distributed in the North Sea and North Atlantic.

Similar species: *Fragilidium subglobosum* is similar in size and shape but has a different plate pattern.

Synonyms:
Gonyaulax tamarensis Lebour, 1925,
Gonyaulax tamarensis var. *excavata* T. Braarud, 1945,
Gonyaulax excavata (Braarud) Balech, 1971,
Gessnerium tamarensis (Lebour) Loeblich III & L. Loeblich, 1979,
Protogonyaulax tamarensis (Lebour) F. J. R. Taylor, 1979,
Protogonyaulax excavata (Braarud) F. J. R. Taylor, 1979,
Alexandrium excavatum (Braarud) Balech & Tangen, 1985.

Literature: Balech 1995, Moestrup & Hansen 1988.

Alexandrium minutum Halim, 1960 Family: Gonyaulacaceae

Season				Trophic mode	Shape	Harmful	Bloom	Resting stage
W	S	S	A	autotrophic	spheroid	PSP toxins	✿	yes

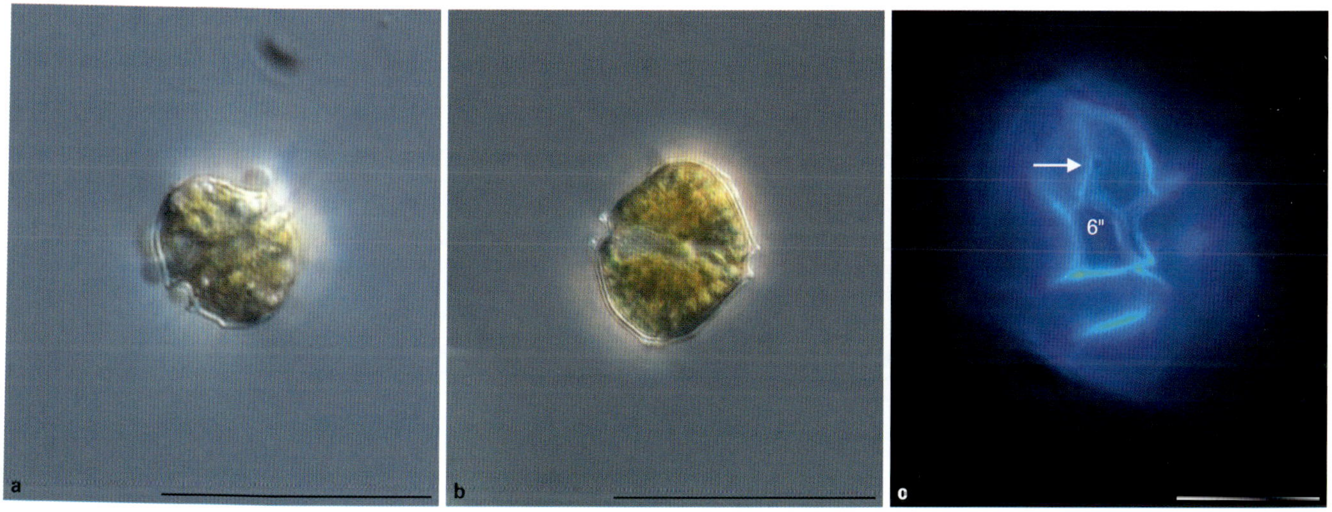

a, b, DIC images of formalin-fixed cells of *Alexandrium minutum*. **c.** Calcofluor stained, formalin-fixed cell showing the first apical plate, ventral pore (arrow) and 6th precingular plate. Scale bars = 50 µm.

Description

Cell outline oval in ventral view. Cells slightly dorso-ventrally compressed. Cingulum wide and descending by about one girdle width. Cingulum bordered by pronounced lists. Sulcus without conspicuous lists. First apical plate relatively narrow with a small ventral pore located nearer to its posterior than the anterior margin. The 6th apical plate higher than wide. The first apical plate is in contact with the apical pore complex, either directly or indirectly, by a thread-like connection.

Size

Diameter: 16-25 µm

Distribution. *Alexandrium minutum* is probably a cosmopolitan species. It has been reported widely, e.g., from the Atlantic and North Sea.

Similar species: *Alexandrium tamarense*, which has a less elongate 6th precingular plate.

Synonyms:
Pyrodinium minutum (Halim) Taylor, 1976,
Alexandrium ibericum E. Balech, 1985,
Alexandrium lusitanicum E. Balech, 1985,
Alexandrium angustitabulatum F. J. R. Taylor, 1995.

Literature: Balech 1995.

Alexandrium ostenfeldii (Paulsen) Balech & Tangen 1985 Family: Gonyaulacaceae

Season				Trophic mode	Shape	Harmful	Bloom	Resting stage
W	S	S	A	autotrophic	rotational ellipsoid	PSP/Spirolide	✿	yes

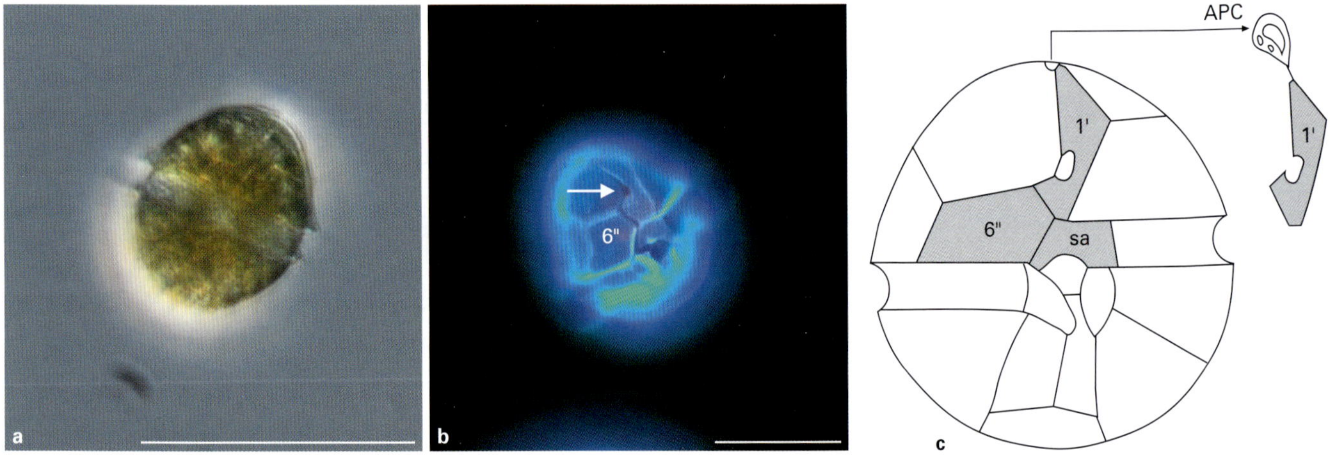

a, Live vegetative cell of *Alexandrium ostenfeldii*. Scale bar = 50 µm.
b, Epifluorescence image (in UV light) showing the first apical plate, the large ventral pore (see arrow) and 6th precingular plate. Scale bar = 20 µm.
c, Schematic drawing of ventral side of the cell and the apical pore complex (APC). Adapted from Balech (1995).

Description

Cells almost spherical, sometimes with a slightly protruding apical pore complex. Cingulum and sulcus shallow and without lists. Cingulum displaced by less than one cingulum width. First apical plate in direct contact with the apical pore complex. Ventral pore large and rounded and positioned at the posterior right margin of the first apical plate. Sixth precingular plate broad. *Alexandrium ostenfeldii* is autotrophic with several red-brown chloroplasts.

Size

Diameter: 35-60 µm,
Length: 40-50 µm.

Distribution. *Alexandrium ostenfeldii* might have a global distribution but is probably often overlooked or misidentified.

Similar species: *Fragilidium subglobosum*: It has a similar outline but a different plate pattern. The theca of the latter easily comes away from the cell body when handled under the microscope.

Synonyms:
Goniodoma ostenfeldii Paulsen, 1904,
Pyrodinium phoneus Woloszynskia & Conrad, 1939,
Goniaulux tamarensis var. *globosa* Braarud, 1945,
Gonyaulax ostenfeldii (Paulsen) Paulsen, 1949,
Heteraulacus ostenfeldii (Paulsen) Loeblich III, 1970,
Gonyaulax globosa (Braarud) Balech, 1971,
Gonyaulax trygvei Parke, 1976,
Protogonyaulax globosa (Braarud) Taylor, 1979,
Gessnerium ostenfeldii (Paulsen) L. Loeblich & Loeblich III, 1979,
Triadinium ostenfeldii (Paulsen) Dodge, 1981.

Literature: Balech 1995, Balech & Tangen 1985.

Protoceratium reticulatum (Claparède & Lachmann) Butschli, 1885 Family: Gonyaulacaceae

Season				Trophic mode	Shape	Harmful	Bloom	Resting stage
W	S	S	A	heterotrophic	sphere-10 %	Yessotoxin	✿	yes

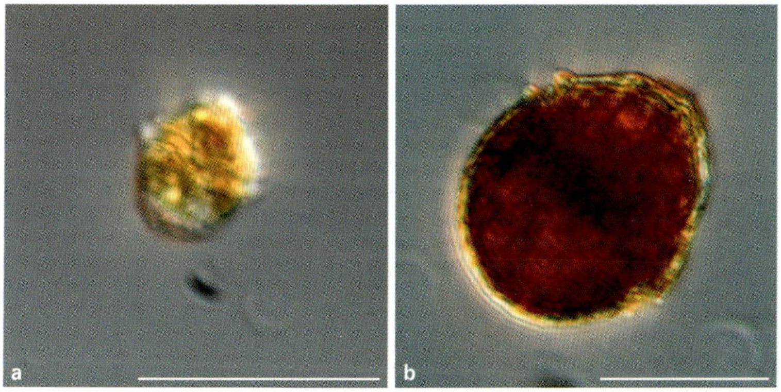

a, Live cell of *Protoceratium reticulatum* from culture. Scale bar = 50 µm. **b,** Lugol-fixed cell. Scale bar = 20 µm. Both images were taken with DIC optics.

Description

Cells polygonal in ventral view. Thecae strongly reticulated resulting in a uneven cell outline in ventral view. Cells pigmented with many dark brown chloroplasts. Cingulum located in a slightly premedian position, descending. Sulcus relatively narrow ending before the antapex. No spines or horns present. The genus *Protoceratium* was re-erected by Dodge 1989 citing the plate formula Po 3' 1a 6" 6c 6s 1p 1"".
This species is by some authors placed in the genus *Gonyaulax* with plate formula: Po, 3', 0a, 6", 6c, 6s, 6"', 0p 2"".

Notes

Protoceratium reticulatum has a characteristic round spiny cyst. The cyst name is *Operculodinium centrocarpum*. The cyst body has about 20-40 slender processes often with bifurcated distal parts.
This species is commonly also referred to as *Gonyaulax grindleyi*, but the genus *Protoceratium* was re-established by Dodge 1989.

Size

Diameter: 24-55 µm

Distribution. *Protoceratium reticulatum* is widely distributed and also regularly occurs in the North Sea. The cyst is a regular and often abundant component of dinocyst assemblages worldwide.

Similar species: none.

Synonyms:
Peridinium reticulatum Claparéde & Lachmann, 1859,
Protoceratium aceros Bergh, 1882.

Literature: Dodge 1989, Hansen et al. 1996/97, Makino et al. 2008, von Stosch 1967.

Heterocapsa rotundata (Lohmann) G. Hansen, 1995

Season				Trophic mode	Shape	Harmful	Bloom	Resting stage
W	S	S	A	autotrophic	2 cones	discolouration	✿ ✿ ✿	no

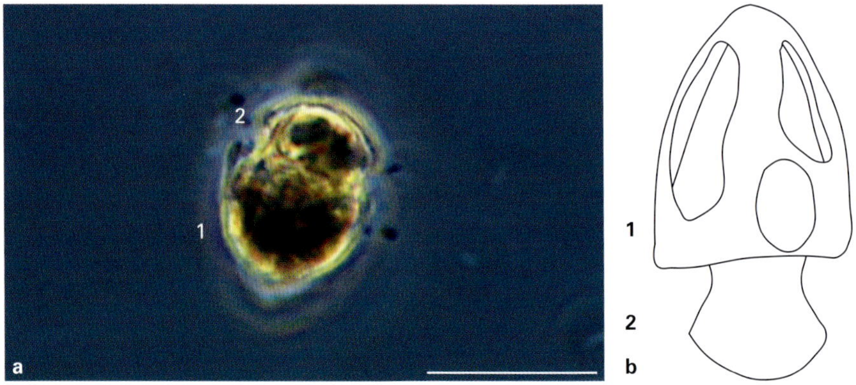

a, Oil immersion image (in phase contrast) of the dinoflagellate *Heterocapsa rotundata.* Scale bar = 10 µm.
b, Schematic drawing of *Heterocapsa rotundata*; 1, epicone, 2, hypocone, redrawn from Dodge 1982

Description

A small thecate species with a much longer epitheca (1) than hypotheca (2). Epitheca cone-shaped and the shorter hypotheca rounded. Thecal plates delicate. Cingulum wide and excavated. Cells pigmented. A large pyrenoid visible in light microscopy.
Because of the delicate plates *Heterocapsa rotundata* has initially been described as a member of the naked genus *Katodinium*.
Plate formula: Po, X, 6', 3a, 7'', 6c, 5s, 5''', 0-1p, 2''''.

Size

Diameter: 6-14 µm
Length: µm 8-17 µm

Distribution. *Heterocapsa rotundata* is a cosmopolitan species. It has been reported from both the North Sea and Baltic. Extensive blooms have been observed along the Norwegian coast, but not yet in the southern North Sea.

Similar species: The species is relatively distinctive but because of its small size it is probably often missed unless oil immersion or epifluorescence microscopy is used.

Synonyms:
Amphidinium rotundatum Lohmann, 1908,
Gymnodinium minutum Lebour, 1925,
Massartia rotundata (Lohmann) Schiller, 1933,
Amphidinium pellucidum Redeke, 1935,
Amphidinium redekei Conrad & Kufferath, 1954,
Massartia rotundatum var. *conradi* Kufferath, 1954,
Katodinium rotundatum (Lohmann) Loeblich III, 1965,
Katodinium minutum Sournia, 1973.

Literature: Hansen 1995.

Heterocapsa triquetra (Ehrenberg) F. Stein, 1883 Peridiniaceae

Season				Trophic mode	Shape	Harmful	Bloom	Resting stage
W	S	S	A	mixotrophic	2 cones	no	✿✿	yes

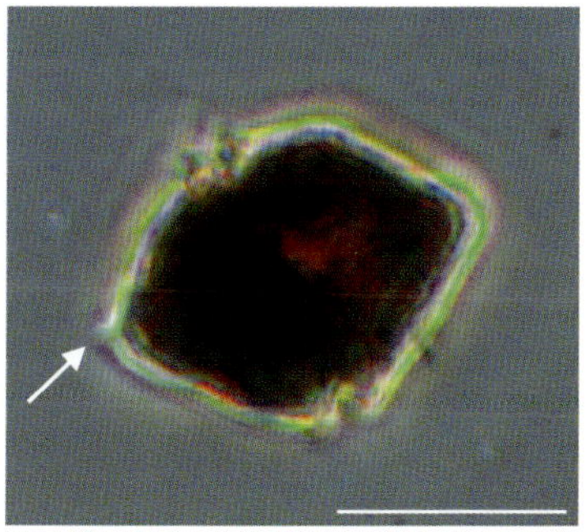

Lugol-fixed single cell of *Heterocapsa triquetra*. The arrow is pointing to the small antapical spine. Scale bar = 20 µm.

Description

A small thecate dinoflagellate. Epitheca and hypotheca conical to rounded in shape, with the hypotheca showing a charac teristic protrusion (see arrow). Theca thin without any spines or ornamentation. Cingulum slightly displaced (descending), wide and deeply excavated. Sulcus hardly discernable. A round pyrenoid present.

Plate formula: Po, x, 5', 3a, 7'', 6c, 5s 5''', 2'''' (according to Hansen & Larsen 1992).

Size

Diameter: 12-18 µm
Length: 18-30 µm

Distribution. *Heterocapsa triquetra* is probably a cosmopolitan species. It is widely distributed, e.g., in the North Sea, Baltic and Atlantic.

Similar species: none.

Synonyms:
Glenodinium triquetrum Ehrenberg, 1840,
Properidinium heterocapsa (Stein) Meunier, 1919,
Peridinium triquetra (Ehrenberg) Lebour, 1925,
Peridinium triquetrum Schiller, 1937.

Literature: Legrand et al. 1998, Morrill & Loeblich III 1981, Olli 1998, Pennick & Clarke 1977, Stein 1883.

Diplopsalis lenticula Bergh, 1882

Family: Protoperidiniaceae

Season				Trophic mode	Shape	Harmful	Bloom	Resting stage
W	S	S	A	heterotrophic	sphere	no		yes

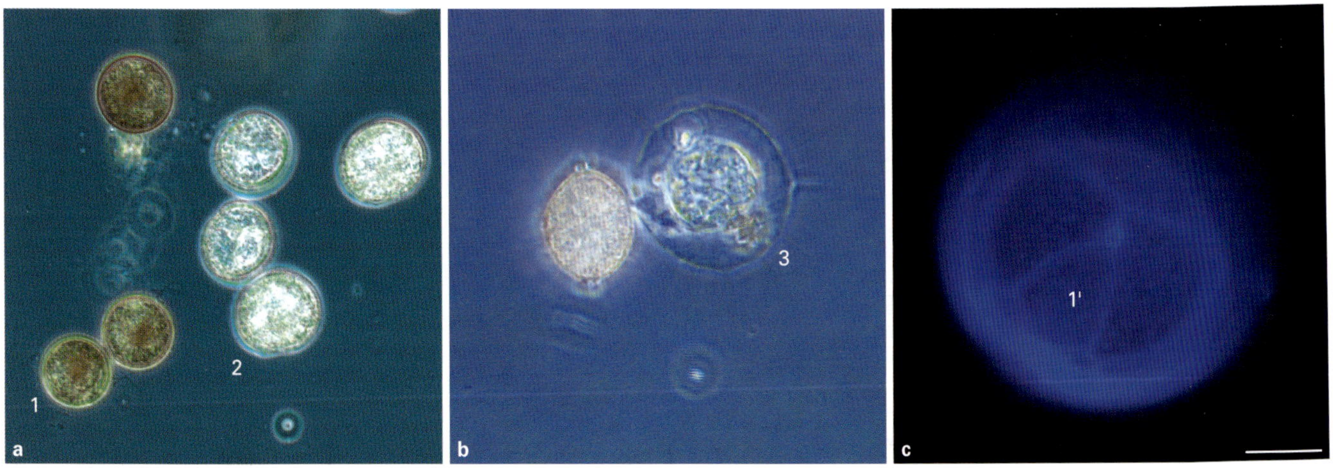

a, Phase contrast image of *Diplopsalis lenticula* cells in culture showing brown cysts (1) and vegetative cells (2).
b, Single cell of *D. lenticula* with everted pallium containing a cell of the diatom genus *Thalassiosira* (3).
c, Epifluorescence image with UV light, showing the first apical plate. Scale bar = 20 µm.

Description

Cells are apically-antapically compressed, usually lying in apical or antapical view on a microscope slide. Plate patterns rarely visible in water mounts. Cells with 3 apical plates and only 1 antapical plate. First apical plate 4-sided and large. Cingulum and sulcus supported by pronounced lists with the left sulcal list extending beyond the antapical cell margin. *Diplopsalis lenticula* is a heterotrophic species feeding with a pallium (3). Cytoplasm in live cells pale pink.
Diplopsalis lenticula forms round, brown cysts (1) with a smooth surface and a large apical archeopyle.

Plate formula: Po, X, 3', 1a, 6'', 4c, 5s, 5''', 1''''.

Size

Diameter: 35-70 µm
Length: 25-50 µm

Distribution. This is a widely distributed species in the North Sea and world-wide.

Similar species: *Diplopsalis lenticula* is a relatively large species but still bears similarities to other members of the *Diplopsalis* group. For reliable identification the plate pattern needs to be examined: In *Diplopsalopsis* Meunier, 1910 for instance there are two antapical plates, whereas there is only one in *Diplopsalis*.

Synonyms:
Peridiniopsis lenticula (Bergh) Starmach, 1974,
Glenodinium lenticula Pouchet, 1883,
Dissodinium lenticulum (Bergh) Loeblich III, 1970.

Literature: Dodge & Hermes, 1981, Dodge & Toriumi 1993, Lebour 1922.

Oblea rotunda (Lebour) Balech ex Sournia, 1973 — Family: Protoperidiniaceae

Season				Trophic mode	Shape	Harmful	Bloom	Resting stage
W	S	S	A	heterotrophic	sphere-10 %	no		yes

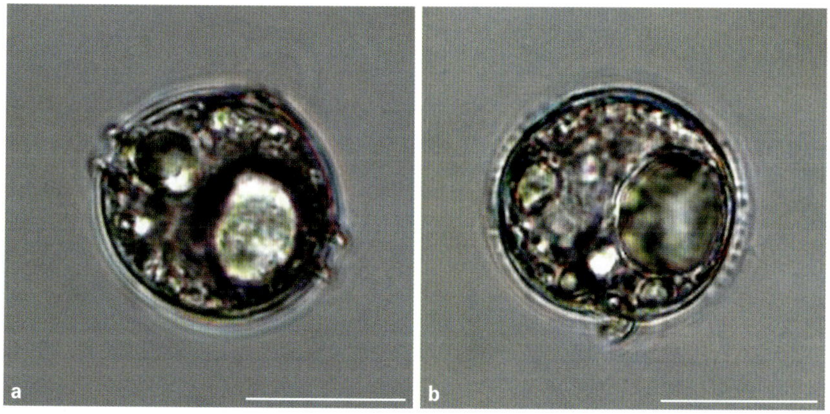

a, DIC image of *Oblea rotunda*. **b,** DIC image of cell in apical view.

Description

Cells globular, only the antapex slightly flattened. Cingulum bordered by lists and situated in a median position. Left side of the sulcus also bordered by a list, extending beyond the antapex of the cell.
Morphology of the first apical (1') and anterior intercalary (1a) plates are diagnostic features. First apical plate broad and five sided (meta). Single intercalary plate large and curving around half of the epitheca touching 5 of 6 precingular plates.
Plate formula: Po, X, 3', 1a, 6'', 4c, 6s, 5''', 2''''.
O. rotunda produces spherical round brown cysts. These cysts are distinguished from other round browns by their small size.

Size

Diameter: 22-34 µm
Length: 22-35 µm

Distribution. *Oblea rotunda* is widely distributed and has been reported, e.g., from the North Sea, English Channel and Mediterranean. It has recently also been described from brackish waters (Chomérat et al. 2004).

Similar species: *Oblea baculifera* Balech ex Loeblich Jr. & Loeblich III, 1966: This species is usually smaller than *O. rotunda*. In *O. rotunda* the first precingular plate is distinctly smaller than the 6th precingular plates.

Synonyms:
Peridiniopsis rotunda Lebour, 1922,
Glenodinium rotundum Schiller, 1937,
Diplopsalis rotunda Wood, 1968,
Diplopsalis rotundata Steidinger & Williams, 1970.

Literature: Chomérat et al. 2004, Lewis 1990, Strom & Buskey 1993.

Preperidinium meunieri (Pavillard) Elbrächter, 1993

Family: Protoperidiniaceae

Season				Trophic mode	Shape	Harmful	Bloom	Resting stage
W	S	S	A	heterotrophic	flattened ellipsoid	no	no	yes

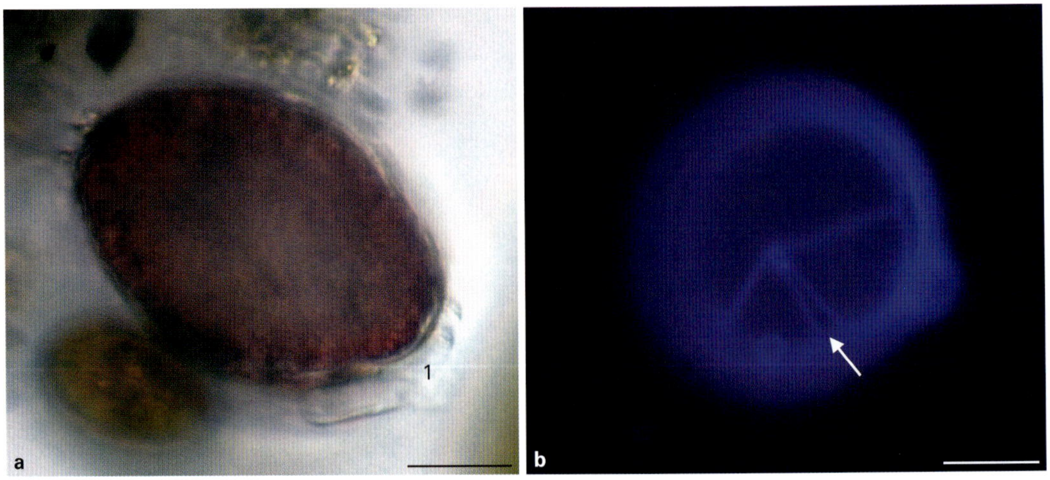

a, Brightfield image of *Preperidinium meunieri* in lateral view showing the sulcal list (1).
b, Epifluorescence image under UV light showing the narrow first apical plate (arrow). Scale bars = 20 μm.

Description

Cell outline oval in ventral view and almost round in apical view. Epitheca with convex sides and a small apical projection. Hypotheca also with convex sides. Girdle in a median position, not excavated but bordered by lists that are supported by spines. Sulcus also bordered by a narrow list appearing to extend beyond the antapex. The list can be seen both in ventral/ dorsal and in apical/antapical view.
First apical plate narrow. First anterior intercalary plate very small with a square outline, located on the left side of the cell.
Hypotheca with one antapical plate.
Plate formula: Po, x, 3', 2a, 7'', 4c, 5s, 5''', 1''''.
Preperidinium produces a round brown cyst. The cyst name is *Dubridinium caperatum*.

Size

Diameter: 28-60 μm

Distribution. *Preperidinium meunieri* has a wide distribution. In European waters it has been reported, e.g., from the Irish, North and Black Seas.

Similar species: Several members of the *Diplopsalis* group are similar in outline. They are distinguished by their plate pattern, but this can often only be visualized using epifluorescence microscopy. The plates are rarely seen in standard Lugol-fixed samples.

Synonyms:
Preperidinium paulseni (Mangin) Mangin,
Diplopsalis lenticula f. *minor* Paulsen, 1907,
Peridinium lenticulum Mangin, 1911,
Peridinium paulsenii Mangin, 1911,
Peridinium meunierii Pavillard, 1912,
Diplopeltopsis minor Pavillard, 1913,
Diplopsalis minor (Paulsen) Lindemann, 1927,
Glenodinium lenticula f. *minor* Schiller, 1937,
Zygabikodinium lenticulatum Loeblich Jr. & Loeblich III, 1970.

Literature: Dodge 1982, Dodge & Hermes 1981, Dodge & Toriumi 1993, Gribble & Anderson 2006.

Protoperidinium (pp. 151-167)

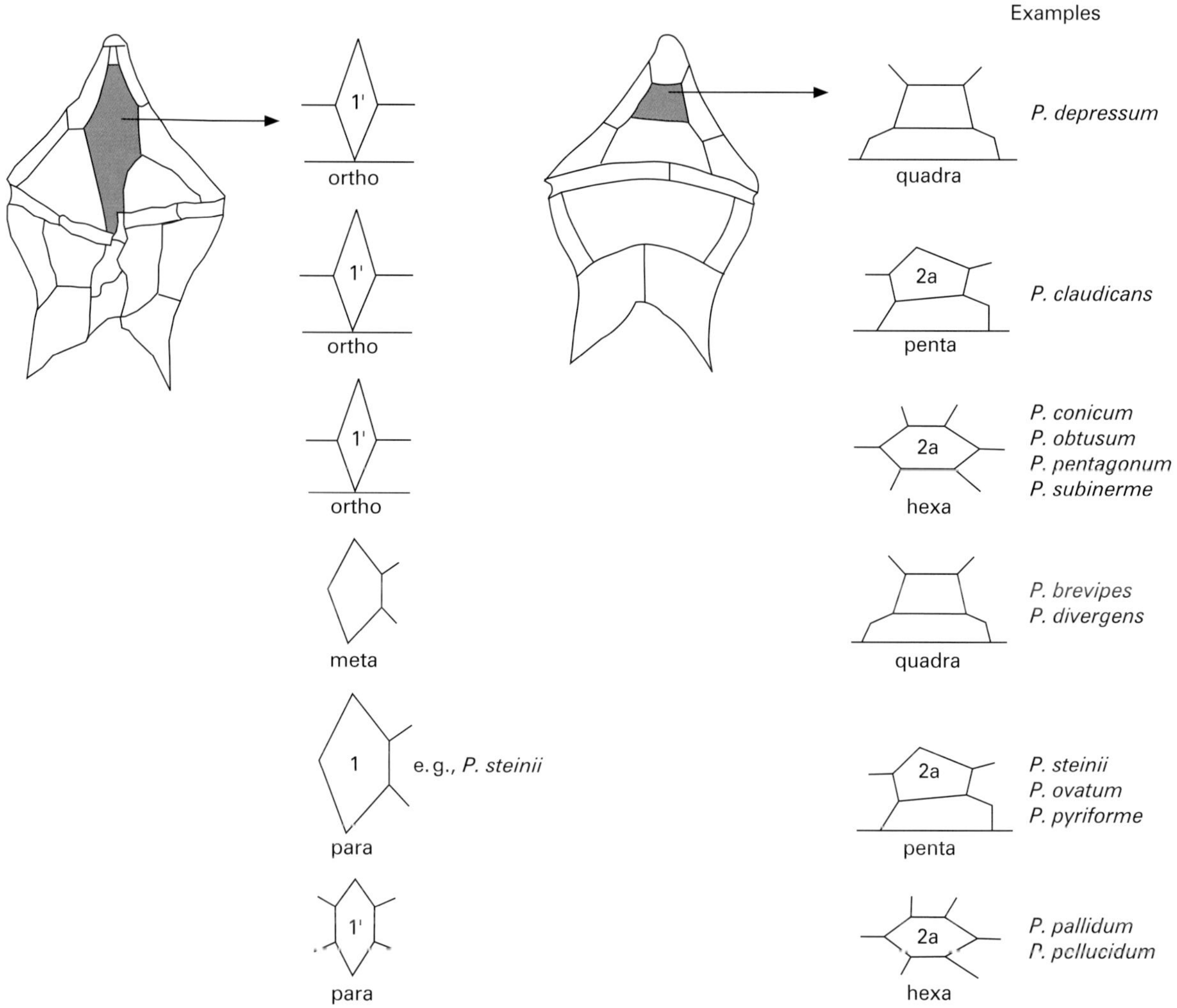

The typical tabulation in the genus *Protoperidinium* is Po, X, 4', 2-3a, 7", 4c, 6s, 5"', 2"". However, the shape of the plates varies considerably between species and can be used for their identification. Two plates are particularly important: The first apical plate (i.e. the first plate in the apical series) and the second anterior intercalary plate. Both plates can be 4, 5 or 6 sided. In the case of the apical plate these are called ortho, meta and para, respectively. The anterior intercalary plates on the other hand are named quadra, penta and hexa. Various combinations of the two plates occur in different *Protoperidinium* species. In addition the length and width of the plates can also vary. This is another character that can be used for species identification.

Protoperidinium claudicans (Paulsen) Balech, 1974   Family: Protoperidiniaceae

Season				Trophic mode	Shape	Harmful	Bloom	Resting stage
W	S	S	A	heterotrophic	cone + half sphere	no	no	yes

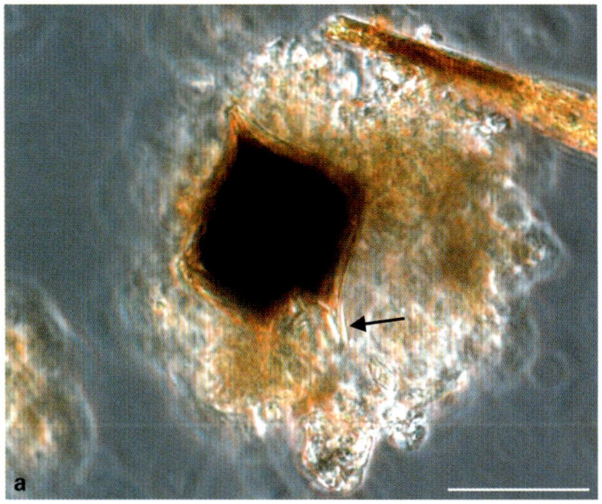

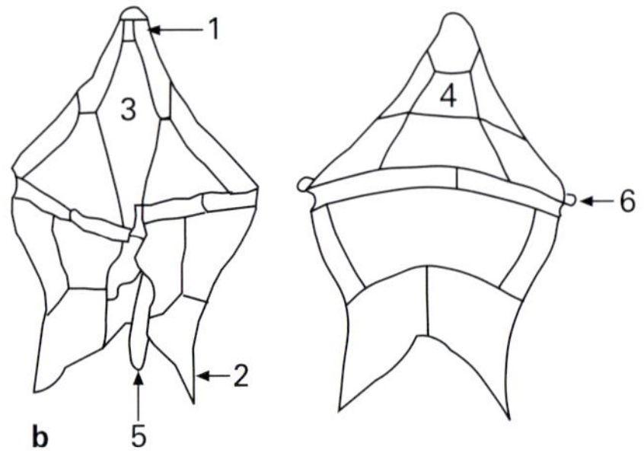

a, Lugol-fixed cell of *Protoperidinium claudicans* with the arrow pointing to the antapical spines. Scale bar = 50 µm (Image courtesy of Martin Löder).
b, General schematic of a *Protoperidinium* cell indicating important features used for species identification: 1, morphology of apical horn; 2, antapical horns; 3,4, Shape of first apical and second anterior intercalary plates respectively; 5, sulcal lists; 6, cingular lists.

Description

Cells strongly dorso-ventrally compressed. Short apical and antapical horns are present. Antapical horns point towards the dorsal side of the cell. Cingulum displaced by at least one cingulum width (descending).
Thecal plates delicate and smooth or weakly reticulated. First apical plate ortho, second intercalary plate penta or rarely hexa.
Protoperidinium claudicans is a cyst producer. The cyst name is *Votadinium spinosum* Reid. Intact cysts have a peridinioid outline and a thin brown cell wall. The cysts are compressed dorso-ventrally.

Size

Diameter: 48-75 µm
Length: 50-105 µm

Distribution. *Protoperidinium claudicans* is probably a cosmopolitan species and has been reported, e.g., from the Atlantic and North Sea and from the Pacific.

Similar species: *Protoperidinium claudicans* can be confused with *P. latidorsale* (P. Dangeard) Balech, 1974 when its 2a plate is hexa.

Synonym:
Peridinium claudicans Paulsen, 1907.

Literature: Marret & Zonneveld 2003, Okolodkov 2005.

Protoperidinium depressum (Bailey) Balech, 1974 — Family: Protoperidiniaceae

Season				Trophic mode	Shape	Harmful	Bloom	Resting stage
W	S	S	A	heterotrophic	2 cones	no	✿✿	yes

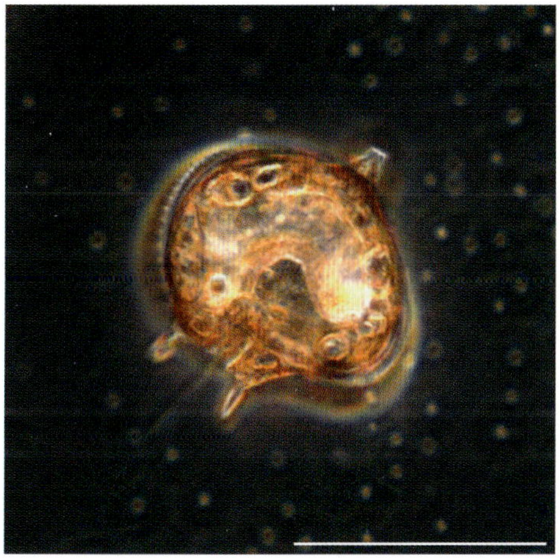

Phase contrast image of *Protoperidinium depressum* from culture. Scale bar = 100 µm.

Description

A large species with long apical and antapical horns. Due to its symmetry (cingulum plane strongly oblique with respect to the longitudinal axis with the dorsal part closer to the cell apex than the venral part), cells often seen in antapical view in settled samples with the apical horn partially obscured. In ventral view outline of the epitheca concave near the apical horn, but convex, i.e. bulging closer to the cingulum. Hypotheca widest near the cingulum. Cingulum descending and the sulcus deeply excavated. First apical plate ortho, second anterior intercalary plate quadra.
Cells normally have a large number of reddish inclusions, sometimes described as 'fat droplets' which are visible in live and Lugol-fixed cells.

Note

This is a heterotrophic species using a pallium to trap and ingest prey organisms. It also shows bioluminescence.

Size

Diameter: 115-144 µm
Length: 116-200 (incl. horns) µm

Distribution. *Protoperidinium depressum* can be abundant and has been reported from Arctic to Antarctic regions and in coastal and oceanic waters.

Similar species: *Protoperidinium divergens*.

Synonym:
Peridinium depressum Bailey, 1854.

Literature: Balech 1974, Carreto 1985.

Protoperidinium pentagonum (Gran) Balech, 1974

Family: Protoperidiniaceae

Season				Trophic mode	Shape	Harmful	Bloom	Resting stage
W	S	S	A	heterotrophic	2 cones-30 %	no	no	yes

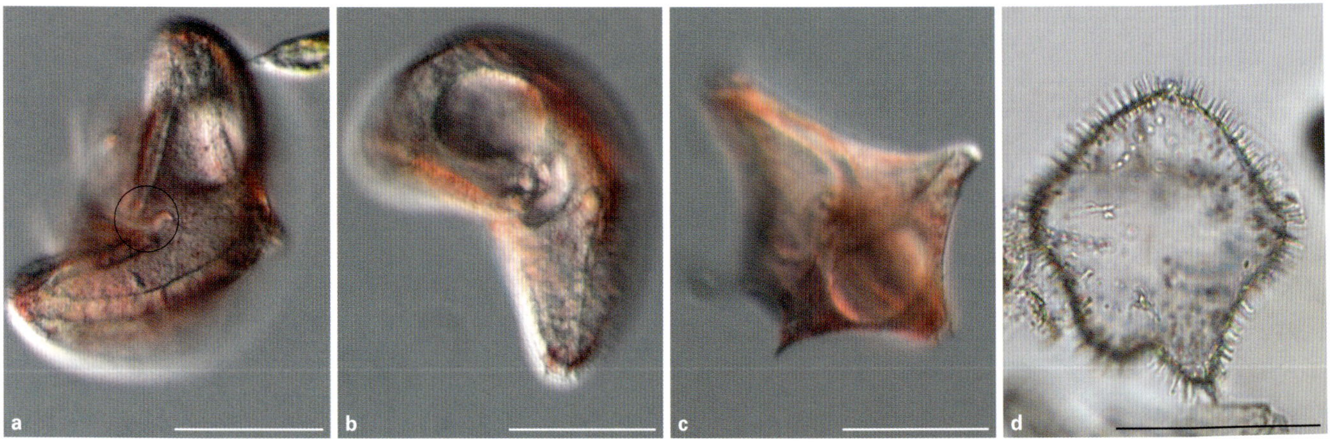

a, Live image of the epitheca of *Protoperidinium pentagonum* in ventral view showing the displaced cingulum (circle). **b,** Ventral view of epitheca with focus on the large vacuole. **c,** Live cell in ventral view. **d,** Spiny resting cyst of *P. pentagonum*. Scale bars = 50 µm.

Description

A large and conspicuous species, cell outline rectangular in ventral view. In apical view, cells strongly dorso-ventrally compressed and kidney shaped. Cingulum descending, displaced by 1-2 times its width. Sulcus slightly invading the epitheca, narrow anteriorly and widening posteriorly. Due to its symmetry *Protoperidinium pentagonum* is rarely seen in ventral or dorsal view in fixed samples. Live cells with many reddish inclusions and vacuoles. This is a heterotrophic species.
The first apical plate is ortho and the second anterior intercalary plate hexa.
Protoperidinium pentagonum is a cyst-producing species. The cyst name is *Trinovantedinium applanatum.* It has a laterally compressed *Peridinium*-like body with an apical horn and two antapical lobes. The cyst wall is pale brown in colour.

Size

Diameter: 75-100 µm
Length: 75-110 µm

Distribution. *Protoperidinium pentagonum* is a cosmopolitan, mainly neritic species. It is also found in brackish waters.

Similar species: *Protoperidinium pentagonum* is a very distinctive species.

Synonyms:
Peridinium pentagonum Gran, 1902,
Peridinium sinuosum Lemmermann, 1905.

Literature: Balech 1974, Hansen & Larsen 1992.

Protoperidinium divergens (Ehrenberg) Balech, 1974 — Family: Protoperidiniaceae

Season				Trophic mode	Shape	Harmful	Bloom	Resting stage
W	S	S	A	heterotrophic	(cone + half sphere)-20 %	no	✿	no

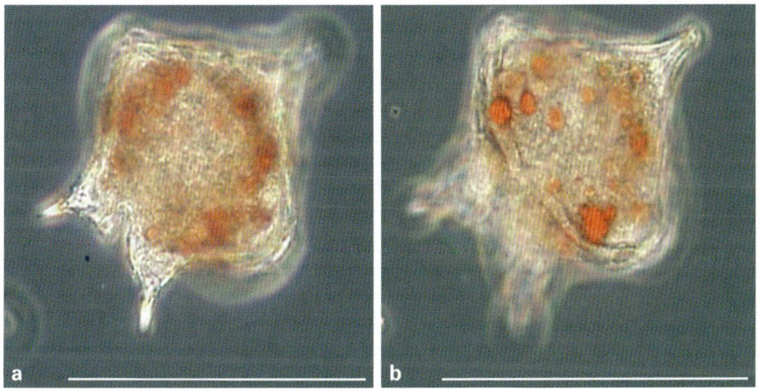

a, b, Lugol-fixed cells of *Protoperidinium divergens*. Images courtesy of Regina Hansen.

Description

Epi- and hypotheca of *Protoperidinium divergens* with concave sides. An apical horn and two slender, diverging antapical horns present. Cingulum located in a median position, circular or slightly ascending. Sulcus straight and bordered by a small list on the left side. The whole theca finely reticulated with additional spines on the hypotheca, particularly the antapical horns.
The first apical plate is meta and the second anterior intercalary plate quadra.

Size

Diameter: 55-65 µm
Length: 75-85 µm

Distribution. *Protoperidinium divergens* is a neritic species. It has been reported from the North Sea, but its worldwide distribution is not well known as it can be misidentified as *P. curtipes* or *P. crassipes.*

Similar species: *Protoperidinium curtipes* (Jorgensen) Balech, 1974 and *Protoperidinium crassipes* (Kofoid) Balech, 1974.

Synonym:
Peridinium divergens Ehrenberg, 1841.

Literature: Jeong 1994, 1994a; Jeong et al. 1997.

Protoperidinium conicum (Gran) Balech, 1974 Family: Protoperidiniaceae

Season				Trophic mode	Shape	Harmful	Bloom	Resting stage
W	S	S	A	heterotrophic	cone + half sphere	no	✿	yes

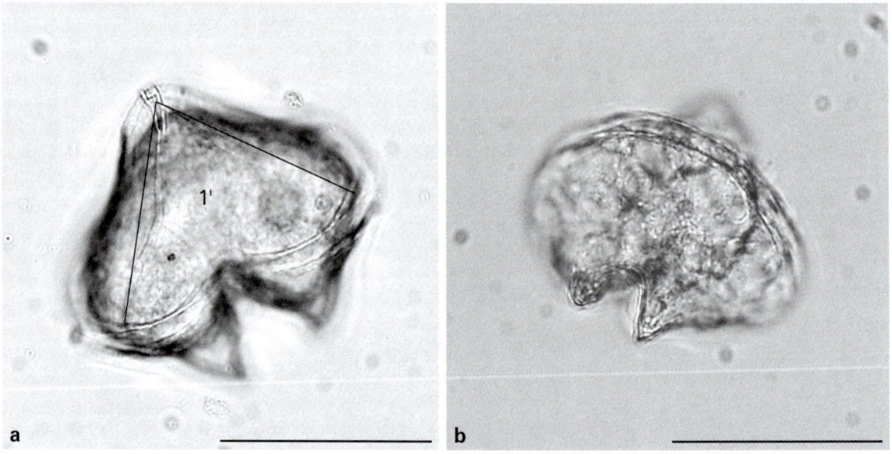

a, Apical part of *Protoperidinium conicum* in ventral view, showing the triangle formed by first apical, first and seventh precingular plates.
b, dorso-antapical view of the cell showing the two horns.
Scale bars = 50 µm. Images courtesy of Regina Hansen.

Description

First apical plate very large and rectangular. The outer sutures of the first and seventh precingular plates aligned with those of the first apical plate to produce two distinct straight lines running from the cell apex to the cingulum and forming a triangle (sometimes visible in light microscopy, see lines in image a). Sulcus deeply excavated and widening posteriorly resulting in two horns. Two short antapical spines present. First apical plate ortho and second anterior intercalary plate hexa.
Protoperidinium conicum forms cysts. The cyst name is *Selenopemphix quanta* (Bradford) Matsuoka. Cysts are subspherical, apically-antapically compressed with long, solid spines that are broadened at their base. These spines occur in rows on the cyst including along the cingular margin.

Size

Diameter: 60-80 µm
Length: 70-90 µm

Distribution. *Protoperidinium conicum* is a cosmopolitan species.

Similar species

Protoperidinium leonis (Pavillard) Balech, 1974: In this species two antapical horns are present on the antapex and it has a more pronounced reticulation. If the apical plates are discernable *P. conicum* can be distinguished from *P. leonis* by the triangle formed by its first apical, first and seventh precingular plates.
Protoperidinium conicoides: This species has a similar plate pattern, but the first apical plate has a characteristic notch, where it meets the sulcus.

Synonyms:

Peridinium divergens var. *conica* Gran, 1900,
Peridinium conicum (Gran) Ostenfeld & Schmidt, 1902.

Literature: Hansen & Larsen 1992, Nehring 1997.

Protoperidinium obtusum (Karsten) Parke & Dodge, 1976 Family: Protoperidiniaceae

Season				Trophic mode	Shape	Harmful	Bloom	Resting stage
W	S	S	A	heterotrophic	cone + half sphere	no	no	yes

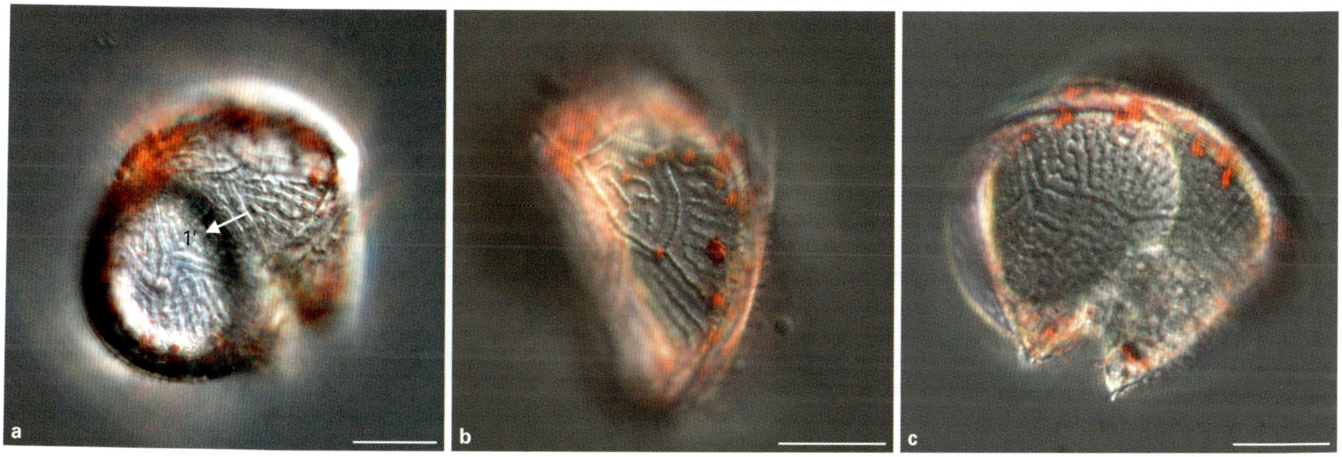

a, DIC image of live cell showing the ventral side with the characteristic first apical plate (arrow).
b, The same cell in lateral view showing the reticulations on some of the apical plates.
c, The same cell in dorso-antapical view. Scale bars = 20 µm.

Description

Epitheca with straight sides in ventral view and a blunt apex without apical horn. Epitheca ornamented with pronounced series of longitudinal ridges (reticulation), running in an apical-antapical direction. Ridges less pronounced on first apical plate. Shape of first apical plate characteristic of the species with the two sides bordering the girdle being much longer than the two sides bordering the apex (image a). Hypotheca ornamented but not in the form of longitudinal striations. Striation patterns clearly visible using light microscopy. Hyptheca terminating in two small antapical horns ending in short spines. First apical plate ortho, second intercalary plate hexa.

Size

Diameter: 50-55 µm
Length: 60-65 µm

Distribution. *Protoperidinium obtusum* has been widely reported from temperate regions, including the North Sea, and tropic regions. But it is absent from the arctic.

Similar species: *Protoperidinium conicum* and *P. leonis*: *P. obtusum* can be distinguished from these two species by the typical reticulation of its theca, which forms striations on the epitheca (except the first apical plate) but much shorter point like structures on the hypotheca.

Synonyms:
Peridinium obtusum Karsten, 1906,
Peridinium divergens var. *obtusum* G. Karsten, 1906,
Peridinium leonis f. *matzenaueri* J. Schillor, 1937.

Literature: Balech 1948, Okolodkov 2005.

Protoperidinium ovatum Pouchet, 1883 Family: Protoperidiniaceae

Season				Trophic mode	Shape	Harmful	Bloom	Resting stage
W	S	S	A	heterotrophic	triaxial ellipsoid	Discolouration	❁	yes

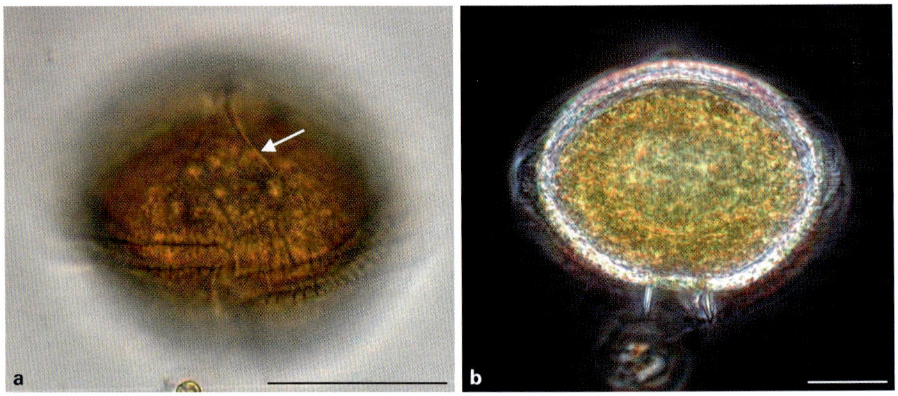

a, Bright field image of a live cell showing the shape of the first apical plate (arrow).
b, Phase contrast image showing the short antapical spines.

Description

Cells compressed in an apical-antapical direction giving them an oval outline in ventral/ dorsal view. Epitheca with a very small apical horn. Hypotheca ending in two short spines. Cell surface covered by reticulations that surround pores in the theca. Cingulum ascending but only weakly displaced and bordered by a broad cingular list supported by spines.
The first apical plate is meta and the second anterior intercalary plate penta. In settled samples, cells are often seen in apical or antapical view.

Size

Diameter: 50-70 µm
Length: 50-88 µm

Distribution. *Protoperidinium ovatum* is widely distributed and it can form blooms.

Similar species: In fixed samples, when plate patterns are obscured, *Protoperidinium ovatum* resembles *Diplopsalis lenticula* when it is seen in apical or antapical view. It is usually possible to distinguish it becuse of the stout spines supporting the cingular list.

Synonyms:
Peridinium ovatum (Pouchet) Schütt, 1895,
Peridinium globulus var. *ovatum* (Pouchet) Schiller, 1937.

Literature: Balech 1976.

Protoperidinium pallidum (Ostenfeld) Balech, 1973 — Family: Protoperidiniaceae

Season				Trophic mode	Shape	Harmful	Bloom	Resting stage
W	S	S	A	heterotrophic	(cone + half sphere)-20 %	no	no	no

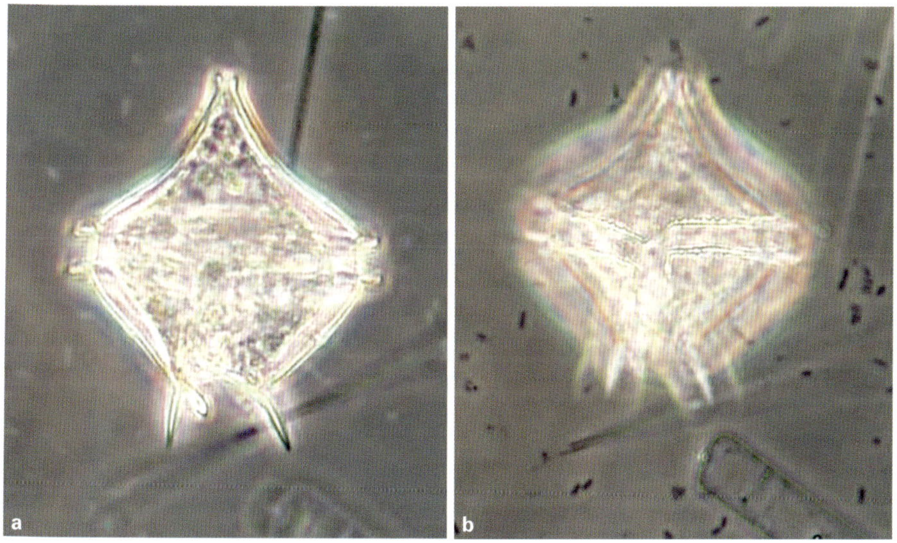

a, b, Images of formalin-fixed *Protoperidinium pallidum* cells (diameter approximately 80 μm). Images courtesy of Regina Hansen.

Description

Protoperidinium pallidum is a large dorso-ventrally compressed species. Epitheca with straight or convex sides when seen in ventral view. Hypotheca also with straight or slightly convex sides and two diverging, winged antapical spines. Cingulum in a median position with no or only very slight displacement (ascending). Sulcus broad and on its left side with a pronounced list extending beyond the cell antapex. The theca is smooth or weakly reticulated. The first apical plate is para. The second anterior intercalary plate is hexa.

Size

Diameter: 65-85 μm
Length: 70-100 μm

Distribution. This species has a worldwide distribution.

Similar species: *Protoperidinium pellucidum*: This species also has a pronounced sulcal list, but is smaller than *P. pallidum*. The first apical plate in *P. pellucidum* is also broader than in *P. pallidum* and the outline of *P. pellucidum* in a transverse section is more rounded.

Synonym:
Peridinium pallidum Ostenfeld, 1899.

Literature: Hansen & Larsen 1992.

Protoperidinium pellucidum Bergh ex Loeblich Jr. & Loeblich III, 1966

Family: Protoperidiniaceae

Season				Trophic mode	Shape	Harmful	Bloom	Resting stage
W	S	S	A	heterotrophic	(cone + half sphere)-25 %	no	✿	yes

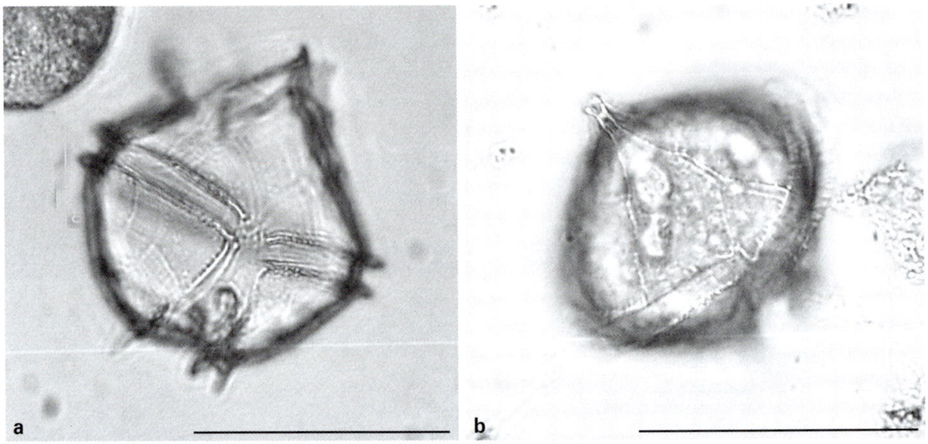

a, Single cell of *Protoperidinium pellucidum* in ventral view.
b, Apical half of the cell in ventral view with the first apical plate just discernable.
Scale bars = 50 μm.

Description

Cells medium-sized with a short apical horn and slightly diverging, antapical spines sometimes bordered by small lists. Theca slightly dorso-ventrally flattened. Epitheca triangular, sometimes with slightly concave sides. Hypotheca with straight to slightly convex sides. Cingulum in a median position. Antapex somewhat incised caused by the posteriorly widening sulcus. Characteristic wing on the left side of the sulcus, extending beyond the antapex, giving the appearance of an additional antapical spine.
First apical plate very broad and para. Second anterior intercalary plate is hexa.

Size

Diameter: 35-70 μm
Length: 40-70 μm

Distribution. This species has a global distribution and has been reported, e.g., from Greenland, the Atlantic, Pacific and Mediterranean. It regularly occurs in the North Sea including Helgoland and Sylt.

Similar species: *Protoperidinium pallidum*: *P. pellucidum* is smaller and circular in cross-section.

Synonym:
Peridinium pellucidum (Bergh) Schütt, 1895.

Literature: Buskey 1997

Protoperidinium subinerme (Paulsen) Loeblich III, 1969 Family: Protoperidiniaceae

Season				Trophic mode	Shape	Harmful	Bloom	Resting stage
W	S	S	A	heterotrophic	(cone + half sphere)-25 %	no	✿	yes

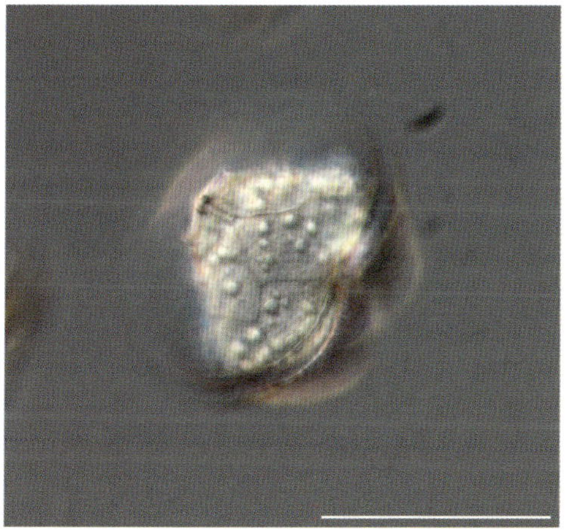

DIC image of the ventral cell of a live cell of *Protoperidinium subinerme*, showing the first apical and adjacent precingular plates. Scale bar = 50 µm.

Description

Cells medium-sized. Almost circular cell outline in apical view. Epitheca with slightly convex sides. Sides of the hypotheca almost straight. Cingulum located in a median position without displacement. Sulcus relatively narrow where it meets the cingulum, but widening posteriorly. Antapex flattened and sometimes bearing short spines.
First apical plate ortho, second intercalary plate hexa.

Protoperidinium subinerme is a cyst former. The cyst name is *Selenopemphix nephroides* Benedeck. These are smooth walled brown cysts with a hexagonal archeopyle. The cysts are slightly dorso-ventrally compressed.

Size

Diameter: 48-60 µm
Length: 50-75 µm

Distribution. *Protoperidinium subinerme* is widely distributed in the North Sea including the islands of Helgoland and Sylt.

Similar species: *Protoperidinium punctulatum*: In *P. punctulatum* the sides of the hypotheca are 'indented' in contrast to *P. subinerme* where they are straight. The surface ornamentation also differs. In *Protoperidinium punctulatum* the theca is punctate giving the cells a rugged outline in ventral or dorsal view. In *P. subinerme* the surface is more reticulate.

Synonyms:
Peridinium subinermis Paulsen, 1904,
Peridinium subinerme Paulsen, 1908.

Literature: Dodge 1982, Wall & Dale 1968.

Protoperidinium pyriforme (Paulsen) Balech, 1974 — Family: Protoperidiniaceae

Season				Trophic mode	Shape	Harmful	Bloom	Resting stage
W	S	S	A	heterotrophic	cone + half sphere	no	no	yes?

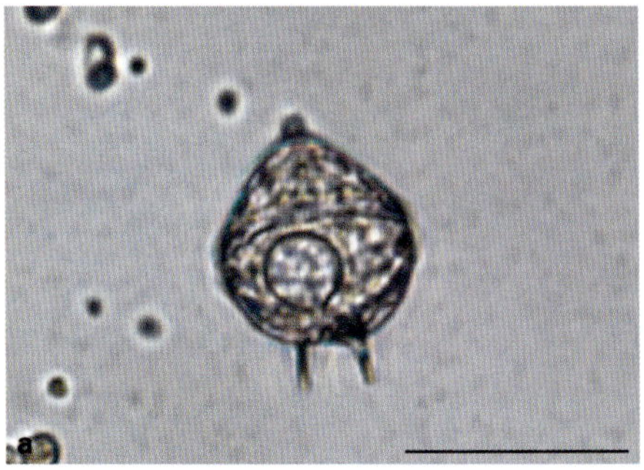

Brightfield micrograph of *Protoperidinium pyriforme* in dorsal view. Scale bar = 50 µm.
Image courtesy of Martin Löder.

Description

Cells small to medium-sized with a conical epitheca. Hypotheca rounded but with a flattened antapex. Theca very finely reticulate. Sutures between plates often extremely wide. Two short, winged antapical spines present, with the left one directed towards the ventral part of the cell. Cingulum circular or very slightly ascending. Sulcus straight and bordered by a list on the left side. Sulcal list continuous with the wing on the left antapical spine.
First apical plate meta, second intercalary plate penta.

Size

Diameter: 32-50 µm
Length: 40-85 µm

Distribution. *Protoperidinium pyriforme* has a very wide distribution. It has been found in many locations in the North Sea and North Atlantic and elsewhere, e.g., in the Mediterranean and Irish Sea.

Similar species: *Protoperidinium steinii.*

Synonyms:
Peridinium steinii var. *pyriformis* Paulsen, 1905,
Peridinium pyriforme Paulsen, 1907.

Literature: Gribble et al. 2007.

Protoperidinium steinii (Jørgensen) Balech, 1974 — Family: Protoperidiniaceae

Season				Trophic mode	Shape	Harmful	Bloom	Resting stage
W	S	S	A	heterotrophic	(cone + half sphere)-25 %	no	✿	yes

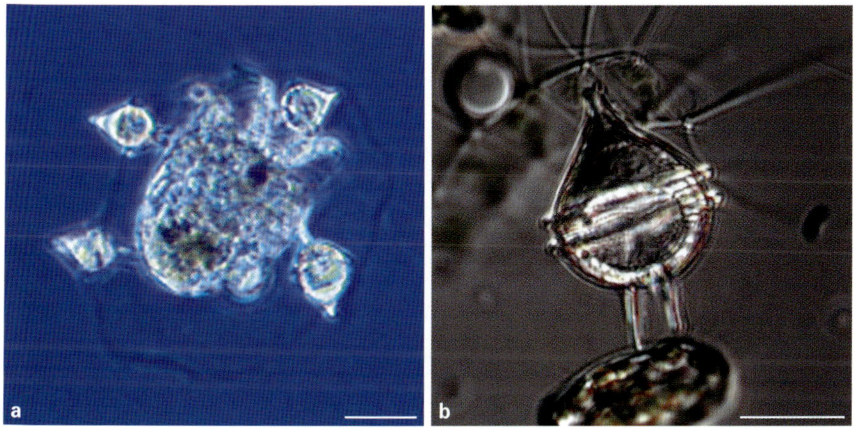

a, A group of *Protoperidinium steinii* cells attached to a nauplius. Scale bar = 50 µm.
b, DIC image of a live cell attached to a diatom cell. Scale bar = 20 µm.

Description

Cells medium-sized. Main cell body rounded. Epitheca extending in a long apical horn and prominent three-winged antapical spines, hypotheca hemispherical. Left side of sulcus usually bordered by a pronounced list. First apical plate meta, second anterior intercalary plate penta. Species heterotrophic and often pink or yellowish in colour, feeding by means of a pallium. In live samples 'group attack' can sometimes be observed. One cell attaches to the prey followed by several more individuals. In culture this species does not grow with immobile prey.

Size

Diameter: 39-60 µm
Length: 22-44 µm

Distribution. *Protoperidinium steinii* has a worldwide distribution.

Similar species: *Protoperidinium cerasus*: The latter species has a more offset apical horn. *Protoperidinium steinii* also resembles *P. pyriforme*. *P. steinii* has a more pronounced apical horn and the three winged spines are distinctive.

Synonyms:
Peridinium michaelis Stein, 1883,
Peridinium steinii Jørgensen, 1889.

Literature: Dodge 1982, Hansen & Larsen 1992, Olsen et al. 2002.

Protoperidinium bipes (Paulsen) Balech, 1974 — Family: Protoperidiniaceae

Season				Trophic mode	Shape	Harmful	Bloom	Resting stage
W	S	S	A	heterotrophic	half cone	no		no?

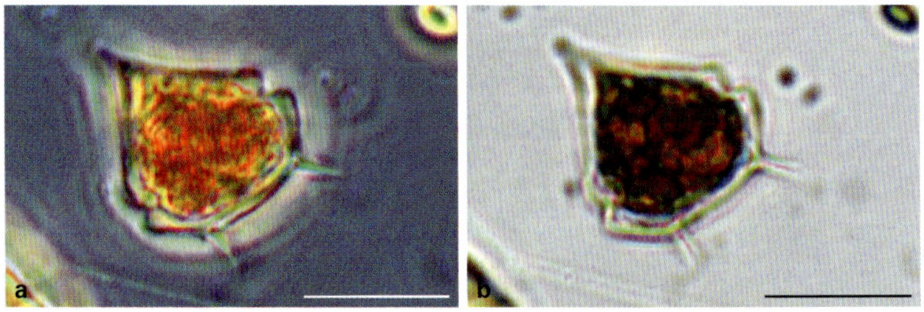

a, Phase contrast image of single cell of *Protoperidinium bipes*.
b, Bright field image of the same cell. Both images were fixed in Lugols iodine. Scale bars = 20 µm.

Description

Cells small. Epitheca, ending in a pronounced apical horn, almost triangular in outline. Sides of the epitheca straight or weakly concave. Cingulum excavated and slightly ascending but without cingular lists. Hypotheca slightly shorter than the epitheca and somewhat asymmetrical with two antapical spines at the point where the sides of the hypotheca meet the antapex. Cells strongly dorso ventrally compressed.
First apical plate ortho. The plate formula of this species is unusual in that it only has 6 precingular plates instead of the usual 7 plates in *Protoperidinium* species.
Plate formula: Po, x, 4', 2a, 6'', 4c, 6s, 5''', 2''''.

Size

Diameter: 15-20 µm
Length: 20-35 µm

Distribution. *Protoperidinium bipes* is a neritic species found in temperate to cold waters. It has also been reported from the Mediterranean.

Similar species: None. This species is a very distinct one.

Synonyms:
Glenodinium bipes Paulsen, 1904,
Peridinium minisculum Pavillard, 1905,
Miniscula bipes Lebour, 1925.

Literature: Jeong et al. 2004.

Protoperidinium brevipes (Paulsen) Balech, 1974 — Family: Protoperidiniaceae

Season				Trophic mode	Shape	Harmful	Bloom	Resting stage
W	S	S	A	heterotrophic	cone + half sphere	no	no	yes

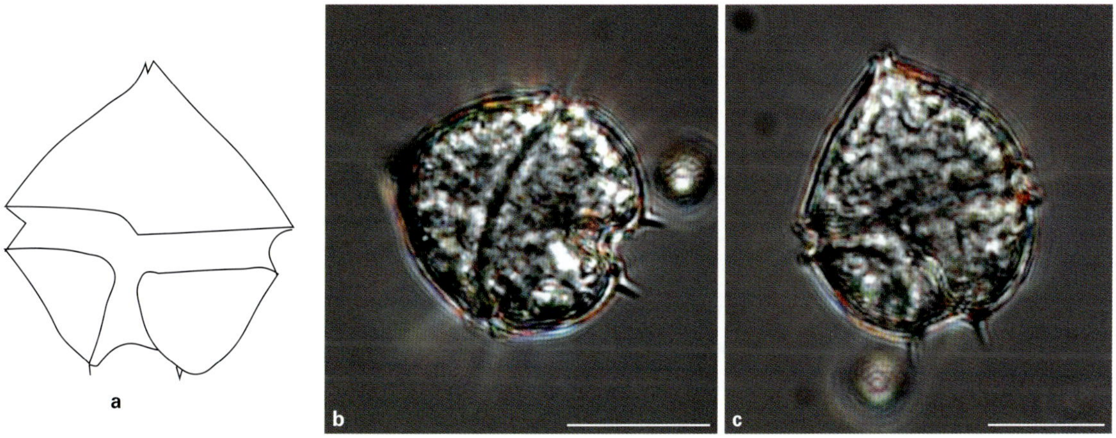

a, Schematic drawing of the thecal outline of *Protoperidinium brevipes* (redrawn from Paulsen, 1908).
b, Phase contrast image of *Protoperidinium brevipes* in ventral view; **c,** the same cell in dorsal view. Scale bar = 20 µm.

Description

Cells small with a roughly triangular epicone. Hypotheca with straight sides terminating posteriorly in two distinct lobes each carrying a short spine. Cingulum slightly displaced (ascending) and very broad. Sulcus excavated and often widening posteriorly. Cells slightly compressed dorsoventrally.
First apical plate large and meta, second anterior intercalary plate quadra and very small.
Protoperidinium brevipes is a heterotrophic species.

Size

Diameter: 20-40 µm
Length: 18-40 µm

Distribution. *Protoperidinium brevipes* is considered a Northern species and has been reported from many Atlantic and North Sea locations.

Similar species: The cell shape is somewhat similar to *Protoperidinium subinerme*, but the latter is a larger species.

Synonyms:
Peridinium brevipes Paulsen, 1908,
Peridinium varicans Paulsen, 1911,
Peridinium incurvum Lindemann, 1924.

Literature: Gribble et al. 2007, Paulsen 1908.

Protoperidinium denticulatum (Gran & Braarud) Balech, 1974 Family: Protoperidiniaceae

Season				Trophic mode	Shape	Harmful	Bloom	Resting stage
W	S	S	A	heterotrophic	flattened ellipsoid	no	no	yes

Pair of live cells of *Protoperidinium denticulatum*. Scale bar = 20 µm..

Description

Cells anteriorly-posteriorly compressed. Epitheca triangular with a short central apical horn with a characteristic apical groove in which the apical pore is located. Hyptotheca trapezoidal in outline. First apical plate ortho, with only two anterior intercalary plates. Cells often occurring as pairs of two cells This species produces round, smooth-walled brown resting cysts.

Size

Length: 30-45 µm
Width: 45-76 µm

Distribution. A North Atlantic species occurring mainly in spring.

Similar species: A distinctive species that can be identified in live as well as Lugol- or formalin-fixed samples.

Synonyms:
Peridinium denticulatum Gran & Braarud, 1935,
Peridinium clavus Abe, 1936.

Literature: Dodge 1982, Hansen & Larsen 1992.

Protoperidinium thorianum (Paulsen) Balech, 1974 — Family: Protoperidiniaceae

Season				Trophic mode	Shape	Harmful	Bloom	Resting stage
W	S	S	A	heterotrophic	rotational ellipsoid - 20 %	no	no	yes

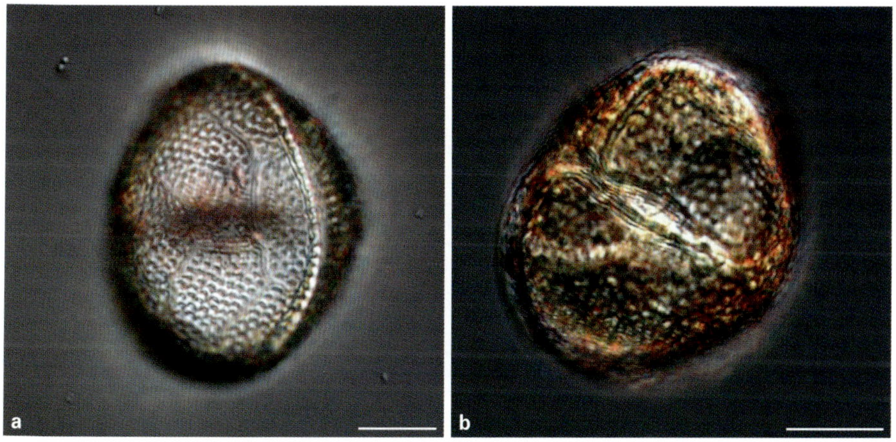

Phase contrast images of *Protoperidinium thorianum*:
a, dorsal view showing the intercalary plates and their thecal ornamentation; **b,** ventral view. Scale bars = 20 µm.

Description

Epitheca more or less triangular and hypotheca rounded but slightly incised by the posterior part of the sulcus. In apical view cells appear rounded.
Cingulum in a premedian position and slightly descending with longitudinal striations throughout the cingular groove. Sulcus can be straight with parallel sides or slightly widening towards the antapex. Theca covered by pits with raised edges giving the cell a rugged outline.
First apical plate is ortho. However, the species is unusual in that it only has two anterior intercalary plates.

Size

Diameter: 55-85 µm
Length: 55-85 µm

Distribution. *Protoperidinium thorianum* is a cosmopolitan species.

Similar species: none.

Synonyms:
Peridinium thorianum Paulsen, 1905,
Properidinium thorianum Meunier, 1919.

Literature: Balech 1974.

Protoperidinium minutum (Kofoid) A. R. Loeblich III, 1970 Family: Protoperidiniaceae

Season				Trophic mode	Shape	Harmful	Bloom	Resting stage
W	S	S	A	heterotrophic	sphere	no		yes

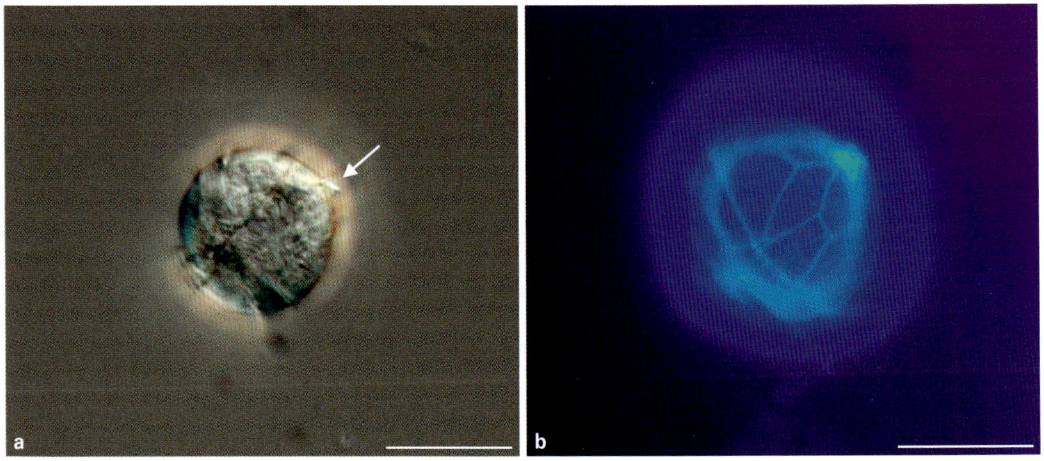

a, DIC image of a live cell of *Protoperidinium minutum* showing the short apical horn (arrow).
b, Epifluorescence image of a calcofluor stained cell excited with UV light.
Scale bars = 20 (a) and 50 µm (b).

Description

Cells globular with a distinct but short apical horn. Antapical horns absent. Theca covered by pores. The girdle is shallow, circular and surrounded by lists. Sulcus also bordered by a list on its left side. The list extends beyond the antapex and terminates in a small spine.
The first apical plate is ortho. There are only two anterior intercalary plates.
Plate formula: Po, x, 4', 2a, 7'', 4c, 6s, 5''', 2''''.
Protoperidinium minutum is a cyst producer (e.g., Ribeiro et al. 2010).

Size

Diameter: 23-56 µm
Length: 23-34 µm

Distribution. *Protoperidinium minutum* has been widely reported from European waters including the North Sea around Helgoland and Sylt.

Similar species: none.

Synonyms:
Peridinium monospinum Paulsen, 1907,
Peridinium minutum Kofoid, 1907.

Literature: Zonneveld & Dale 1994.

Scrippsiella cf. *trochoidea* Family: Peridiniaceae

Season				Trophic mode	Shape	Harmful	Bloom	Resting stage
W	S	S	A	autotrophic	cone + half sphere	no	✿✿✿	yes

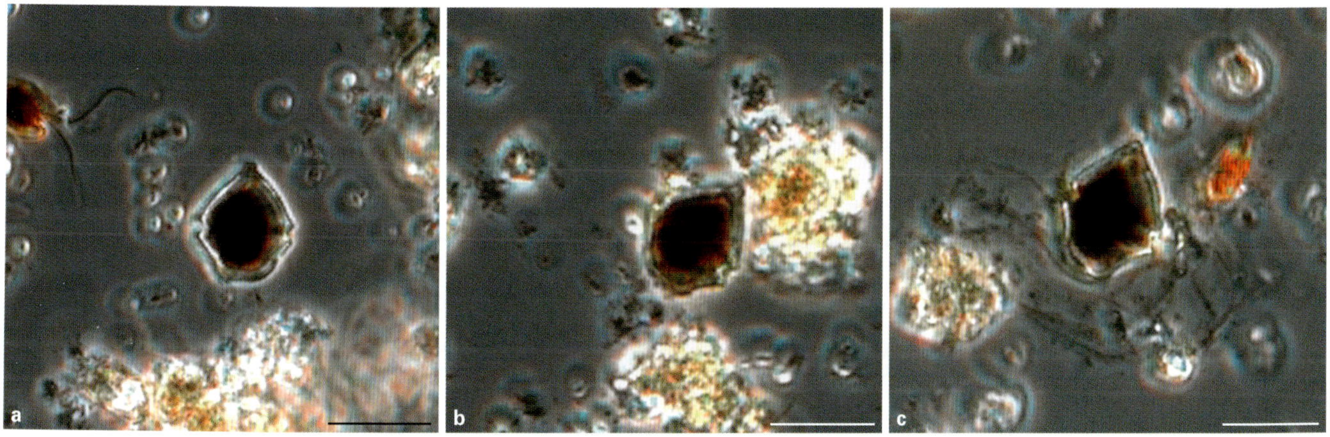

a-c, Lugol-fixed cells of *Scrippsiella* cf. *trochoidea*. Scale bars = 20 µm. Images courtesy of Martin Löder.

Description

Cells small. Cell outline round in apical view. Epitheca is roughly conical in ventral view. Hypotheca slightly smaller than the epitheca with a rounded antapex without spines or horns. Cingulum excavated and without cingular lists (narrow lists are present according to Dodge 1982). Sulcus wide and excavated, terminating before reaching the antapex. Cells pigmented and appearing very dark brown in Lugol-fixed samples. Nucleus located in the cell centre.
As the cells shown here were only identified from Lugol preserved cells from light microscopy we only named the species as 'cf. *trochoidea*', meaning that the species looks like *Scrippsiella trochoidea*. Only more detailed examination of the plate tabulation, particularly the cingulum, however will facilitate an unambiguous identification.
Plate formula: 4', 3a, 7'', 6c, 5s, 5''', 2''''.
Scrippsiella trochoidea produces oval calcareous cysts.

Size

Diameter: 20-23 µm
Length: 16-36 µm

Distribution. *Scrippsiella* cf. *trochoidea* is probably cosmopolitan, but it is likely to have been misidentified many times. Although this species is known to form very dense blooms these are not common in the Southern North Sea including the island of Helgoland.

Similar species:

Pentapharsodinium dalei Indelicato & Loeblich III 1986: This species is very similar to *Scrippsiella trochoidea* in cell outline, but it only has 5 cingular plates as opposed to 6 in *S. trochoidea* and has also been found in the waters around Helgoland.
S. trochoidea also resembles *S. minima*. The latter is smaller and has a different sulcal plate pattern.

Synonyms for *S. trochoidea*:

Glenodinium trochoideum Stein, 1883,
Glenodinium acuminatum Jörgensen, 1899,
Peridinium faeroense Paulsen, 1905,
Peridinium trochoideum (Stein) Lemmermann, 1910,
Scrippsiella faeroense (Paulsen) Balech & Soares, 1967,
Scrippsiella faeroense Dickensheets & Cox, 1971.

Literature: Balech 1990, Dodge 1982, Montresor et al. 2003.

Dissodinium Klebs in Pascher, 1916 Family: Pyrocystaceae

Season				Trophic mode	Shape	Harmful	Bloom	Resting stage
W	S	S	A	parasitic	sphere	no	no	yes

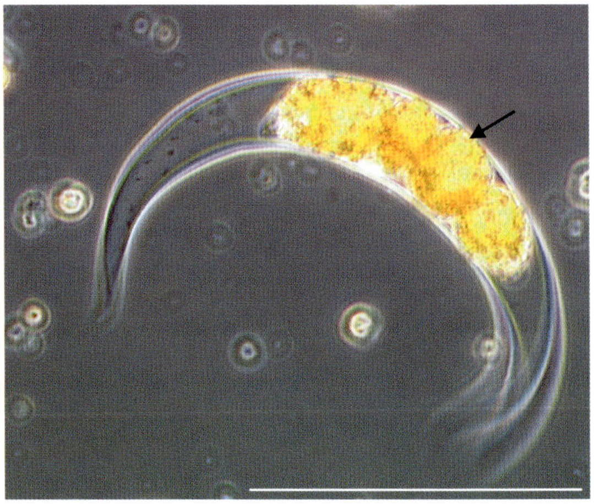

Lugol-fixed secondary cyst of *Dissodinium* sp. Scale bar = 100 µm.

Description

Species in this genus have a complex life cycle involving several morphologically very different stages. The primary cyst is a bladder like, globular structure with a large central vacuole and a few chloroplasts. Inside this cyst, several crescent-shaped secondary cysts are formed and then released. These cysts are the most conspicuous stage in the plankton. Inside the secondary cyst in turn several dinospores are formed (arrow). In some species these are thecate in others gymnoinioid. The number of biflagellate stages per cyst also varies between species.
These biflagellate stages can be pigmented but upon release from the cyst they become parasitic, infecting copepod eggs.

Size

Length: 100-140 µm (secondary cyst stage); 10-20 µm (biflagellate stage)

Distribution. They are probably widely distributed in the North Sea but due to the taxonomic confusion surrounding the genera *Dissodinium* and *Pyrocystis* J. Murray ex Haeckel, 1890 in the past, misidentification might be common in older records, especially since in preserved samples often only the lunate stages are conspicuous.

Similar species: The genera *Pyrocystis* and *Dissodinium* have traditionally been separated among other life-cycle characteristics, by the fact that the lunate stages of *Pyrocystis* species only contain two zooids but the number of zooids in *Dissodinium* is higher. However, several *Dissodinium* species have now been transferred to the genus of *Pyrocystis* and this group might need further revision.

Literature: Elbrächter & Drebes 1978, Taylor 1972.

Pyrophacus horologium Stein, 1883 — Family: Pyrophacaceae

Season				Trophic mode	Shape	Harmful	Bloom	Resting stage
W	S	S	A	autotrophic	ellipsoid - 20 %	no		yes

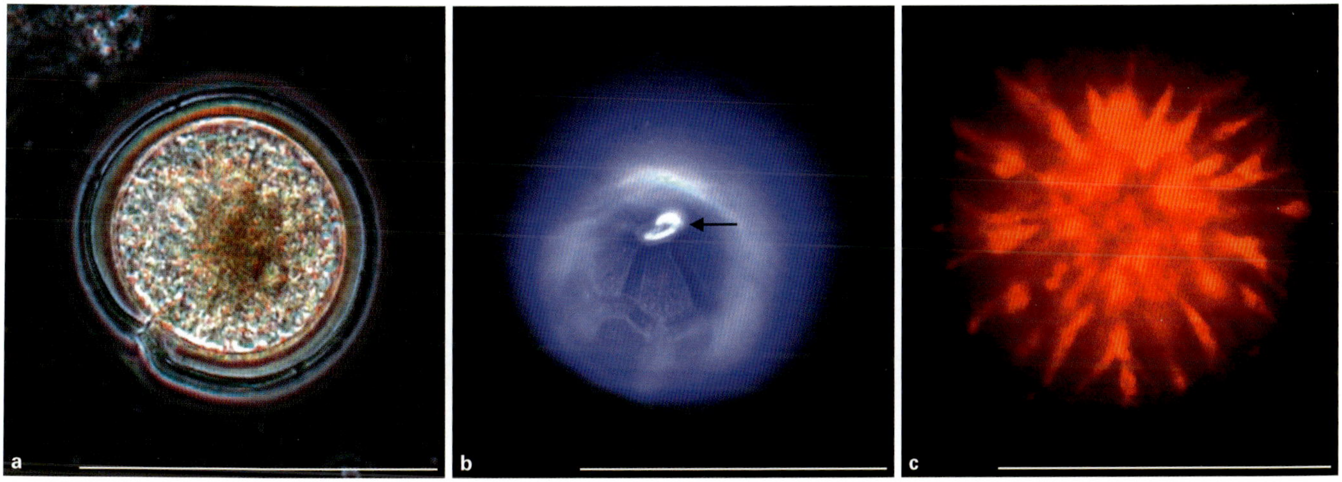

a, Phase contrast image of a live cell of *Pyrophacus horologium* in apical view.
b, c, Epifluorescence images of a cell in apical view, showing the plate patterns including the apical pore (**b**, see arrow) and the autofluorescence of the chloroplast (**c**). Scale bars = 100 µm.

Description

Cells strongly apically-antapically compressed. Epitheca and hypotheca of similar height and width. Cingulum slightly descending and not bordered by lists. Because of its symmetry cells of this species are always seen in apical view (the outline resembling large *Diplopsalis* cells). Cells clearly pigmented.
The plate numbers and patterns are sometimes visible in water mounts but are variable. The thecae readily disintegrate in fixed material.
Plate formula (highly variable): 5-6', 0-1a, 7-10'', 9c, 8-10''', 3-5''''.
Pyrophacus horologium is a cyst producer.

Size

Diameter: 35-136 µm
Length: 32-125 µm

Distribution. *Pyrophacus horologium* is a cosmopolitan species in cold temperate to tropical waters. It has been reported from the North Sea and often occurs in coastal waters. It has also been found in the waters around Helgoland.

Similar species: When seen in Lugol-fixed samples in which the plates are not discernable, it might be mistaken for a large *Diplopsalis* cell.

Synonyms: none.

Literature: Wall & Dale 1971.

Fragilidium subglobosum (von Stosch) Loeblich III, 1965 — Family: Pyrophacaceae

Season				Trophic mode	Shape	Harmful	Bloom	Resting stage
W	S	S	A	mixotrophic	sphere	no	✿	no

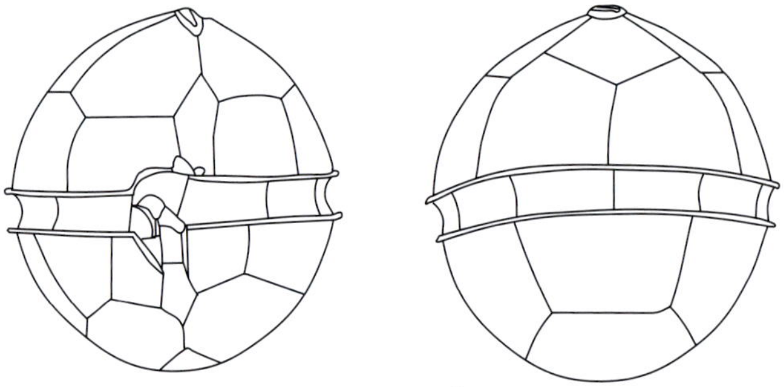

a Ventral side **b** Dorsal side

Schematic of *Fragilidium subglobosum* in ventral (**a**) and dorsal (**b**) view.
Both figures have been redrawn from von Stosch 1969.

Description

A medium-sized thecate species with a rounded cell outline when viewed in ventral view. The theca is very delicate and not easily seen in unstained watermounts. No thecal ornamentation is visible using light microscopy. The apical pore plate is elongate with a comma shaped slit. It is surrounded by four apical plates of similar size. No intercalary plates are present.
The first precingular plate is similar in shape and position to the first apical plate in *Alexandrium*, but there is not, as in some *Alexandrium* species, an apical pore. Cells are pigmented with many elongate chloroplasts.
According to von Stosch (1969) the plate formula is Po 4', 0a 9'', 10c, 7s + a transitional plate, 7''', 3''''. Steidinger and Tangen (1997) on the other hand ste Po, cp, 4-5', 0a, 7-9'', 9-11c, 6-8s, 7-8''', 1p, 2'''', interpreting one of Stosch's antapical plates as a posterior intercalary plate

Size

Diameter: 35-50 µm

Distribution. The distribution of *Fragilidium subglobosum* is poorly known as it is probably often misidentified.

Similar species: *Fragilidium subglobosum* resembles species of *Alexandrium* in general cell outline. But it has a very different plate pattern. Also, when manipulated the theca comes away from the cell contents very easily.

Synonym:
Helgolandinium subglobosum von Stosch, 1969.

Literature: Hansen & Nielsen 1997, Skovgaard 1996, Skovgaard, Hansen & Stoecker 2000, von Stosch 1969.

Marine Flagellates

Short summaries are given of the major flagellate groups with more detailed accounts for some common and/or harmful species

Prymnesiophyceae Hibbert, 1976

Species of the class Prymnesiophyceae (often also referred to as Haptophyceae) are regularly found in the open North Sea and coastal waters, sometimes forming dense blooms. Groups with very different life cycles are combined in this group. In genera such as *Chrysochromulina* and *Prymnesium* cells are motile and are covered by minute scales but look like naked cells when viewed under light microscope. Motile cells have two flagella and a haptonema which in the light microscope sometimes appears as the "third" flagellum but is a special organelle, possibly involved in prey capture.

The second group, the coccolithophorids, can have motile flagellate stages but also non-motile stages in their life-cycle (Klaveness 1972, 1972a). The best known stage are the coccolith bearing cells, which are covered by conspicuous calcareous plates, which often occur detached from the cell surface in fixed samples. Especially in fixed samples it is often difficult to determine the genus let alone the species due to the change in cell size and form, or the loss of scales or plates as the result of chemical preservation. We here briefly describe two genera. *Phaeocystis*, the most important genus of this class for North Sea coastal waters, is described separately. A more detailed description of the Prymnesiophyceae is found in Moestrup & Thomsen 2003.

Chrysochromulina sp. Lackey, 1939

Family: Prymnesiaceae

Description. Cell asymmetrical, shape variable, from spherical to elliptical. Cells with a coiling haptonema, extremely long in some species (up to 18 × cell length). Cells mostly with two chloroplasts, sometimes one. Cell surface covered by scales. Sometimes several types of scales occurring in the same species. Haptonema mostly longer than the cell, varying from 1 to 1.5 to ca. 18 × cell length depending on the species (Figure). The two flagella range from 1 to ca. 3 × cell length. *Chrysochromulina* species known to be mixotrophic.

Note. Electron microscopy should be used to identify species, as structural differences in the minute scales covering the body surface have to be examined to discriminate between individual species.

Distribution. Most *Chrysochromulina* species seem to be distributed throughout the world's oceans and coastal areas, up to 40 species can co-occur. The peak season lasts from April to August in the Skagerrak and from May to October in the Baltic Sea. Blooms of the toxic species *Chrysochromulina polylepis* were only observed in the Skagerrak and Kattegat; however the species has a world-wide distribution. In North Sea coastal waters species of this genus occur mainly during summer.

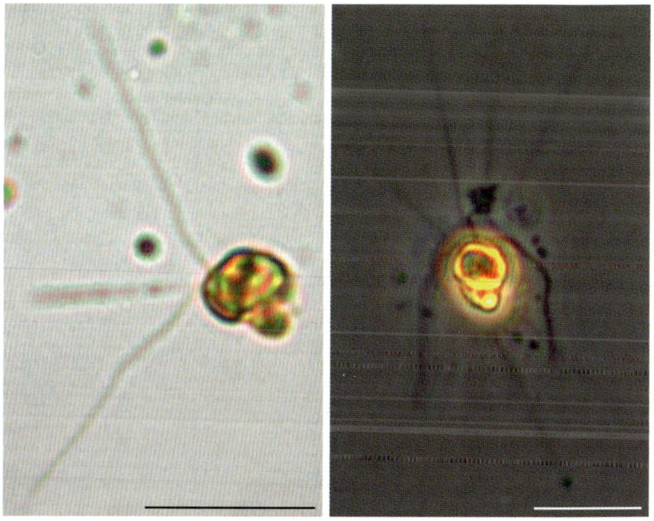

Two different species of *Chrysochromulina* in Lugol-fixed samples. Scale bars = 10 µm.
Left image taken in bright field and right image in phase contrast optics.

Prymnesium sp. Massart, 1920

Family: Prymnesiaceae

Description. Species in this genus usually ellipsoid, sometimes slightly asymmetrical. Haptonema shorter than in *Chrysochromulina* species, approximately one fourth to one third of cell length depending on the species. All species with two chloroplasts.

Note. *Prymnesium parvum* N. Carter, 1937 is the most common *Prymnesium* species. Cell length varying from 6 to 18 μm. The two flagella are longer than the cell, from 1 to approximately 2 × the cell length. *Prymnesium parvum* causes fish kills, destroying red blood cells.

Distribution. *Prymnesium* species seem to be restricted to temperate and tropical regions. In temperate regions (such as the North Sea) the species mainly occurs during summer, in warmer regions throughout the year.

Emiliania huxleyi (Lohmann) W. W. Hay & H. Mohler, 1967

Family: Noëlaerhabdaceae

Description. *Emiliania huxleyi* is probably the best known coccolithophorid species. Characteristic of this group are the coccoliths. These are plates of calcium carbonate (as calcite) which cover the cells of one of its life-cycle stages. *Emiliana huxleyi* has a life cycle with three cell types: 1, non-motile spheres with coccoliths; these are oval placoliths and cover the whole cell; 2, non-motile completely naked cells (without any coccoliths) and 3. motile cells with flagella. The coccosphere is very characteristic; however the naked cells and especially the motile cells with flagella can be mistaken for other non-calcified flagellate species. Coccoliths consist of partially interlocked placoliths, approximately 3.5 μm long and 3 μm wide, variable in size and form even on one coccosphere. Calcification varies with environmental calcite concentrations.

Size. Non-motile coccosphere: 5-10 μm diameter.

Distribution. This species is cosmopolitan and can be found throughout the year.

Synonyms:
Pontosphaera huxleyi Lohmann, 1902,
Hymenomonas huxleyi (Lohmann) Kamptner, 1930,
Coccolithus huxleyi (Lohmann) Kamptner, 1943,
Gephyrocapsa huxleyi (Lohmann) Reinhardt, 1972.

Literature: Jordan et al. 2005, Marchant et al. 2005, Throndsen 1997.

Phaeocystis sp. Lagerheim, 1893

Family: Phaeocystaceae

Season				Trophic mode	Shape	Harmful	Bloom	Resting stage
W	S	S	A	autotrophic	spherical to ellipsoid	yes	✿✿✿	no

Two species of this genus are found in North Sea and adjacent waters. They can be distinguished mainly by their colony form. However this is often destroyed in fixed samples. *Phaeocystis* species can form very dense blooms and produce mucilage, which when the bloom dies off, often washes up on beaches in large quantities causing foam.

Phaeocystis globosa **Scherffel, 1899**

Description. Motile stage: spherical cells (two chloroplasts) with two flagella (longer than cell length) and a shorter haptonema; three types, differing in size, macro-, meso and microflagellated cells. Nonmotile stage: cells (two chloroplasts) without flagella and haptonema either as single cells (Figure), however usually embedded in gelatinous mucilage of mucopolysaccharides forming large buoyant colonies of thousands of cells; smaller colonies spherical, older ones elongate.

Note. In Lugol-fixed samples the cells take on a characteristic 'butterfly shape'.

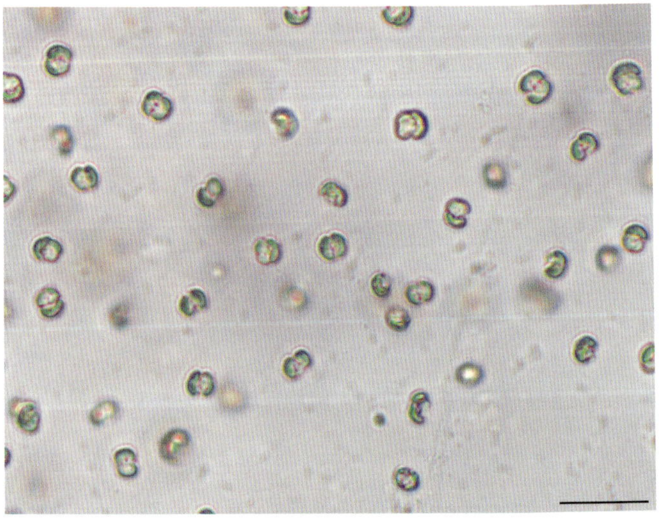

Bright field image of part of a Lugol-fixed mucilage colony of *Phaeocystis globosa* with the non-motile cells. Scale bar = 20 μm.

Size. Motile stage: up to 5 µm. Size of non-motile cells up to 7 µm. Cells in colonies evenly distributed (difference to *Phaeocystis pouchetii*); colony size up to 2 mm, in calm conditions to ca. 8-10 mm.

Distribution. *Phaeocystis globosa* occurs in temperate and tropical waters of both hemispheres, both in coastal and oceanic waters. It can be found in the North Sea from late winter to summer with the main biomass peak usually during late spring or early summer. Extensive blooms of the species can cause beach fouling when cells die off and mortality in marine organisms due to oxygen depletion in the water.

Synonyms: none.

Similar species and literature: see *Phaeocystis pouchetii*.

Phaeocystis pouchetii (Harriot) Lagerheim, 1896

Description. Motile stage: As in *Phaeocystis globosa* spherical cells with two flagella (ca. 1,5 × cell length) and a shorter haptonema. There are at least two cell types, differing in size. In all, 4 cell types have been reported. Nonmotile Cells without flagella and haptonema are embedded in gelatinous mucilage of muco-polysaccharides forming buoyant colonies of hundreds of cells, the shape of the colony is irregularly lobed due to the more delicate nature of the mucilage compared to that of *P. globosa*. In both stages cells with two chloroplasts. Cells in colonies are not evenly distributed as in *P. globosa*, but with 4 cells forming a square, in between larger cell free areas; colony size up to 1 mm, not larger.

Size
Length of motile cells: 4,5-8 µm.
Diameter of non-motile cells: 5 µm.

Distribution. This species occurs in cold waters (Arctic) of the northern hemisphere, both in coastal and oceanic waters throughout the year often forming massive blooms in spring. It has been found also in the North Sea in winter and spring. Like *Phaeocystis globosa* it can cause mortality after massive blooms due to oxygen depletion in the water when blooms die off. The species has also been reported as toxic to cod larvae forming a polyunsaturated aldehyde (Stabell et al. 1999, Hansen & Eilertsen 2007).

Synonym: *Tetraspora pouchetii* Hariot, 1892.

Literature: Baumann et al. 1994, Hansen & Eilertsen 2007, Jordan et al. 2005, Medlin & Zingone 2007, Stabell et al. 1999.

Raphidophyceae M. Chadefaud ex P. C. Silva, 1980

Season				Trophic mode	Shape	Harmful	Bloom	Resting stage
W	S	S	A	autotrophic	ellipsoid to pyriform	yes	✿✿✿	yes

Description. Cells relatively large compared with other flagellate groups and containing many chloroplasts and a large conspicuous rounded nucleus.
Some members of the class Raphidophyceae can cause fish kills resulting from toxins. Damage to fish gills during raphidophyte blooms have also been reported. Raphidophytes can form blooms in the North Sea area from late spring onwards.

Note: Raphidophytes are typically very fragile in fixed samples.

Chattonella sp. B. Biecheler, 1936

Family: Chattonellaceae

Description. Cells with an ellipsoid to pyriform shape, body asymmetric somewhat flattened. Cells with two flagella: a forward pointing flimmer flagellum which pulls the cell and a back pointing flagellum which seems to be less active, both inserting in a shallow ventral groove. *Chattonella* species have many discoid small chloroplasts.

Size
Length: cell 8-50 µm
Forward pointing flagellum: approximately equal to the cell length
Backward pointing flagellum: 1.5-2 × the cell length

Distribution. Species have a global distribution are mainly found in warmer regions from June to September when temperature exceeds 20 °C. But they can occur in the southern North Sea from May. During the rest of the year they are attached as cysts to solid surfaces such as diatom frustules and sand grains.

Literature: Underdahl 1989.

Fibrocapsa japonica S. Toriumi & H. Takano, 1973

Family: Chattonellaceae

Description. Shape ovoid. Two flagella (as typical for the group): a forward pointing one and a back pointing flagellum, both inserting anteriorly. Chloroplasts numerous, slightly flattened and densely packed. Characteristic features: Trichocysts in the posterior part of the cell.

Size. Cell length 20-30 µm, the trailing flagellum is approximately 1 1/4 the cell length, the swimming flagellum almost one cell length.

Distribution: It has probably a worldwide distribution in temperate waters. Since the 1990s it is also observed in the North Sea mainly during summer building up a moderate to high biomass.

Literature: Throndsen 1997.

Cryptophyceae Fritsch in G. S. West & Fritsch, 1927

Season				Trophic mode	Shape	Harmful	Bloom	Resting stage
W	S	S	A	autotrophic	ellipsoid	no	no	no (usually)

Description. Most cryptophycean species ellipsoid to drop-shaped in outline with a typical furrow or depression. An ejectile organelle (ejectosome), typical of cryptophytes and some prasinophytes, associated with the furrow. Two flagella arising from the end of the furrow. Some genera heterotrophic, i.e. without chloroplasts.

Distribution. Cryptophycean species can be found in nearly every phytoplankton sample throughout the year, especially in coastal environments. Main appearance is during late spring and summer.

Cryptomonas sp. C. G. Ehrenberg, 1832

Family: Cryptomonadaceae

Description. Cells usually with an ovoid shape. Cells with a furrow and sack-like gullet with ejectosomes and two flagella which are 3/4 of cell length. Two chloroplasts per cell coloured brown, yellow-brown or yellow-green, each with a pyrenoid (Figure).

Size. 15-40 µm.

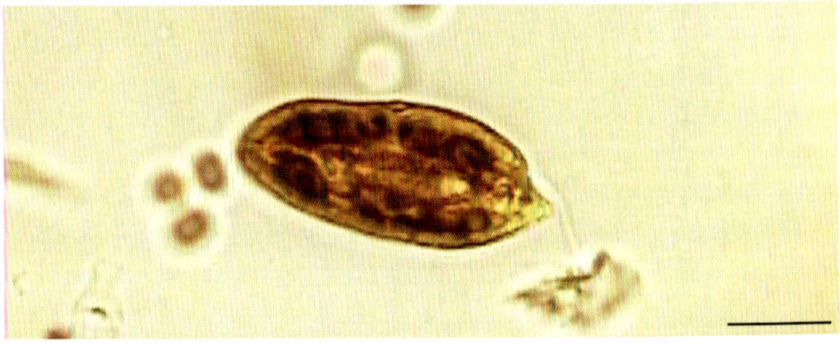

Bright field image of a Lugol-fixed *Cryptomonas* cell. Scale bar = 10 µm. Picture courtesy of Gabriele Krauß.

Leucocryptos marina (Braarud) Butcher, 1967

Family: Cryptomonadaceae

Description. Cells with a characteristic pyriform shape and two flagella exceeding the cell in length. *Leucocryptos* species without chloroplasts. Characteristic feature: Two ventral rows of ejectosomes (5 to 18 in each row), without true gullet or furrow (Figure).

Size. 15 to 18 µm, flagella ca. 10 µm.

Synonyms:
Bodo marina Braarud, 1935,
Chilomonas marina (Braarud) Halldal, 1953.

Literature: Clay et al. 1999, Throndsen 1997.

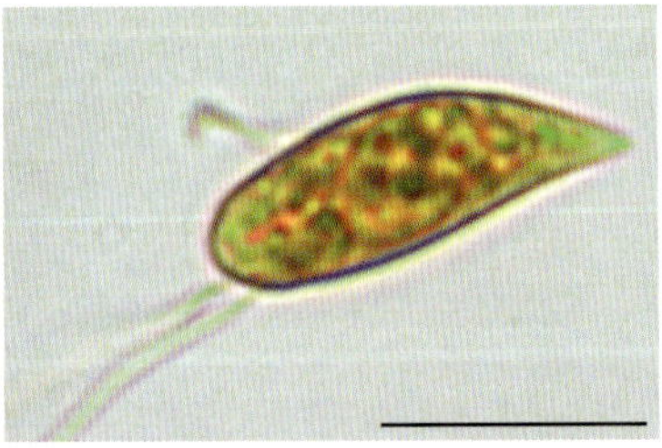

Bright field image of a Lugol-fixed *Leucocryptos marina* (Braarud) Butcher, 1967 cell. Scale bar 10 µm. Picture courtesy of Susanne Busch.

Hemiselmis sp. Parke, 1949

Family: Hemiselmidaceae

Description. Cells of a more or less ovoid shape. Two flagella inserted in a transverse gullet and of the same length as the cell, no furrow. Flagella may be unequal depending on the species. One green, blue-green, red or brown chloroplast per cell (Figure).

Size. 4-8.5 µm, flagella as long as the cell or slightly shorter.

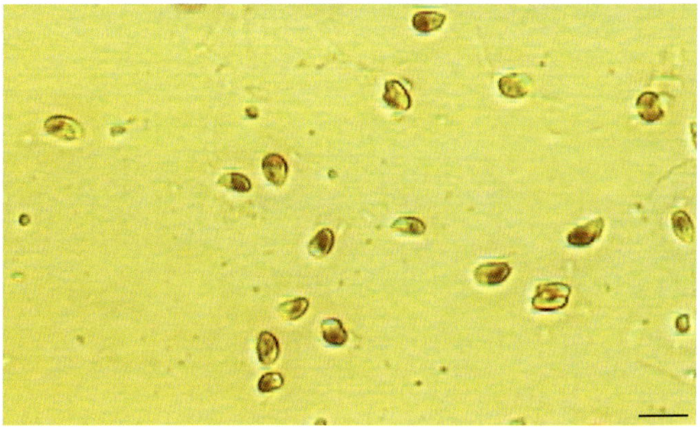

Bright field image of Lugol-fixed *Hemiselmis virescens* Droop, 1955 cells. Scale bar = 10 µm. Picture courtesy of Gabriele Krauß.

Rhodomonas sp. Karsten, 1898

Family: Pyrenomonadaceae

Description. Cells with a more or less ovoid shape. Two flagella, 1/2 to 3/4 of cell length with a short furrow and inserting in a tubular gullet. Single chloroplast bilobed, coloured red or red-brown, with a single pyrenoid (Figure).

Size. 10-30 µm, flagella ca. 10 µm.

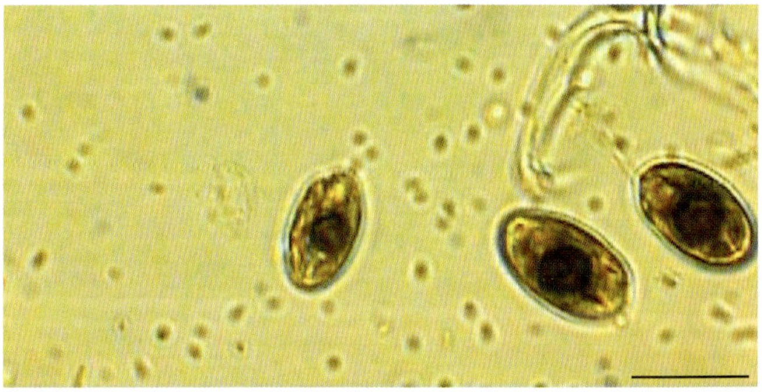

Bright field image of Lugol-fixed *Rhodomonas salina* (Wislouch) D. R. A. Hill & R. Wetherbee 1989 cells. Scale bar = 10 µm. Picture courtesy of Gabriele Krauß.

Teleaulax sp. Hill, 1991

Family: Geminigeraceae

Description. Cells with an acute anterior and posterior. Two flagella, each 3/4 of cell length. A single long chloroplast coloured red-brown or orange, with one pyrenoid.

Size. 12-15 µm, flagella ca. 10 µm (Figure).

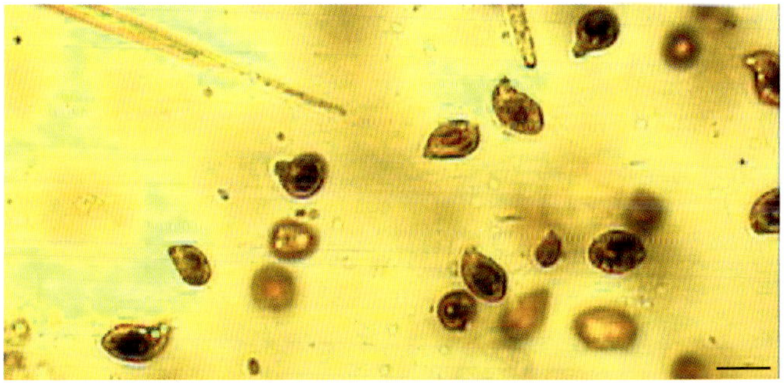

Bright field image of Lugol-fixed *Teleaulax acuta* (Butcher) D. R. A. Hill, 1991cells. Scale bar = 10 µm. Picture from Gabriele Krauß.

Euglenophyceae Schoenichen, 1925

Season				Trophic mode	Shape	Harmful	Bloom	Resting stage
W	S	S	A	autotrophic	ellipsoid	no	no	no

***Eutreptiella* sp.** da Cunha, 1913

Family: Eutreptiaceae

Description. Cells with an elongate sometimes ovoid outline and with a canal at its anterior end. Two sometimes four (*Eutreptiella braarudii* Throndsen, 1969) subequal heterodynamic flagella emerging from this canal. Cells containing a single or many chloroplasts, always of greenish colour. Some species with an eyespot (Figure).

Size. Length: 17-90 (115) µm

Distribution. *Eutreptiella* species appear in coastal and oceanic environments, in temperate and cold water. In the North Sea they appear mainly in late spring and autumn.

Literature: Throndsen 1997.

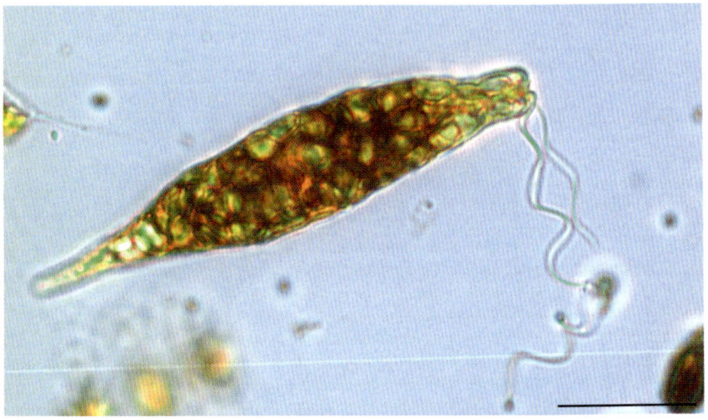

Bright field image of a Lugol-fixed *Eutreptiella* sp. cell. Scale bar = 20 µm. Picture courtesy of Susanne Busch.

Prasinophyceae T. Christensen ex Ø. Moestrup & J. Throndsen, 1988

Season				Trophic mode	Shape	Harmful	Bloom	Resting stage
W	S	S	A	autotrophic	sphere, ellipsoid or conic	no	no	no

Description. Species of this class with a quadrangular or ellipsoid (bilaterally compressed) cell shape and with one, two, four, eight or sixteen flagella. Cell either naked or with organic scales covering cell body and flagella. One or two chloroplasts present. Eyespots, if present, located within the chloroplast. Two examples are shown here. For more detailed accounts see Throndsen 1997.

Note. The Prasinophyceae are also an important component of the picoplankton i.e. cells smaller than 2-3 µm in size. The best known example is *Micromonas pusilla.* The identification of most species is only possible by examining the body scales with the use of electron microscopy (scanning electron and transmission electron microscopy)

Distribution. Many prasinophytes probably have a very wide distribution and the larger species are regularly found in samples from the southern and central North Sea. However, particularly with the smaller species, identification in routine monitoring samples is almost impossible and species are probably often misidentified. We, therefore, provide no detailed account of the distribution of particular groups here.

***Tetraselmis* sp.** Stein, 1878

Family: Chlorodendraceae

Description. Cell outline ovoid or heart shaped and with four flagella. Some species with a more or less pronounced four-lobed furrow. Cells compressed or slightly compressed. Some species possessing a large posterior or median eyespot coloured red-orange. Most species with large pyrenoids.

Size. Cell length: mostly in the range between 6-17 µm.

Pyramimonas sp. Schmarda, 1849

Family: Pyramimonadaceae

Description. Cells with quadrilateral symmetry, and with usually four or sometimes eight flagella; one recently found species apparently with 16 flagella. One cup-shaped chloroplast per cell divided anteriorly into four or eight lobes, a single pyrenoid, located posteriorly. Some species have one or two eye spots, location of eyespot varying with species (Figure).

Size. Cell length of the different species ranges from 4-6 up to 16-35 µm; flagella are as long as the cell or 2 × cell length.

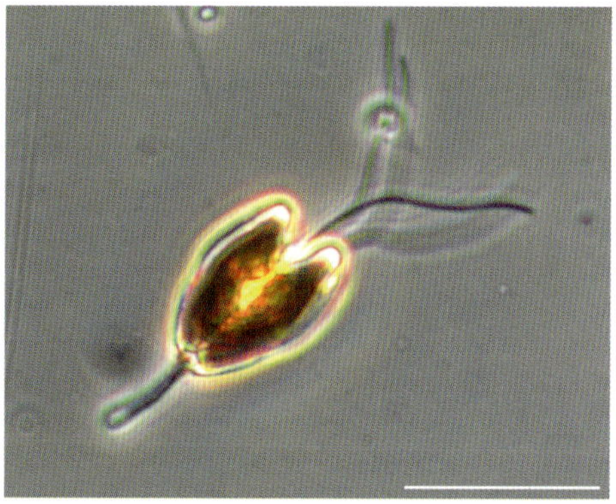

Phase contrast image of a Lugol-fixed *Pyramimonas* cell. Scale bar = 20 µm. Picture courtesy of Susanne Busch.

Dictyochophyceae P. C. Silva, 1980

Season				Trophic mode	Shape	Harmful	Bloom	Resting stage
W	S	S	A	autotrophic	sphere or pyriform	no		no

Description. Cells unicellular, with several different life stages: a stage with a silica skeleton and one flagellum, an amoeboid stage and several more flagellate, naked cells (Moestrup & Thomsen 1990).

Dictyocha speculum Ehrenberg, 1837

Family: Dictyochaceae

Description. Cells with skeleton, uniflagellate. Silica skeleton consisting of two differently sized hexagonal rings with six long spines emerging from the outer ring, shorter spines present on the inner ring. Cell body naked with numerous pseudopodia. Cells with many yellow-brown chloroplasts. Nucleus in a central position within the cells (Figures).

Size. Cell length varying from 19 to 35 µm (plus spines ca. 60 µm), flagellum as long as the cell or slightly shorter.

Distribution. The species appears in coastal and oceanic environments in cold and temperate water. In the North Sea it can be found throughout the year, mainly during late winter and spring

Similar species: none.

Synonyms:
Distephanus speculum (Ehrenberg) Haeckel, 1887,
Cannopilus calyptra Haeckel, 1887.

Literature: Throndsen 1997.

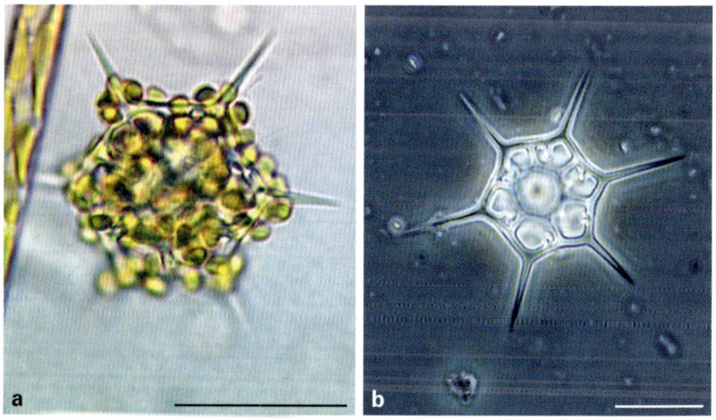

Phase contrast images of *Dictyocha speculum*. **a,** live, intact cell; **b,** skeleton. Scale bar = 20 µm.

Dictyocha fibula Ehrenberg, 1837

Description. Cells with internal silica skeleton with four protruding spines and four "windows". Cell body naked with numerous pseudopodia. The single flagellum of the same length as the cell. Cell containing many yellow-brown chloroplasts (Figure).

Size. Cell length varies from 10 to 45 µm (plus spines up to ca. 90 µm), flagellum as long as the cell or slightly shorter.

Distribution. This species occasionally appears in oceanic environments throughout the year.

Synonym: none.

Literature: Throndsen 1997.

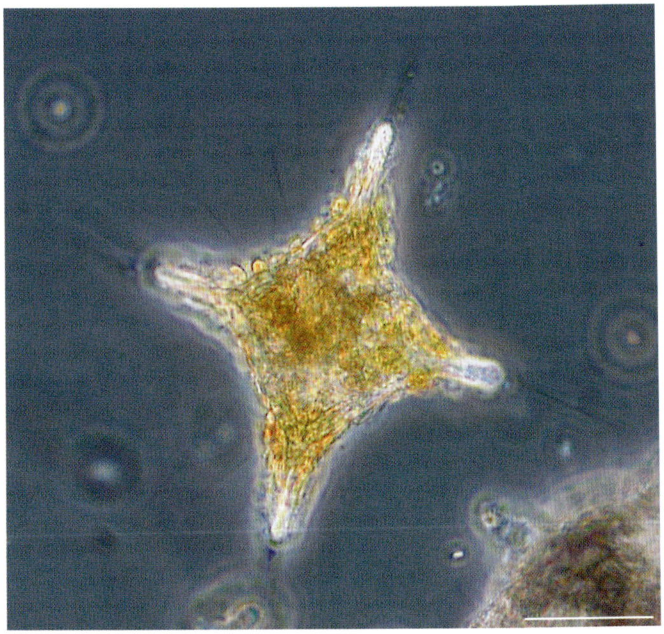

Phase contrast image of a live cell of *Dictyocha fibula* with pseudopodia. Scale bar = 20 µm.

Ciliates

Like dinoflagellates, ciliates are an important component of the microzooplankton. They are capable of feeding on a considerable range of prey items from bacteria to large diatoms. Here we give some examples of the most common species that can be identified in Lugol-fixed samples.

Laboea strobila Lohmann, 1908

Laboea strobila is a very characteristic species as it has a row of cilia (girdle kinety) forming five whorls around the body. The oral cavity is narrow with distinctive oral cilia extending deeply into the oral groove. Several rounded macronuclei are scattered throughout the body. Cell outline strongly conical.

Length: 45-115 µm

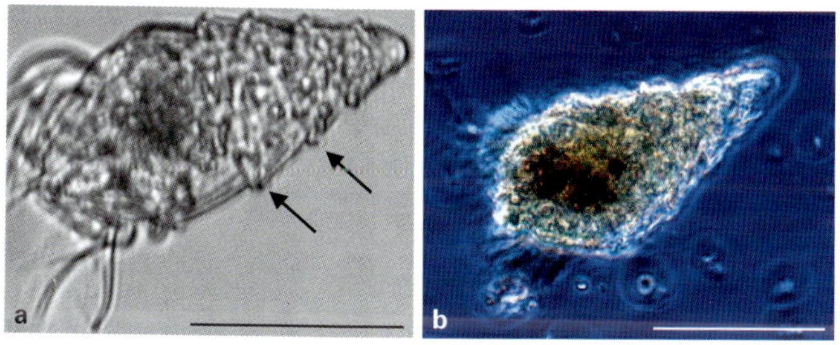

Laboea strobila. **a,** Flowcam image; courtesy of Florian Hantzsche; **b,** Lugol-fixed cell. Scale bars = 50 µm.

Myrionecta rubra (Lohmann) Jankowski, 1976

Myrionecta rubra has an ovoid cell outline with a sub-median constriction. Two macronuclei are located in a central position. The cytoplasm has a reddish colour due to the presence of cryptophyte pigments. A ring of cilia is located in a median position. Oral tentacles can be present.

Length: 10-100 µm

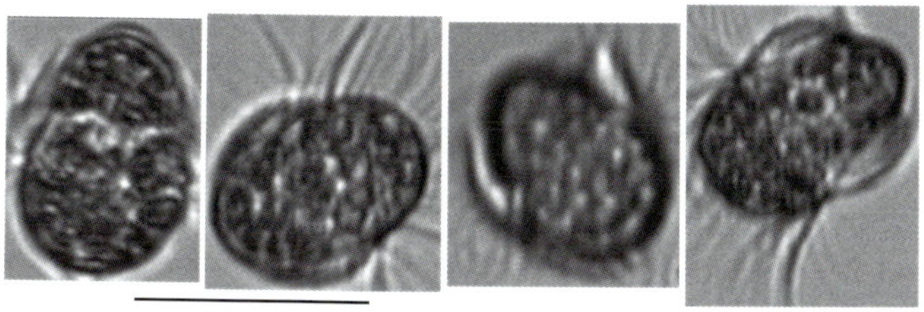

Series of FlowCAM images of *Myrionecta rubra*. Images courtesy of Florian Hantzsche.

Lohmanniella oviformis Leegaard, 1915

Cells are small and non-pigmented. The oral cavity is non-centric. One kidney shaped macronucleus is present. 3-5 monokinetids are located at the posterior pole of the cell.

Length: 10-25 µm

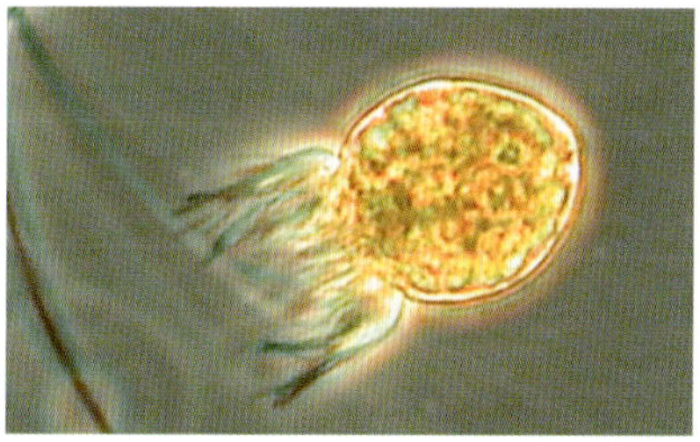

Lugol-fixed cell of *Lohmanniella oviformis*. Image courtesy of Nicole Aberle-Malzahn.

Tiarina fusus (Claparède & Lachmann) Bergh, 1880

Spindle-shaped body, pointed posteriorly. The anterior end is also pointed or slightly flattened. The anteriorly positioned oral cavity is surrounded by a ring of cilia (17 rows of kinetosomes).
Tiarina fusus can be very abundant even causing a reddish-brown water discolouration. At high densities, it can have a considerable grazing impact (Jeong et al. 2002).

Length: 60-90 µm

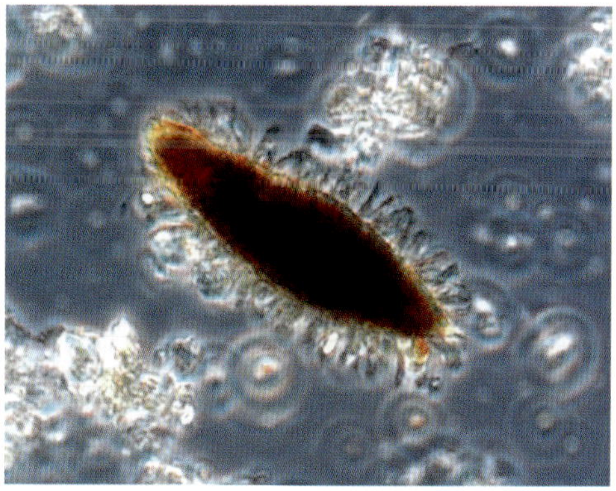

Lugol-fixed cell of *Tiarina fusus*. Image courtesy of Martin Loeder.

Glossary

Antapex: The posterior end of a dinoflagellate cell.

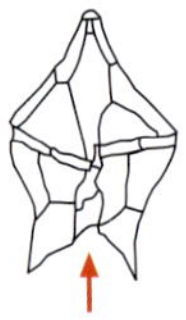

Aperture (in diatoms): The space between adjacent cells in a chain, e.g., in the genera *Eucampia* and *Chaetoceros.* The apertures between cells are seen in girdle view.

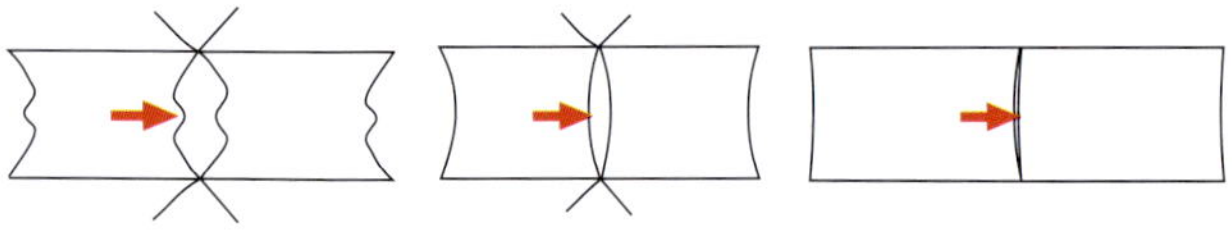

Apex: The anterior end of a dinoflagellate cell.

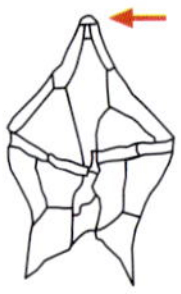

Apical axis: In diatoms generally: the cell diameter; in bipolar diatoms: the axis between the two poles of the frustule (i.e. the broadest part of the cell).

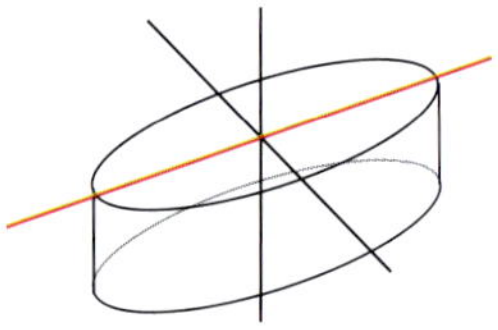

Apical groove: A small groove running around the apex or in a straight line from the apex towards the cingulum of the cell in many unarmored dinoflagellates (Gymnodiniales). It is an important morphological feature for distinguishing genera of unarmoured dinoflagellates.

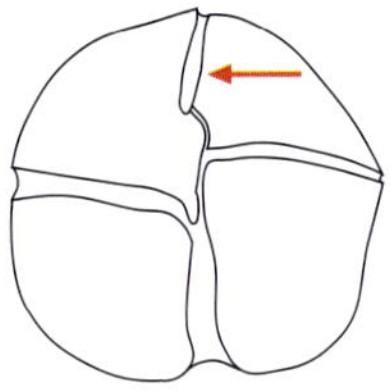

Apical pore complex (APC): One or more pores located in a special pore plate at the apex of an armoured dinoflagellate cell, e.g., in the genus *Alexandrium*.

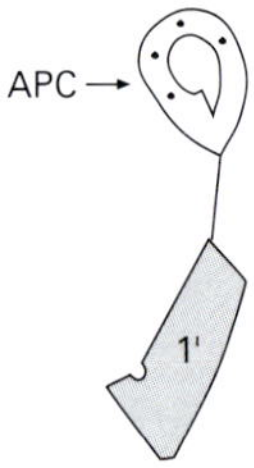

Apical pore field: An area of small pores (usually in longitudinal rows) at the apices of pennate diatom valves. They are similar to the ocelli found in centric diatoms.

Archeopyle: The aperture in a dinoflagellate cyst wall from which the new cell has hatched. The shape and position of the archeopyle is an important diagnostic feature in the identification of dinoflagellate cysts.

Areola: Pores through the valve wall in diatoms. The pores on the valve face (arrow in schematic) form characteristic patterns that together with other features are used in the identification of a species.

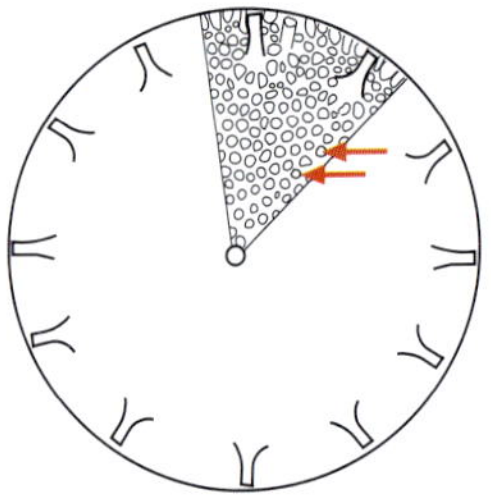

Areolation pattern: The pattern of areolation on the valve face and/or mantle in diatoms. The areolation patterns on the valve face of many centric diatoms are diagnostic features for distinguishing different species, e.g., in the genera *Coscinodiscus* and *Thalassiosira*. The main patterns are: **a**, curved tangential; **b**, straight tangential; **c**, fasciculated; **d**, fasciculated (curvatulus type); **e**, radial.

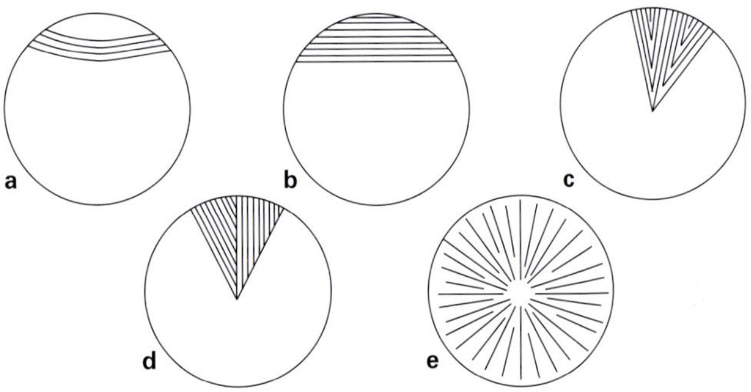

Athecate (= naked): A dinoflagellate without cellulose plates. Athecate dinoflagellates are often a lot more variable in shape than are thecate dinoflagellates.

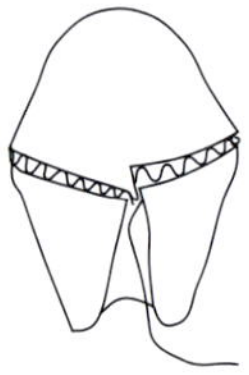

ATP (Adenosine triphosphate): A carrier of chemical energy in living organisms. The molecule contains a chain of three phosphate groups. The breakage of the bonds between these groups releases energy needed, e.g., for physical movement. The molecule with two and one phosphate group are termed adenosine diphosphate and monophosphate respectively. Using energy from the ingestion of food the triphosphate molecule can be restored again.

Autotrophy: Synthesis of complex organic compounds from inorganic molecules using energy from light or inorganic chemical reactions.

Auxospore (in diatoms): A special cell that is usually the result of sexual reproduction. The auxospore is larger than the cells it originates from thereby helping to restore size (cell size in a diatom population decreases as the result of asexual division).

Basal part: A term referring to the diatom genus *Chaetoceros*: It describes that part of an inner seta between the point where it arises from the valve face and the point of fusion with the seta from the neighbouring cell.

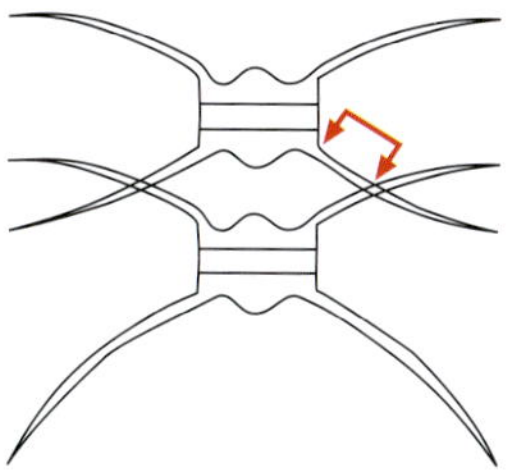

Bilabiate process: A process with an external and internal tube. The internal part is expanded at its distal end. The tip of the swollen section of the internal part of the process is closed, while its sides have a slit-like opening.

Bipolar: Term used for centric diatoms with two poles with areolae radiating from the valve centre or central part of valve margin towards the poles (e.g., *Odontella*).

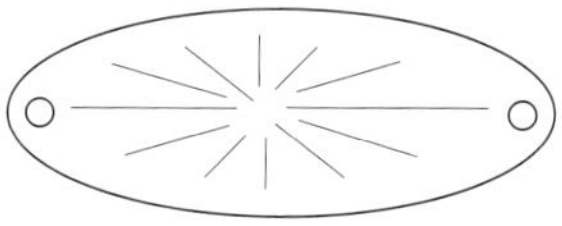

Canal raphe: A raphe system in diatoms: A canal, set-off from the remainder of the frustule running along the inner side of the raphe.

Central annulus: An area in diatoms, often in the centre of the valve face, with a different surface ornamentation from the rest of the valve. The area is surrounded by a thickened rim.

Central hyaline area: Area in the centre of a valve, not exhibiting any areolae. An example is the diatom *Coscinodiscus wailesii.*

Central interspace: A term used for pennate diatoms: It is a central area of solid silica in pennate diatoms, e.g., in the genus *Pseudo-nitzschia.*

Central rosette: An area in the centre of the valve face in centric diatoms with areolae (usually less than 10) larger than on the rest of the valve.

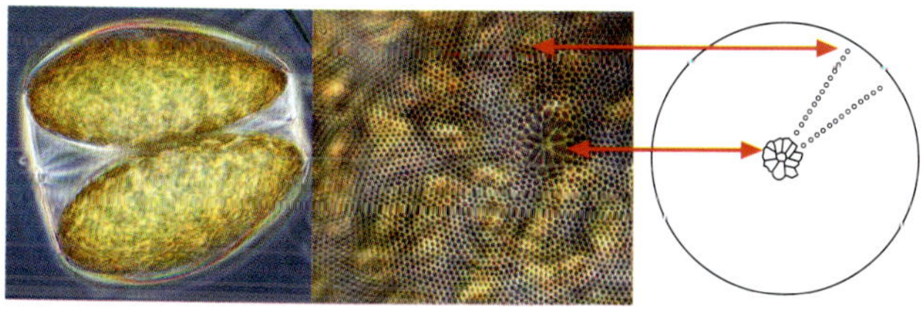

Centric diatoms: Diatoms in which the ornamentation on the valve face is arranged in relation to a central point (e.g., a central process, areola or hyaline area).

Chloroplast: The organelle in phototrophic organisms, which contains the photosynthetic pigments. Their number and shape can be a diagnostic feature.

Cingulum (= Girdle).
In diatoms: The part of the frustule between the epivalve and the hypovalve, consisting of the epicingulum and the hypocingulum. In *Chaetoceros*, the girdle is composed of numerous bands, but these are almost always too lightly silicified to see in the light microscope without special preparation (e.g., naphrax mounts).
In dinoflagellates. The cingulum is a furrow containing the transverse flagellum. This furrow encircles the cell once or several times. In the latter case, the girdle is twisted around the cell (torsion).

Clasper: Membranous structures on the valve ends of cells in the genus *Rhizosolenia*. When *Rhizosolenia* cells form chains the clasper holds the otarium of the adjacent cell.

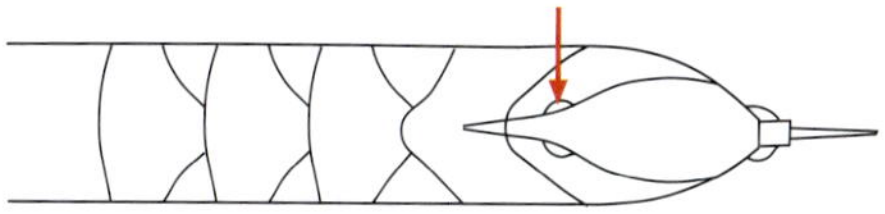

Classification: The grouping and categorization of organisms into a hierarchical system based, in its simplest form on shared morphological characteristics.

Coccoid: With a round shape, like a ball.

Contiguous area: The area of overlap between adjacent cells in a chain of the diatom *Rhizosolenia*.

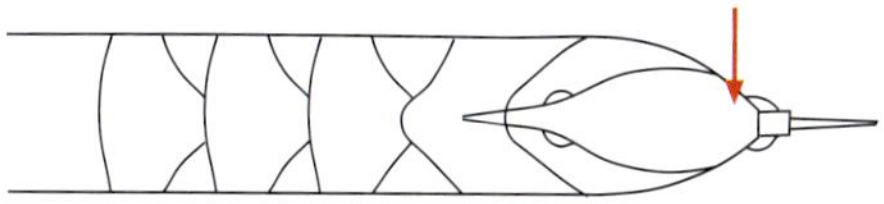

Costa (in diatoms): a thickened line of silica, e.g., in the diatom genus *Rhabdonema.*

Costate ocellus: These occur, e.g., in the genus *Cerataulina*. They are similar to simple ocelli but with siliceous ribs between rows of pores.

Cribrum: A velum (thin layer of silica covering an areola) perforated by small pores. In *Coscinodiscus* the cribra are visible when looking at the outside of the valve. In *Thalassiosira* the cribra are internal.

Cytoplasm: The material bound within the plasma membrane. In eukaryotic organisms the cytoplasm also contains the cell organelles, e.g., the mitochondria.

Cytostome: The cell mouth in ciliates.

Desmokont: A dinoflagellate cell type in which two dissimilar flagella emerge from the anterior part of the cell, e.g., in the genus *Prorocentrum*. This morphological type does not have a cingulum or sulcus.

Diagnosis: A Description of an organism outlining the unique features of the organism that are needed to distinguish it from other species. Morphological characteristics of value in identifying a species (and distinguishing it from others) are called 'diagnostic'.

Differential interference contrast (DIC): A microscopy technique to produce high contrast images. DIC images typically have a relief-like shadow cast appearance. The technique is similar to phase contrast but without producing the distinct halo around cells that is typical for phase contrast images. DIC is well suited, e.g., for the visualisation of surface structures on diatom valves or dinoflagellate thecae. The use of DIC requires special prisms and a beam splitter.

Dinokaryon: A nucleus with permanently condensed chromosomes making them easily visible. This type of nucleus is typical of the dinoflagellates.

Dinokont: The cell type in dinoflagellates in which the two flagella are inserted on the ventral side of the cell. One flagellum is located in the cingulum (transverse flagellum) and the other in the sulcus (longitudinal flagellum).

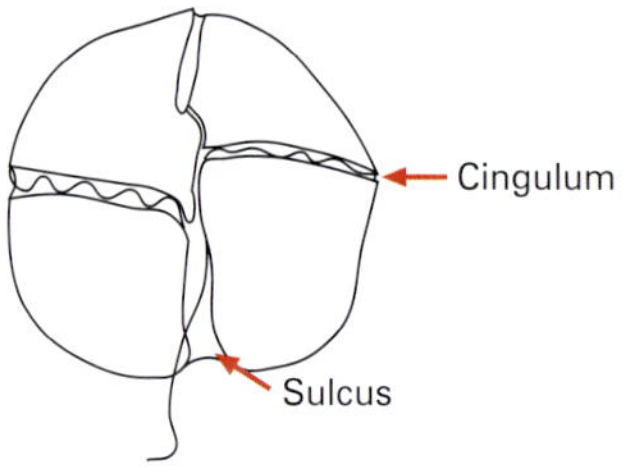

Diploid: Vegetative cells with two sets of chromosomes (2N), e.g., the diatoms.

Displacement: A term for a cingulum the two ends of which do not meet. Displacement is high for instance in species of the genus *Gonyaulax*. The displacement is called descending when the portion on the left side of the cell (in ventral view) lies below that on the right side. In the opposite case the cingulum is ascending.

Distal: The part of a structure furthest away from the point of attachment to a substratum, organism etc.

DNA (desoxyribonucleic acid): The carrier of genetic information within an organism's cells. The DNA is made up of genes which contain the 'instructions' for the building of proteins.

Dorsal: In dinoflagellates: that side of the cell opposite the side containing the sulcus.

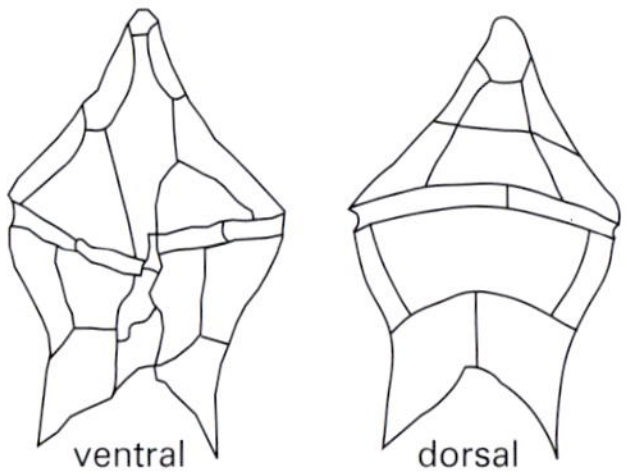

Ejectosome: An ejectile organelle. It is found in cryptophytes and also some prasinophytes.

Epicone: The part of the cell above the cingulum in unarmoured dinoflagellates.

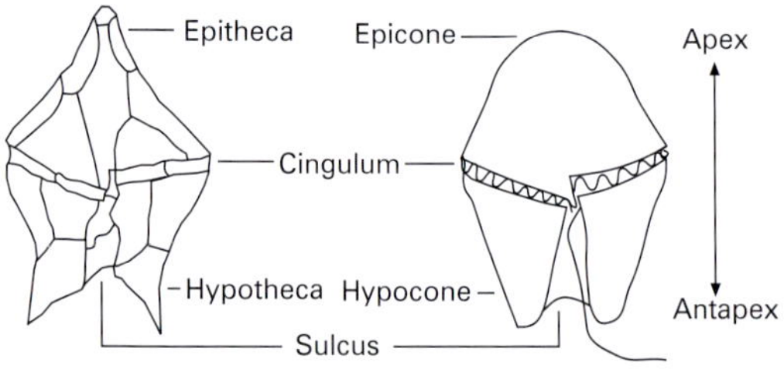

Epitheca: The part of the cell above the cingulum in armoured dinoflagellates (see epicone).

Euphotic zone: The topmost layer of a waterbody in which light penetration is sufficient for net primary production to be possible. The depth of the euphotic zone depends on the degree of turbidity of the water.

Euryhaline: Term referring to organisms that can tolerate a wide range of salinities.

Eurytherm: Term referring to organisms that can tolerate a wide range of temperatures.

Fibula: A bridge of silica between portions of the valve on either side of the raphe in pennate diatoms.

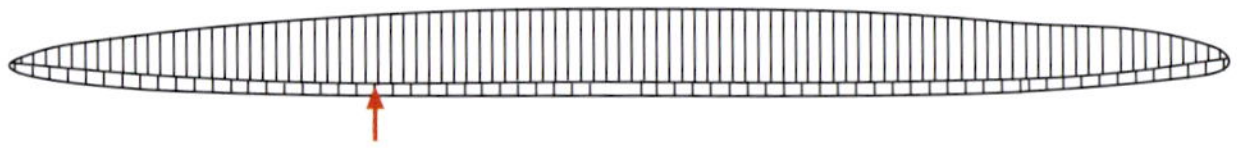

Fimbrate: Fringed, e.g., the marginal ridge in the diatom *Ditylum brightwellii*.

Fixation: The procedure of treating samples of live material with chemicals to immobilize cells and facilitate their long-term preservation. Common fixatives are Lugols iodine solution and formalin.

Flagellum (pl. Flagella): Hair-like swimming organelles in dinoflagellates and other small flagellates.

Flimmer flagellum: Flagellar with short tubular hairs arranged in (two) rows along the flagellum.

Focal plane: The area in a camera where light is focused. An object on a slide or in a settling chamber has a three dimensional structure not all of which can be in focus at the same time. When the fine or coarse focus are adjusted on the microscope different areas of the specimen can be focused on. This is important when trying to identify plankton. Cells with a convex or concave surface with different types of processes for instance need to be viewed in different focal planes to see all important morphological features. An example is the genus *Coscinodiscus* where one needs to adjust the microscope focus when looking at marginal processes and the central rosette.

Food web: The foodweb is an extension of the food chain concept and summarizes the relationships between the members of the different trophic levels in the ecosystem.

Foot pole: The thickened valve end in some pennate diatom species such as *Asterionellopsis* and *Asteroplanus*. It is the footpole that connects individual cells into chains or colonies.

Frustule: The siliceous part of the wall in a diatom cell.

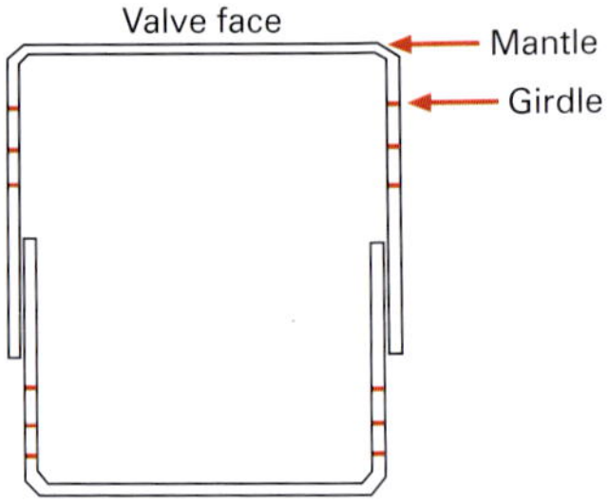

Girdle: See cingulum.

Girdle bands (in diatoms): A series of bands constituting the girdle (= cingulum) between the epi- and hypovalve.

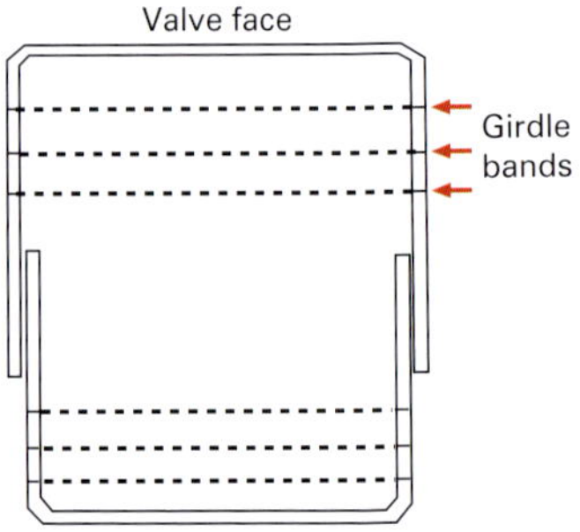

Girdle view: Viewing a diatom along the transapical (broad girdle view) or apical axis (narrow girdle view), at a right angle to the valve face.

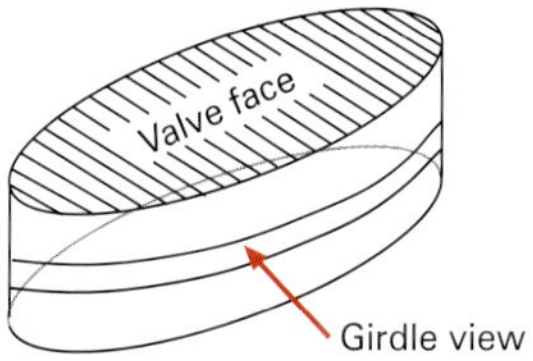

Gullet: A sac-like invagination of the cell surface in cryptophyceans

Haploid: Vegetative or gametic cells with one set of chromosomes (denoted as 1N). Vegetative dinoflagellate cells are 1N.

Haptonema: Flagellum-like organelle characteristic of the Haptophyta. It can coil and are not used for swimming but for attachment and capture of food particles. The haptonema has a different microtubular structure to flagella.

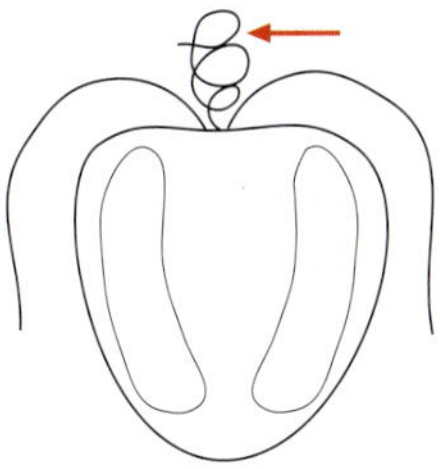

Heterodynamic: Flagella with different (independent) beating patterns, e.g., one pulling and one pushing.

Heterotrophy: A type of nutrition where organic material has to be taken up to obtain energy and the materials for biosynthesis.

Heterovalvate: Diatom frustules in which the two valves look dissimilar, e.g., one valve with a concave and one with a convex valve.

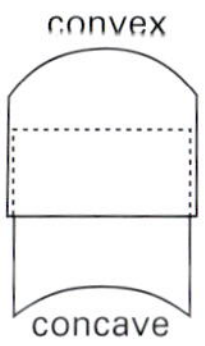

Horn: A hollow structure protruding from a valve or theca: This structure occurs commonly in diatoms (e.g., the genus *Odontella*) and in armoured dinoflagellates (e.g., in *Ceratium* and some *Protoperidinium* species), where it is an extension of the thecal plates.

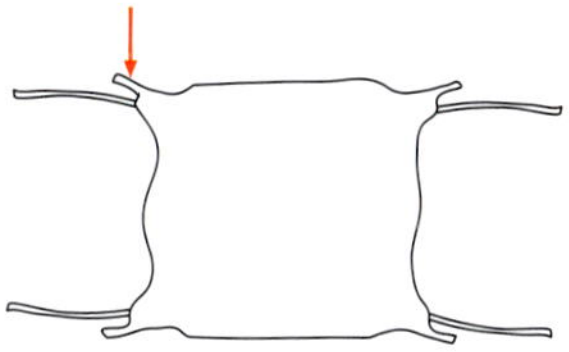

Hypocone: The part of the cell below the cingulum in unarmoured dinoflagellates.

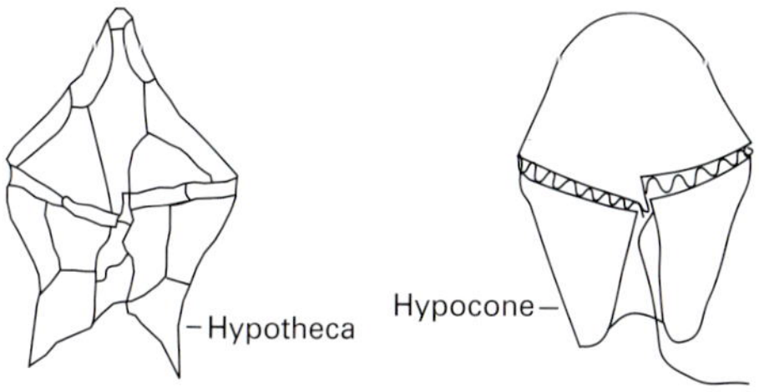

Hypotheca: The part of the cell below the cingulum in armoured dinoflagellates.

Interspace: In pennate diatoms: The space between two fibulae.

Interstria: In pennate diatoms, e.g., in the genus *Pseudo-nitzschia* the lines of non areolated silica located between the striae.

Isogamy: A form of sexual reproduction in which both male and female gametes are of the same size and appearance. They can, therefore, not be distinguished visually as male or female. The gametes can be flagellate or non-flagellate. In many pennate diatoms for instance sexual reproduction involves isogamous non-flagellate sperm.

Isovalvate: The two valves of a diatom frustule look the same (see also heterovalvate).

Kinetid: Appendages in ciliates, consisting of a basal body and associated structures. These can be ciliated or non-ciliated. The kinetids often form distinct patterns around or within the oral cavity or on the body (somatic kinetids). Their arrangement on the cell is used in species identification.

Kinety: A row of kinetids.

Labiate process: A type of process found in many diatom genera, e.g., *Thalassiosira* and *Coscinodiscus*. The labiate process is a tube through the valve wall with an opening on the outer and inner valve surface. On the inside of the valve the opening of the tube is surrounded by two lips the shape of which is used in the identification of the species. (Adapted from Hasle & Syvertsen, 1997).

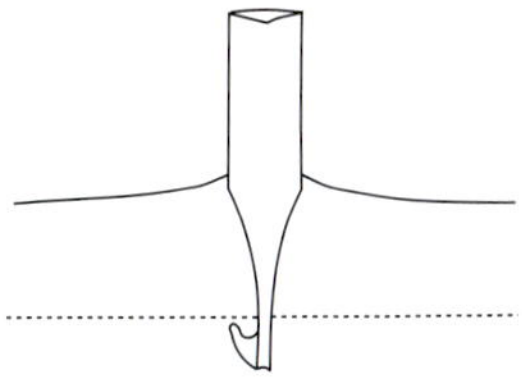

Lanceolate: Lance shaped.

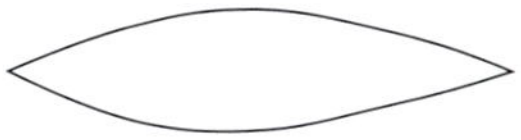

Lenticular: Lentil shaped.

Ligulate: Shaped like a strap or tongue.

List: Extensions of the thecal plates in dinoflagellates. They are usually most pronounced around the cingulum and sulcus (e.g., in the genera *Dinophysis*, *Diplopsalis* and *Protoperidinium*)

Littoral: Pertaining to the shore.

Longitudinal flagellum: The flagellum in dinoflagellates that is housed in the sulcus.

Lugol: An iodine-based solution used for the preservation of phytoplankton samples.

Macronucleus: The macronucleus is one of two types of nuclei found in ciliates. It is involved in protein synthesis but not in meiotic divisions.

Macrorimoportula: These are found in the genus *Coscinodiscus*. Within the marginal ring of labiate processes two distinctly larger labiate processes, the macrorimoportulae, are located. The angle at which they are positioned to each other is a diagnostic feature.

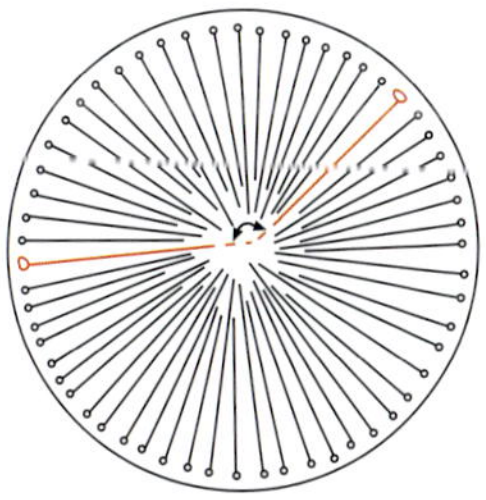

Marginal ridge (in diatoms): A bridge between the valve face and the valve mantle. It can be continuous or interrupted perforated or solid (especially in the Lithodesmiaceae).

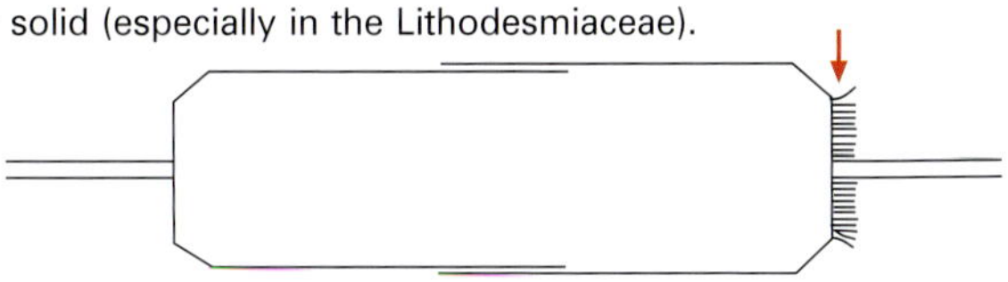

Meiosis (reduction division): Nuclear division leading to the production of reproductive cells (gametes).

Micronucleus: One of the two types of nuclei in ciliates. The micronucleus is involved in sexual reproduction.

Mitosis: Nuclear division that results in two daughter cells each with the same number of chromosomes as the mother cell.

Mixotrophy: A phototrophic organism that complements its diet by the uptake of organic material.

Morphology: The shape and form of an organism or part of an organism.

Motile: Able to move.

Mucilage: A thick, sticky substance produced by some animals and plants (including some diatoms) often to attach to a substratum or to trap prey.

Naphrax: A substance used to mount and permanently fix cleaned phytoplankton samples on microscope slides. It not only preserves the sample but also facilitates the better resolution of structures on the diatom valve when viewing the sample under the microscope. Naphrax contains the chemical Toluol as a solvent. Toluol is toxic. Therefore, Naphrax should be used under a fume hood.

Nematocyst: A large ejectile organelle in the dinoflagellate families Polykrikaceae and Warnowiaceae. It is used to capture prey.

Neritic: The zone from the low water mark to the edge of the continental shelf.

Oblique: Sloping.

Occluded process: In the diatom family Thalassiosiraceae, a process on the valve face. It has a hollow external tube but does not penetrate through the valve wall. An example is the *Thalassiosira punctigera.*

Oceanic: In the open sea.

Ocellus

In diatoms: An area of areolae on the valve face that is surrounded by a rim of silica and forms an area discrete from the rest of the valve. Ocelli are formed, e.g., in the genera *Eucampia*, *Asteroplanus,* and *Odontella*.

In dinoflagellates: A pigmented eye spot. Ocelli are found in the dinoflagellate *Nematodinium armatum* for instance.

Oil immersion objective: A high magnification lens to be used with immersion oil between the lens and the slide or chamber. The oil facilitates the formation of a sharp image. Immersion objectives for use in water are also available. These are used for viewing live samples, e.g., in a petri dish.

Oogamy: Form of sexual reproduction in which the male and female gametes are of unequal size with the female gametes being large and non-motile and the male gametes small and flagellate.

Oral: A term referring to the mouth region of a ciliate.

Organic compounds: Chemical compounds that contain carbon. Examples are sugars. A few simple compounds such as carbon dioxide (CO_2) and methane (CH_4) are traditionally considered to be inorganic despite the presence of carbon.

Otarium: Winglike structures found associated with the spines of cells in the genus *Rhizosolenia.* The exact shape and positioning of these structures is used in the identification of *Rhizosolenia* species.

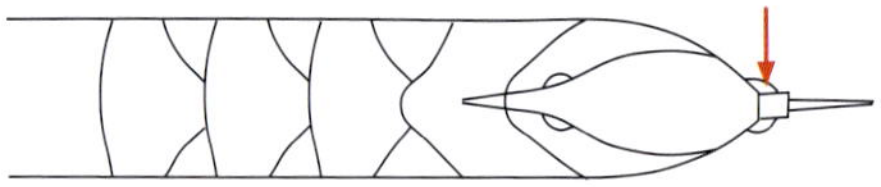

Pennate diatom: Diatoms that are in contrast to centric diatoms bilaterally symmetrical.

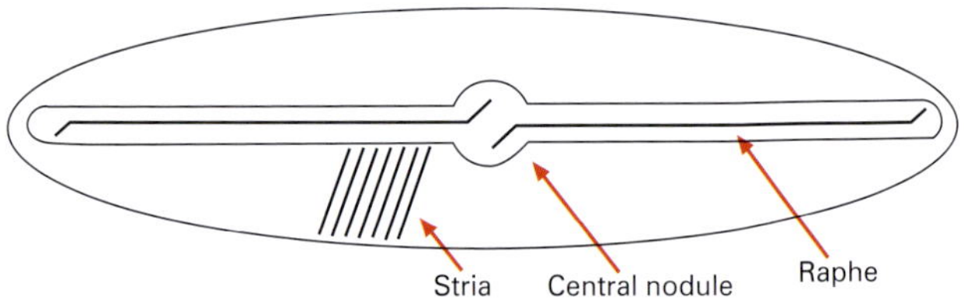

Periflagellar area: The area of thecal plates near the cell apex in the dinoflagellate genus *Prorocentrum*. These harbour the pores from which the flagella emerge.

Pervalvar axis: The pervalvar axis is the longitudinal axis through the centre point of the two valves in a diatom cell. It describes the cell height of a diatom cell.

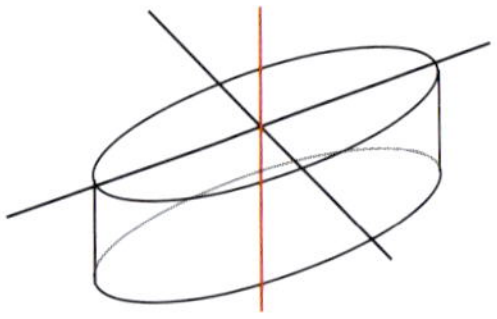

Phagotrophy: A special form of food uptake, e.g., in the dinoflagellates. Food particles are taken up as whole cells and digested within special food vacuoles in the predator.

Photosynthesis: The use of the energy from sunlight to transform two simple compounds (carbon dioxide and water) into sugars. The sun's energy is captured by photosynthetic pigments arranged within the chloroplasts inside the plant cell.

Phytoplankton: The term refers to photosynthesizing (photoautotrophic) plankton, which comprises mainly diatoms, green algae, small flagellates such as cryptohytes, and dinoflagellates. Note: the latter group also contains heterotrophic species and an increasing number of studies reveal dinoflagellates capable of mixotrophy.

Pigments: Coloured compounds, which absorb light of different wavelengths. The major plant pigment is chlorophyll. Chlorophyll absorbs yellow and blue light (green is not absorbed which is why leaves have a green colour). Plants also have further pigments (accessory pigments) which absorb light at additional wavelengths.

Pilus: Long, hairlike structures arising from the poles of diatom valves in the family Cymatosiraceae.

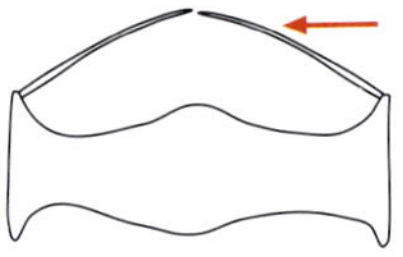

Placolith: A coccolith composed of an upper and a lower shield separated by a tube.

Plankton: Organisms that drift, rather than actively swim, in the water as they are either incapable of movement or cannot generate sufficient movement to withstand the direction of the currents.

Plasmolysis: The contraction of cell contents due to water loss or as in plankton due to shock (e.g., violent shaking of a sample).

Polykinetid: A kinetid composed of more than two kinetosomes (the basal structures from which the cilia arise). These can occur, e.g., in the oral region of the ciliate.

Poroids: In the genus *Pseudo-nitzschia*: rows of small pores in the striae on the valve face. The number of rows of poroids and their shape are important for the identification of species.

Primary production: The term refers to the production of organic materials from carbon dioxide as the result of photo- or chemosynthesis. Primary production is usually expressed as mass per unit area and time, e.g., g $C/m^2 \cdot a$).

Proboscis: A tube terminating in a blunt tip. A proboscis is typical for the genus *Proboscia.*

Protist: Unicellular, eukaryotic organisms, existing either as independent cells or, if they are colonial, show a colonial habit without differentiation into different types of tissues (e.g., ciliates, different types of flagellates, and algae).

Proximal: The part of a structure closest the point of attachment to a substratum, organism etc.

Pseudocellus (in diatoms): Similar to an ocellus but without being surrounded by a rim of silica.

Pseudonodulus: A marginal to submarginal structure (always only one per valve) in the diatom family Hemidiscaceae. With the light optical microscope it becomes evident as a hole or an area covered by densely packed smaller areolae near the valve margin.

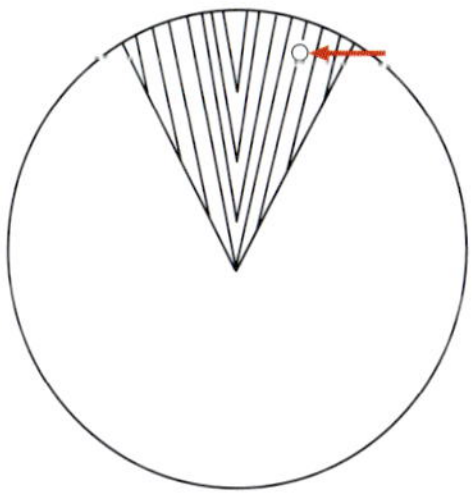

Pseudopodium: Temporary outgrowth of the cell of some protozoans (amoebae) used for feeding and locomotion.

Pyrenoid: A specialized area of the algal chloroplast that is the centre of carbon dioxide fixation. They contain high levels of RubisCo (the enzyme catalyzing the first step in carbon fixation).

Raphe: In pennate diatoms: a longitudinal slit running along the apical axis (e.g., in *Pseudo-nitzschia*) or around the cell margin (e.g., in *Surirella*)

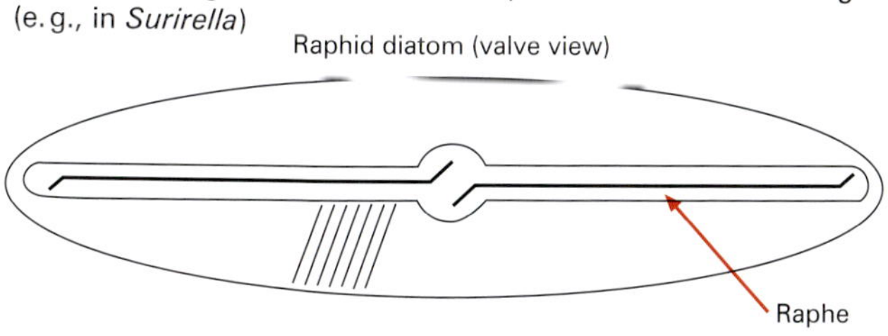

Resting cyst (Hypnozygote): Dormant stage of dinoflagellates that germinates under favourable environmental conditions to produce swimming cells. It is formed by the fusion of two motile gametes.

Reticulations: Ridges over the surface of armoured dinoflagellates. They can be a diagnostic feature.

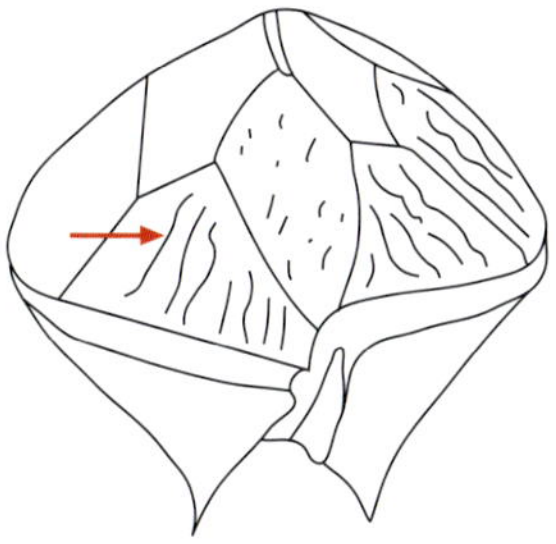

Sensu lato: In a broad sense.

Sensu stricto: In a narrow sense.

Septum (in diatoms): An internal plate projecting from the girdle band into the interior of the frustule.

Solitary: Living as single cells.

Somatic: Referring to the body of a ciliate.

Spine: A solid process protruding from a valve or theca.

Sternum (= Pseudoraphe): An elongate structure in pennate diatoms that is only sparsely areolated or not areolated at all.

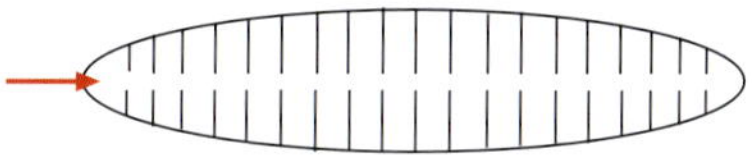

Striation

In dinoflagellates: Structures on the cell surface appearing as longitudinal lines or ridges.

In diatoms: Lines of pores on the valve face or mantle.

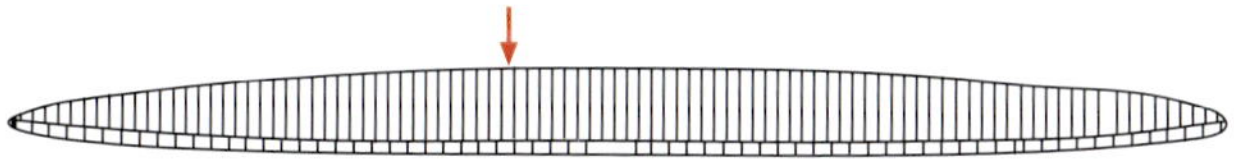

Strutted process: A process in the Thalassiosirales, which penetrates the valve wall. Internally it is supported by struts and satellite pores. Externally it may have a tube of varying length. While the internal structures can only be visualized using Scanning electron microscopy, the external tubes of the strutted processes are often very conspicuous. An example is *Thalassiosira nordenskioeldii* (Graphic adapted from Hasle & Syvertsen, 1997).

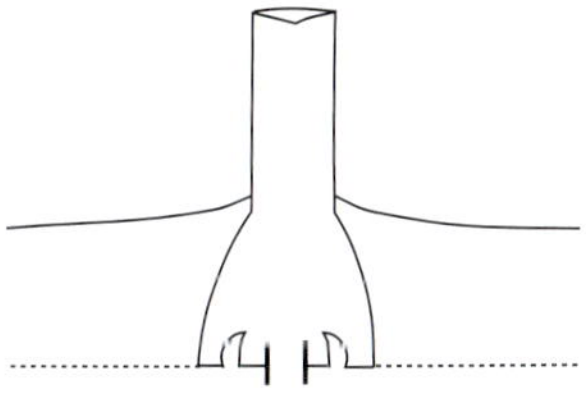

Sulcus: The groove in dinokont dinoflagellates, which houses the longitudinal flagellum. It usually runs from the cell antapex towards the cingulum (sometimes it extends beyond the cingulum into the epitheca or epicone)

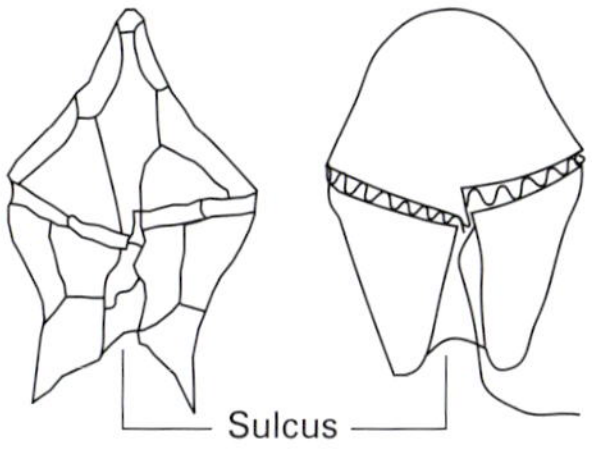

Sutures (in thecate dinoflagellates). The boundaries between cells. These are thicker than the plates themselves. In cells stained with calcofluor the stronger staining of these thickened boundaries facilitates the visualization of plate patterns

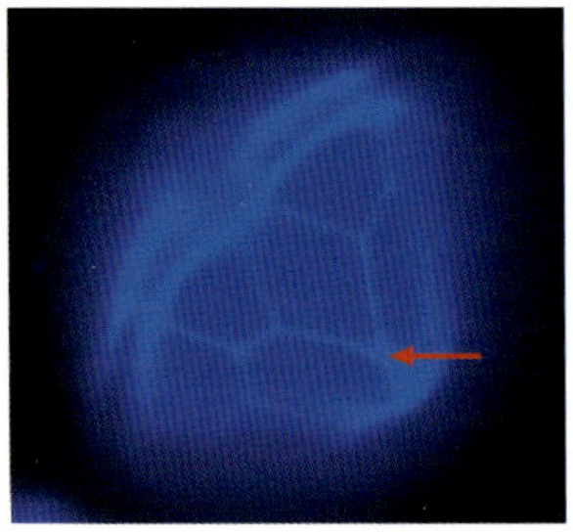

Synonym: Although a species can only have one valid name, several different names have often been assigned in the past, e. g., because the same species has originally been described independently by two different authors or because a species is 're-classified'. This can be the case if re-examination of a specimen reveals new morphological characteristics. This is often the case when new microscope techniques become available.

Tabulation (of thecal plates): For easier comparison of plate patterns in dinoflagellates the thecal plates are named according to their location on the cell: apical, intercalary, precingular, cingular, postcingular, posterior intercalary antapical and sulcal. These have a standardized designation, here shown for the genus *Protoperidinium* (Apical pore plate = Po).

Po	X	Apical	Anterior intercalary	Precingular	Cingular
x	x	4'	2-3a	7''	4c

Sulcal	Postcingular	Posterior intercalary	Antapical
6s	5'''	0p	2''''

The plate pattern is a highly diagnostic feature, e. g., in the genus *Protoperidinium*.

Taxonomy: The process of naming a species so as to reflect its systematic position, i. e. it's relatedness to other species. To name a species its different morphological features will be described (and compared with those of already known species).

Torsion: Twisting along the longitudinal axis of an organism or a chain as in the diatom *Helicotheca tamesis*.

Transapical axis: The depth of the valve in diatoms In bipolar diatoms this refers to the narrowest part of the valve, while the apical axis describes the bradest part.

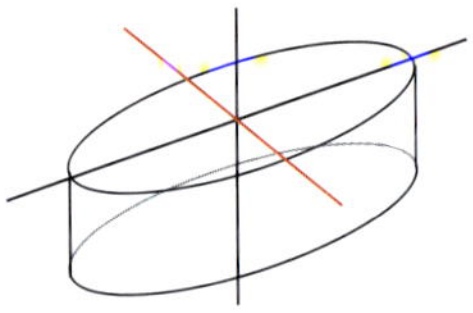

Transapical plane (in diatoms): The plane perpendicular to the apical axis.

Transverse flagellum: The flagellum in dinoflagellates that is located in the cingulum.

Trichocyst pore: Pores in the theca of dinoflagellates, which can be involved in the excretion of mucus. The pore number and pattern on the theca can be variable in some species but in others, e. g., of the genus *Prorocentrum* they are a diagnostic feature.

Trophic level: An ecological term describing the position of a species in a food chain or food web (i. e. what does the organism eat and who eats the organism in question). The lowest trophic level (the base of the foodweb) is formed by autotrophic organisms that use the sun's energy to produce complex organic compounds. They are called primary producers. Primary producers are eaten by the primary consumers (plant eaters) and these in turn are eaten by secondary consumers.

Tychopelagic: A species that is found mainly in the benthos but can be found in the water column in turbulent conditions, e. g., after a storm.

Valve face: The view of a diatom when looking at the upright valve from above.

Valve mantle: The part of the diatom frustule that lies between valve face and girdle.

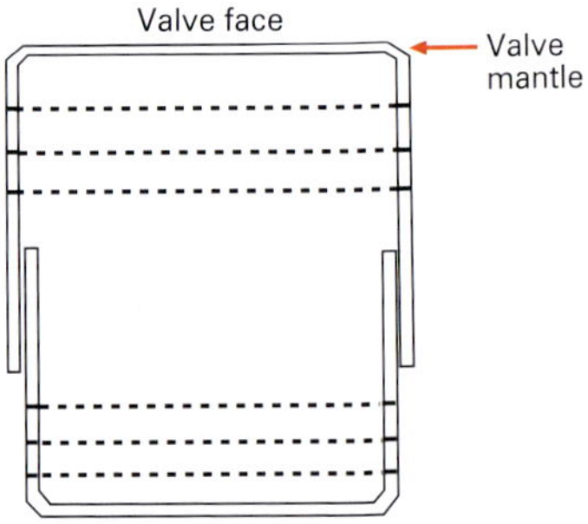

Ventral: In dinoflagellates that side of the cell that shows the ventral groove (sulcus).

Zooplankton: Animals that drift, rather than swim, in the water, incapable of generating sufficient movement to withstand the currents. This term encompasses a large size range of organisms from large jellyfish to small crustacean animals such as copepods and different larval stages.

Zygote: The cell resulting from the fusion of two gametes. In dinoflagellates, there are two types the motile planozygote and the non-motile hypnozygote. In diatoms the zygote (auxospore) is usually the stage in the life cycle which restores cells to a maximum size after a series of vegetative divisions.

General phytoplankton references

Hoppenrath, M., M. Elbrächter & G. Drebes, (2009). Marine Phytoplankton: Selected microplankton species from the North Sea around Helgoland and Sylt. – Kleine Senckenberg-Reihe 49, 264 pp.; Stuttgart (E. Schweizerbart'sche Verlagsbuchhandlung, Nägele u. Obermiller).

Lee, J. J., G. F. Leedale & P. Bradbury (eds.) (2000). An illustrated guide to the protozoa. – 1430 pp. (two volumes); Lawrence, Kansas (Society of Protozoologists).

Reynolds, C. S. (2006). "The Ecology of Phytoplankton". – 535 pp.; London, UK (Cambridge University Press).

Scott, F. J. & H. J. Marchant (eds.) (2005). Antarctic Marine Protists. – 579 pp.; Canberra and Hobart (Australian Biological Resource Study and Australian Antarctic Division).

Taylor, F. J. R. (1987). The biology of dinoflagellates. – 705 pp.; Oxford (Blackwell Scientific Publications).

Throndsen, J., G. R. Hasle & K. Tangen (2007). Phytoplankton of Norwegian coastal waters. – 343 pp.; Oslo (Almater Forlag AS 2007).

Tomas, C. R. (ed.) (1997). Identifying marine phytoplankton. – 858 pp.; San Diego (Academic Press).

Valiela, I. (1995). Marine Ecological Processes. – 686 pp.; Berlin (Springer).

Taxonomic references

Amato, A., L. Orsini, D. D'Alelio & M. Montresor (2005). Life cycle, size reduction patterns, and ultrastructure of the pennate planktonic diatom *Pseudo-nitzschia delicatissima* (Bacillariophyceae). – Journal of Phycology **41** (3): 542-556.

Andrews, G. W. (1975). Taxonomy and stratigrapic occurrence of the marine diatom genus *Rhaphoneis*. – Beiheft zur Nova Hedwigia: 193-227.

– (1977). Morphology and stratigraphic significance of *Delphineis*, a new marine diatom genus. – Beiheft zur Nova Hedwigia **54**: 243-260.

– (1981). Revision of the diatom Genus *Delphineis* and morphology of *Delphineis surirella* (Ehrenberg) G. W. Andrews, n. comb. – Pp. 81-92 in: Ross R. (ed.) Sixth symposium on recent and fossil diatoms. Koenigstein (Koeltz Scientific Books).

Arrigo, K. R. (2005). Marine microorganisms and global nutrient cycles. – Nature **437** (7057): 349-355.

Azam, F., T. Fenchel, J. G. Field, J. S. Gray, L. A. Meyer-Reill & F. Thingstad (1983). The ecological role of water-column microbes in the sea. – Marine Ecology Progress Series **10**: 257-263.

Balech, E. (1948). Etudes de quelques especes de *Peridinium*, souvent confondues. – Hydrobiologia **1** (4): 390-409.

(1974). El genero *Protoperidinium* (*Peridinium* Ehrenberg, 1931, partim). – Museo Argentino de Ciencias naturales 'Bernadino Rivadavia' e Instituto Nacional de investigaciones de las ciencias naturales, Revistas, Hidrobiologia **4**: 1-79.

– (1976) Sur quelques *Protoperidinium* (Dinoflagellata) du Golfe du Lion. – Vie et Milieu **26** (1): 27-46.

– (1990). Four new dinoflagellates. – Helgolaender Meeresuntersuchungen **44** (3-4): 387-396.

– (1995). The genus *Alexandrium* Halim (Dinoflagellata). – 151 pp.; Cork, Ireland (Sherkin Island Marine Station, Sherkin Island, Co).

Balech, E. & K. Tangen (1985). Morphology and taxonomy of toxic species in the Tamarensis group (Dinophyceae): *Alexandrium excavatum* (Braarud) comb. nov. and *Alexandrium ostenfeldii* (Paulsen) comb. nov. – Sarsia **70** (4): 333-343.

Baumann, M. E. M., C. Lancelot, F. P. Brandini, E. Sakshaug & D. M. John, (1994). The taxonomic identity of the cosmopolitan Prymnesiophyte *Phaeocystis*: A morphological and ecophysiological approach. – Journal of Marine Systems **5** (1): 5-22.

Berland, B. R., S. Y. Maestrini & D. Grzebyk (1995). Observations on possible life cycle stages of the dinoflagellates *Dinophysis* cf. *acuminata*, *Dinophysis acuta* and *Dinophysis pavillardi*. – Aquatic Microbial Ecology **9** (2): 183-189.

Boalch, G. (1975). The Lauder species of the diatom genus *Bacteriastrum* Shadbolt. – Beiheft zur Nova Hedwigia **53**: 185-192.

Boalch, G. T. (1971). The typification of the diatom species *Coscinodiscus concinnus* W. M. Smith and *Coscinodiscus granii* Gough. – Journal of the Marine Biological Association of the United Kingdom **51** (3), 685-695.

Boalch, G. T. & D. S. Harbour (1977). Observations on the structure of a planktonic *Pleurosigma*. – Beiheft zur Nova Hedwigia **54**: 275-280.

Brooks, M. (1975a). Studies of the genus *Coscinodiscus* 3. Light transmission and scanning electron microscopy of *C. granii* Gough. – Botanica Marina **18** (1): 29-39.

– (1975b). Studies on the genus *Coscinodiscus* 1. Light transmission and scanning electron microscopy of *C. concinnus* Wm. Smith. – Botanica Marina **18** (1): 1-14.

– (1975c). Studies on the genus *Coscinodiscus* 2. Light, transmission and scanning electron microscopy of *C. asteromphalus* Ehr. – Botanica Marina **18** (1): 15-27.

Bursa, A. (1959). The genus *Prorocentrum* Ehrenberg. Morphodynamics, protoplasmatic structures and taxonomy. – Canadian Journal of Botany **37** (1). 1-31.

Buskey, E. J. (1995). Growth and bioluminescence of *Noctiluca scintillans* on varying algal diets. – Journal of Plankton Research **17** (1): 29-40.

– (1997). Behavioural components of feeding selectivitiy of the heterotrophic dinoflagellate *Protoperidinium pellucidum*. – Marine Ecology Progress Series **153**: 77-89.

Carpenter, E. J., S. Janson, R. Boie, F. Pollehne & J. Chang (1995). The dinoflagellate *Dinophysis norvegica*: Biological and ecological observations in the Baltic Sea. – European Journal of Phycology **30** (1): 1-9.

Carreto, J. I. (1985). A new ketocarotenoid from the heterotrophic dinoflagellate *Protoperidinium depressum* (Bayley) Balech, 1974. – Journal of Plankton Research **7** (3): 421-423.

Carvalho, W. F., S. Minnhagen & E. Graneli (2008). *Dinophysis norvegica* (Dinophyceae), more a predator than a producer?. – Harmful Algae **7** (2): 174-183.

Chepurnov, V. A., D. G. Mann, K. Sabbe, K. Vannerum, G. Casteleyn, E. Verleyen, L. Peperzak & W. Vyverman (2005). Sexual reproduction, mating system, chloroplast dynamics and abrupt cell size reduction in *Pseudo-nitzschia pungens* from the North Sea. – European Journal of Phycology **40** (4): 379-395.

Chepurnov, V. A., D. G. Mann, P. von Dassow, E. V. Armbrust,K. Sabbe & Dasseville, R. (2006). Oogamous reproduction, with two-step auxosporulation, in the centric diatom (Bacillariophyta). – Journal of Phycology, **42**(4): 845-858.

Chomérat, N., A. Couté, S. Fayolle, G. Mascarell & A. Cazaubon (2004). Morphology and ecology of *Oblea rotunda* (Diplosalidaceae, Dinophyceae) from a new habitat: a brackish and hypertrophic ecosystem, the Etang de Bolmon (South of France). – European Journal of Phycology **39**(3): 317-326.

Clay, B. L., P. Kugrens & R. Lee (1999). A revised classification of Cryptophyta. – Botanical Journal of the Linnean Society **131**(2): 131-151.

Cloern, J. E. (1996). Phytoplankton bloom dynamics in coastal ecosystems: A review with some general lessons from sustained investigation of San Francisco Bay, California. – Reviews of Geophysics **34**(2): 127-168.

Cohen-Fernandez, E. J., E. M. Del Castillo, I. H. Salgado Ugarte & F. F. Pedroche (2006). Contribution of external morphology in solving a species complex: The case of *Prorocentrum micans*, *Prorocentrum gracile* and *Prorocentrum sigmoides* (Dinoflagellata) from the Mexican coast. – Phycological Research **54**(4): 330-340.

Cox, E. & P. C. Reid (2004). Generic relationships within the Naviculinae: A preliminary cladistic analysis. – Proceedings of the 17th International diatom symposium, Ottawa, Canada, pp. 49-62.

Crawford, R. M. (1979). Taxonomy and frustular structure of the marine centric diatom *Paralia sulcata*. – Journal of Phycology **15**(2): 200-210.

– (1977). The taxonomy and classification of the diatom genus *Melosira* C. Ag. II. *M. moniliformis*. – Phycologia **16**(3): 277-285.

Crawford, R. M. & C. Gardner (1997). The transfer of *Asterionellopsis kariana* to the new genus *Asteroplanus* (Bacillariophyceae). – Nova Hedwigia **65**(1-4): 47-57.

Daugbjerg, N., G. Hansen, J. Larsen & Ø. Moestrup (2000). Phylogeny of the major genera of dinoflagellates based on ultrastructure and partial LSU rDNA sequence data, including the erection of three new genera of unarmoured dinoflagellates. – Phycologia **39**(4): 302-317.

Davis, C. O., J. T. Hollibaugh, D. L. R. Seibert, W. H. Thomas & P. J. Harris (1980). Formation of resting spores by *Leptocylindrus danicus* (Bacillariophyceae) in a controlled ecosystem. – Journal of Phycology **16**(2): 296-302.

Dodge, J. (1982). Marine dinoflagellates of the British Isles. – London, Her Majesty's Stationary Office: 303.

Dodge, J. D. (1981). Three new generic names in the Dinophyceae: *Herdmania*, *Sclerodinium* and *Triadinium* to replace *Heteraulacus* and *Goniodoma*. – British Phycological Journal **16**(3): 273-280.

– (1989). Some revisions of the family Gonyaulacaceae (Dinophyceae). – Botanica Marina **32**(4): 275-298.

Dodge, J. D. & H. Hermes (1981). A revision of the *Diplopsalis* group of dinoflagellates (Dinophyceae) based on material from the British Isles. – Botanical Journal of the Linnean Society **83**(1): 15-26.

Dodge, J. D. & S. Toriumi (1993). A taxonomic revision of the *Diplopsalis* group (Dinophyceae). – Botanica Marina **36**(2): 137-147.

Drebes, G. (1972). The life history of the centric diatom *Bacteriastrum hyalinum* Lauder. – Beiheft zur Nova Hedwigia **39**: 95-110.

Dring, M. (1991). The Biology of Marine Plants. – 208 pp.; Cambridge (Cambridge University Press).

Dujardin, F. (1841). Histoire naturelle des Zoophytes, infusoires, comprenant la physiologie et la classification de ces animaux et la maniere de les etudies a l'aide du microscope. – 684 pp.; Paris, Roret (Nouvelles suites a Buffon).

Durbin, E. G. (1974). Studies on the autecology of the marine diatom *Thalassiosira nordenskioeldii* Cleve. I. The influence of daylength, light intensity and temperature on growth. – Journal of Phycology **10**(2): 220-225.

– (1977). Studies on the autecology of the marine diatom *Thalassiosira nordenskioeldii*. II. The influence of cell size on growth rate, carbon, nitrogen, chlorophyll a and silica content. – Journal of Phycology **13**(2): 150-155.

– (1978). Aspects of the biology of resting spores of *Thalassiosira nordenskioeldii* and *Detonula confervacea*. – Marine Biology **45**(1): 31-37.

Dürselen, C. D. & H. J. Rick (1999). Spatial and temporal distribution of two new phytoplankton diatom species in the German Bight in the period 1988 and 1996. – Sarsia **84**(5-6): 367-377.

Edlund, M. B. & E. F. Stoermer (1997). Ecological, evolutionary and systematic significance of diatom life histories. – Journal of Phycology **33**(6): 897-918.

Edvardsen, B., K. Shalchian-Tabrizi, K. Jakobsen, S., L. K. Medlin, E. Dahl, S. Brubak & E. Paasche (2003). Genetic variability and and molecular phylogeny of *Dinophysis* species (Dinophyceae) from Norwegian waters inferred from single cell analyses of rRDNA. – Journal of Phycology **39**(2): 395-408.

Elbrächter, M. (1996). Toxische Algen in der Nordsee. – Deutsche Hydrographische Zeitschrift. Supplement **4-6**: 37-44.

Elbrächter, M. & G. Drebes (1978). Life cycles, phylogeny, and taxonomy of *Dissodinium* and *Pyrocystis* (Dinophyta). – Helgoländer Wissenschaftliche Meeresuntersuchungen **31**(3): 347-366.

Elbrächter, M. & E. Schnepf (1996). *Gymnodinium chlorophorum*, a new green bloom forming dinoflagellate (Gymnodiniales, Dinophyceae) with a vestigial prasinophyte endosymbiont. – Phycologia **35**(5): 381-393.

Ellegaard, M., N. Daugbjerg, A. Rochon, J. Lewis & I. Harding (2003). Morphological and LSU rDNA sequence variation within *Gonyaulax spinifera-Spiniferites* group (Dinophyceae) and proposal of *G. elongata* comb. nov. and *G. membranacea* comb. nov. – Phycologia **42**(2): 151-164.

Ewart, C. S., M. K. Meyers, E. R. Wallner, D. J. McGillicuddy & C. A. Carlson (2008). – Microbial dynamics in cyclonic and anticyclonic mode-water eddies in the northwestern Sargasso Sea. – Deep Sea Research II **55**(10-13): 1334-1347.

Faust, M. A., J. Larsen & Ø. Moestrup (1999). Potentially Toxic Phytoplankton: 3. Genus *Prorocentrum* (Dinophyceae). – ICES Identification leaflets for plankton **184**: 1-24.

Fehling, J., D. H. Green, K. Davidson, C. J. Bolch & S. S. Bates (2004). Domoic acid production by *Pseudo-nitzschia seriata* (Bacillariophyceae) in Scottish waters. – Journal of Phycology **40**(4): 622-630.

Fernandes, L. F., L. Zehnder-Alves & J. C. Bassfeld (2001). The recently established diatom *Coscinodiscus wailesii* (Coscinodiscales, Bacillariophyta) in Brazilian waters. I: Remarks on morphology and distribution. – Phycological Research **49**(2): 89-96.

Figueroa, R. I. & I. Bravo (2005). Sexual reproduction and two different encystment strategies of *Lingulodinium polyedrum* (Dinophyceae) in culture. – Journal of Phycology **41**(2): 370-379.

Figueroa, R. I., I. Bravo & E. Garcés (2005). Effects of nutritional factors and different parental crosses on the encystment and excystment of *Alexandrium catenella* (Dinophyceae) in culture. – Phycologia **44**(6): 658-670.

Figueroa, R. I., I. Bravo & E. Garcés (2008). The siginificance of sexual versus asexual cyst formation in the life cycle of the noxious dinoflagellate *Alexandrium peruvianum*. – Harmful Algae **7**(5): 653-663.

Francis, D. (1967). On the eyespot of the dinoflagellate, *Nematodinium*. – Journal of Experimental Biology **47**: 495-501.

French III, F. W. & P. E. Hargraves (1985). Spore formation in the life cycles of the diatoms *Chaetoceros diadema* and *Leptocylindrus danicus*. – Journal of Phycology **21** (3): 477-483.

– (1986). Population dynamics of the spore forming diatom *Leptocylindrus danicus* in Narragansett Bay, Rhode Island. – Journal of Phycology **22** (4): 411-420.

Fritz, L. & R. E. Triemer (1985). A rapid and simple technique utilizing Calcofluor White M2R for the visualization of dinoflagellate thecal plates. – Journal of Phycology **21** (4): 662-664.

Fryxell, G. A. & G. R. Hasle (1972). *Thalassiosira eccentrica* (Ehrenb.) and Cleve, T. symmetrica sp. nov. and some related centric diatoms. – Journal of Phycology **8** (4): 297-317.

– (1977). The genus *Thalassiosira*: Some species with a modified ring of central strutted processes. – Beiheft zur Nova Hedwigia **54**: 67-89.

Genovesi-Giunti, B. & A. Vaquer (2006). The benthic resting cysts: A key actor in harmful dinoflagellate blooms – a review. – Vie et Milieu **56** (4): 327-337.

Gifford, D. J. & D. A. Caron (2000). Sampling, preservation, enumeration and biomass of marine protozooplankton. – Pp. 193-221 in Harris, R. P., P. H. Wiebe, J. Lenz, H. R. Skjoldal & M. Huntley (eds.): ICES Zooplankton Methodology Manual. London (Academic Press).

Gollasch, S., E. Macdonald, S. Belson, H. Botnen, J. T. Christensen, J. P. Hamer, G. Houvenaghel, A. Jelmert, I. Lucas, D. Masson, T. McCollin, S. Olenin, I. Persson, A. Wetsteyn & T. Wittling (2002). Life in Ballast tanks. – Pp. 217-231 Leppäkoski, E., S. Gollasch & S. Olenin (eds.): Invasive aquatic species of Europe: distribution, impacts and management. Dordrecht (Kluwer).

Gomez, F. (2008). Phytoplankton invasions: Comments on the validity of categorizing the non-indigenous dinoflagellates and diatoms in European Seas. – Marine Pollution Bulletin **56**: 620-628.

Granéli, E. & J. T. Turner (2006). Ecology of Harmful Algae. – 413 pp.; Berlin Heidelberg New York (Springer).

Gribble, K., E. & D. M. Anderson (2006). Molecular phylogeny of the heterotrophic dinoflagellates, *Protoperidinium*, *Diplopsalis* and *Preperidinium* (Dinophyceae), inferred from large subunit rDNA – Journal of Phycology **42** (5): 1081-1095.

Gribble, K., E., G. Nolan & D. M. Anderson (2007). Biodiversity, biogeography and potential impact of *Protoperidinium* spp. – Journal of Plankton Research **29** (11): 931-947

Grosjean, P., M. Picheral, C. Warembourg & G. Gorsky (2004). Enumeration, measurement and identification of net zooplankton samples using the ZOOSCAN digital imaging system. – ICES Journal of Marine Science **61**: 518-525.

Hallegraeff, G. M. & I. A. N. Lucas (1988). The marine dinoflagellate genus *Dinophysis* (Dinophyceae): photosynthetic, neritic and non-photosynthetic, oceanic species. – Phycologia **27** (1): 25-42.

Hallegraeff, G. M., D. A. Anderson & A. Cembella (eds.) (2003). Manual on harmful marine microalgae. – 793 pp.; Paris (UNESCO Publishing).

Hansen, E. & H. C. Eilertsen (2007). Do polyunsaturated aldehydes produced by *Phaeocystis pouchetii* (Hariot) Lagerheim influence diatom growth during the spring bloom in Northern Norway? – Journal of Plankton Research **29** (1): 87-96.

Hansen, G. (1993). Dimorphic individuals of *Dinophysis acuta* and *D. norvegica* (Dinophyceae) from Danish waters. – Phycologia **32**: 73-75.

– (1995). Analysis of the thecal plate pattern in the dinoflagellate *Heterocapsa rotundata* (Lohmann) comb. nov. (= *Katodinium rotundatum*) (Lohmann) Loeblich). – Phycologia **34** (2): 166-170.

Hansen, G. & N. Daugbjerg (2004). Ultrastructure of *Gyrodinium spirale*, the type species of *Gyrodinium* (Dinophyceae) including a phylogeny of *G. dominans*, *G. rubrum* and *G. spirale* deduced from partial LSU rDNA sequences. – Protist **155** (3): 271-294.

Hansen, G. & J. Larsen (1992). Dinoflagellater i danske farvande. Havforskning fra Miljøstyrelsen. – 331 pp.; Copenhagen (H. A. Thomsen, ed.).

Hansen, G., O. Moestrup & K. R. Roberts (1996/97). Light and electron microscopical observations on *Protoceratium reticulatum* (Dinophyceae). – Archiv für Protistenkunde **147**: 381-391.

Hansen, G., N. Daugbjerg & P. Henriksen (2000). Comparative study of *Gymnodinium mikimotoi* and *Gymnodinium aureolum*, comb. nov. (= *Gyrodinium aureolum*) based on morphology, pigment composition, and molecular data. – Journal of Phycology **36** (2): 394-410.

Hansen, P. J. (1992). Prey size selection, feeding rates and growth dynamics of heterotrophic dinoflagellates with special emphasis on *Gyrodinium spirale*. – Marine Biology **114** (2): 327-334.

Hansen, P. J. & T. G. Nielsen (1997). Mixotrophic feeding of *Fragilidium subglobosum* (Dinophyceae) on three species of *Ceratium*: effects of prey concentration, prey species and light intensity. – Marine Ecology Progress Series **147** (2): 187-196.

Hargraves, P. E. (1972). Studies on marine plankton diatoms I *Chaetoceros diadema* (Ehr.) Gran: life cycle, structural morphology and regional dist. – Phycologia **II** (3/4): 247-257.

– (1982). Resting spores formation in the marine diatom *Ditylum brightwellii* (West) Grun. ex V. H. – Pp. 33-46 in Mann, D. G. (ed.): 7th Diatom symposium; Königstein (Koeltz Scientific Publications).

– (1990). "Studies on marine planktonic diatoms. V. Morphology and distribution of *Leptocylindrus minimus* Gran." – Beiheft zur Nova Hedwigia **100**: 47-60.

– (1979). Studies on marine plankton diatoms, IV. Morphology of *Chaetoceros* resting spores. – Beiheft zur Nova Hedwigia. **64**: 99-120.

Hasle, G. R. (1973a). Morphology and Taxonomy of *Skeletonema costatum*, Bacillariophyceae. – Norwegian Journal of Botany **20**: 109-137.

– (1973b). Some marine plankton genera of the diatom family Thalassiosiraceae. – Beiheft zur Nova Hedwigia **45**: 1-49.

– (1975). Some living marine species of the diatom family Rhizosoleniaceae. – Beiheft zur Nova Hedwigia **53**: 99-140.

– (1976). The biogeography of some marine planktonic diatoms. – Deep Sea Research **23**: 319-338.

– (1978). Some *Thalassiosira* species with one central process (Bacillariophyceae). – Norwegian Journal of Botany **25**: 77-110.

– (1980). Examination of *Thalassiosira* type material: *T. minima* and *T. delicatula* (Bacillariophyceae). – Norwegian Journal of Botany **27**: 167-173.

– (1983). *Thalassiosira punctigera* (Castr.) comb. nov., a widely distributed marine planktonic diatom. – Nordic Journal of Botany **3**: 593-608.

– (1994). *Pseudo-nitzschia* as a genus distinct from *Nitzschia* (Bacillariphyceae). – Journal of Phycology **30** (6): 1036-1039.

– (1995). *Pseudo-nitzschia pungens* and *Pseudo-nitzschia multiseries* (Bacillariphyceae); nomenclatural history, morphology and distribution. – Journal of Phycology **31** (3): 428-435.

– (2001). The marine planktonic family Thalassionemataceae: morphology, taxonomy and distribution. – Diatom Research **16** (1): 1-82.

Hasle, G. R. & C. B. Lange (1992). Morphology and distribution of *Coscinodiscus* species from the Oslofjord, Norway, and the Skagerrak, North Atlantic. – Diatom Research **7** (1): 37-68.

Hasle, G. R. & P. A. Sims (1986). The diatom genus *Coscinodiscus* Ehrenb.: *C. argus* Ehrenb. und *C. radiatus* Ehrenb. – Botanica Marina **29**(4): 305-318.

Hasle, G. R. & E. E. Syvertsen (1980). The diatom genus *Cerataulina*: morphology and taxonomy. – Bacillaria **3**: 79-113.

– (1997). Marine Diatoms. – Pp. 5-386 in Tomas, C. R. (ed.): Identifying marine phytoplankton. San Diego (Academic Press).

Hasle, G. R., C. B. Lange & E. E. Syvertsen (1996). A review of *Pseudo-nitzschia*, with special reference to the Skagerrak, North Atlantic, and adjacent waters. – Helgoländer Meeresuntersuchungen **50**(2): 131-175.

Hasle, G. R., H. A. von Stosch & E. E. Syvertsen (1983). *Cymatosiraceae*, a new diatom family. – Bacillaria **6**: 9-156.

Heimdal, B. R. (1971). Vegetative cells and resting spores of *Thalassiosira constricta* Gaarder (Bacillariophyceae). – Norwegian Journal of Botany **18**: 153-159

– (1974). Further observations on the resting spores of *Thalassiosira constricta* (Bacillariophyceae). – Norwegian Journal of Botany **21**: 303-307

Hendey, N. I. (1964). An Introductory Account of the Smaller Algae of British Coastal Waters Pt. V, Bacillariophyceae. – xii + 317 pp.; London (Her Majesty's Stationery Office).

– (1974). A revised checklist of British marine diatoms. – Journal of the Marine Biological Association of the United Kingdom **54**(2): 277-300.

Hendey, N. F., D. H. Cushing & G. W. Ripley (1954). III Electron microscope studies of diatoms. – Journal of the Royal Microbiological society **74**(1): 22-34.

Hernández-Becerril, D. U. (1991). Note on the morphology of *Chaetoceros didymus* and *C. protuberans*, with some considerations on their taxonomy. – Diatom Research **6**(2): 289-297.

Hernández-Becerril, D. U. (2000). Morfologia y taxonomia de algunas especies de diatomeas del genero Coscinodiscus de las costas del Pacifico Mexicano (Morphology and taxonomy of some diatom species of the genus *Coscinodiscus* from the Mexican Pacific Ocean coasts). – Revista De Biologia Tropical **48**(1): 7-18.

Hernández-Becerril, D. U. & M. E. Meave del Castillo (1997). *Neocalyptrella*, gen. nov., a new name to replace *Calyptrella* Hernandez-Becerril and Meave. – Phycologia **36**(4): 329.

Hoban, M. A. (1983). Biddulphiod diatoms: II The morphology and systematics of the Pseudocellate species, *Biddulphia biddulphiana* (Smith) Boyer, *B. alternans* (Bailey) Van Heurck, and *Trigonium arcticum*. – Botanica Marina **26**(6): 271-284.

Honsell, G. & L. Talarico (2004). *Gymnodinium chlorophorum* (Dinophyceae) in the Adriatic: electron microscopial observations. – Botanica Marina **47**(2): 152-166.

Hoppenrath, M., B. Beszteri, G. Drebes, H. Halliger, J. E. E. Van Beusekom, S. Janisch & K. H. Wiltshire (2007). *Thalassiosira* species (Bacillariophyceae, Thalassiosirales) in the North Sea at Helgoland (German Bight) and Sylt (North Frisian Wadden Sea) – a first approach to assessing diversity. – European Journal of Phycology **42**(3): 271-288.

Horner, R. A. (2002). A taxonomic guide to some common marine phytoplankton. – 195 pp.; Bristol (Biopress Ltd.).

Hulburt E.M. (1957). The taxonomy of unarmoured Dinophyceae of shallow embayments on Cape Cod, Massachusetts. – Biol. Bull. mar. biol. Lab., Woods Hole **112**: 196-219.

Irigoien, X., K. J. Flynn & R. P. Harris (2005). Phytoplankton blooms: a loophole in microzooplankton grazing impact? – Journal of Plankton Research **27**(4): 313-321.

Jacobsen, A., A. Larsen, J. Martinez-Martinez, P. G. Verity & M. E. Frischer (2007). Susceptibility of colonies and colonial cells of *Phaeocystis pouchetii* (Haptophyta) to viral infection. – Aquatic Microbial Ecology **48**(2): 105-112.

Jacobson, D. M. & D. M. Anderson (1986). Thecate heterotrophic dinoflagellates: Feeding behaviour and mechanisms. – Journal of Phycology **22**(3): 249-258.

Jacobson, D. M. & R. A. Anderson (1994). The discovery of mixotrophy in photosynthetic species of *Dinophysis* (Dinophyceae): light and electron microscopical observations of food vacuoles in *Dinphysis acuminata*, *D. norvegica* and two heterotrophic dinophysoid dinoflagellates. – Phycologia **33**(2): 97-110.

Jahn, R. & A. M. Schmid (2007). Revision of the brackisch-freshwater diatom genus *Bacillaria* Gmelin (Bacillariophyta) with the description of a new variety and two new species. – European Journal of Phycology **42**(3): 295-312.

Jansen, S. (2008). Copepods grazing on *Coscinodiscus wailesii*: a question of size? – Helgoland Marine Research **62**(3): 251-255.

– (2004). Molecular evidence that plastids in the toxin-producing dinoflagellate genus *Dinophysis* originate from the free-living cryptophyte *Teleaulax amphioxa*. – Environental Microbiology **6**: 1102-1106.

Jeong, H. J. (1994). Predation by the heterotrophic dinoflagellate *Protoperidinium* cf. *divergens* on copepod eggs and nearly naupliar stages. – Marine Ecology Progress Series **114**: 203-208.

– (1994a). Predation effects of the calanoid copepod *Acartia tonsa* on a population of the heterotrophic dinoflagellate *Protoperidinium* cf. *divergens* in the presence of co-occurring red-tide dinoflagellate prey. – Marine Ecology Progress Series **111**: 87-97.

– (1999). The ecological roles of heterotrophic dinoflagellates in marine planktonic community. – Journal of Eukaryotic Microbiology **46**(4): 390-396.

Jeong, H. J., C. W. Lee, W. H. Yih & J. S. Kim (1997). *Fragilidium* cf. *mexicanum*, a thecate mixotrophic dinoflagellate which is prey for and a predator on co-occurring thecate heterotrophic dinoflagellate *Protoperidinium* cf. *divergens*. – Marine Ecology Progress Series **151**: 299-305.

Jeong, H. J., J. Y. Yoon, J. S. Kim, Y. D. Yoo & K. A. Seong (2002). Growth and grazing rates of the prostomatid ciliate *Tiarina fusus* on red-tide and toxic algae. – Aquatic Microbial Ecology **28**(3): 289-297.

Jeong, H. J., Y. D. Yoo, S. T. Kim & N. S. Kang (2004). Feeding by the heterotrophic dinoflagellate *Protoperidinium bipes* on the diatom *Skeletonema costatum*. – Aquatic Microbial Ecology **36**(2): 171-179.

Jeong, H. J., Y. D. Yoo, J. S. L. Kim, Kim T. H., J. H. Kim, N. S. Kang & W. H. Yih (2004). Mixotrophy in the phototrophic harmful alga *Cochlodinium polykrikoides* (Dinophycean): Prey species, the effect of prey concentration and grazing impact. – Journal of Eukaryotic Microbiology **51**(5): 563-569.

Jeong, H. J., Y. D. Yoo, J. Y. Park, J. Y. Song, S. T. Kim, S. H. Lee, K. Y. Kim & W. H. Yih (2005). Feeding by phagotrophic red-tide dinoflagellates: five species newly revealed and six species previously known to be mixotrophic. – Aquatic Microbial Ecology **40**(2): 133-150.

Jordan, R. W., L. Cros & J. R. Young (2005). A revised classification scheme for living haptophytes. – Micropaleontology **50** (suppl. 1): 55-79.

Jordan, R. W., R. Ligowski, E. M. Noethig & J. Priddle (1991). The diatom genus *Proboscia* in Antarctic waters. – Diatom Research **6**(1): 63-78.

Jørgensen, F. M., S. Murray & N. Daugbjerg (2004). *Amphidinium* revisited. I. Redefinition of *Amphidinium* (Dinophyceae) based on cladistic and molecular phylogenetic analyses. – Journal of Phycology **40**(2): 351-365.

Kaczmarska, I., M. Beaton, A. C. Benoit & L. K. Medlin (2006). Molecular phylogeny of selected members of the order Thalassiosirales (Bacillariophyta) and evolution of the fultoportula. – Journal of Phycology **42**(1): 121-138.

Kiørboe, T. & J. Titelman (1998). Feeding, prey selection and prey encounter mechanisms in the heterotrophic dinoflagellate *Noctiluca scintillans*. – Journal of Plankton Research **20** (8): 1615-1636.

Klaveness, D. (1972). *Coccolithus huxleyi* (Lohm.) Kamptn. II. The flagellate cell, aberrant cell types, vegetative propagation and life cycles. – British Phycological Journal **7** (3): 309-318.

– (1972a). *Coccolithus huxleyi* (Lohmann) Kamptner. 1. Morphological investigations on the vegetative cell and the process of coccolith formation. – Protistologica **8**: 335-346.

Koester, J. A., S. H. Brawley, L. Karp-Boss & D. G. Mann (2007). Sexual reproduction in the marine centric diatom *Ditylum brightwelli* (Bacillariophyta). – European Journal of Phycology **42** (4): 351-366.

Kofoid, C. A. & O. Swezy (1921). The free-living unarmored dinoflagellata. – Memoirs of the University of California **5**: 562 pp.

Krock, B., U. Tillmann, U. John & A. D. Cembella (2000). Characterization of azaspiracids in plankton size-fractions and isolation of azaspiracid-producing dinoflagellate from the North Sea. – Harmful Algae **8** (2): 254-263.

Kühn, S. F., G. Klein, H. Halliger, P. Hargraves & L. K. Medlin (2006). A new diatom, *Mediopyxis helysia* gen. nov. and sp. nov. (Mediophyceae) from the North Sea and the Gulf of Maine as determined from morphological and phylogenetic characteristics. – Beiheft zur Nova Hedwigia **130**: 307-324.

Larsen, J. & Ø. Moestrup (eds.) (1992). Potentially toxic phytoplankton. 2. Genus *Dinophysis*. – 12 pp; ICES Identification leaflet for plankton; Copenhagen (International Council for the exploration of the sea).

Lebour, M. V. (1922). Plymouth peridinians. I. *Diplopsalis lenticula* and its relatives. – Journal of the Marine Biological Association of India **12** (4): 795-812.

– (1925). Dinoflagellates of Northern Seas. – 250 pp.; Plymouth (Biological Association of the United Kingdom).

Le Gall, F., F. Rigaut-Jalabert, D. Marie, L. Garczarek, M. Viprey, A. Gobet & D. Vaulot (2008). Picoplankton diversity in the South-East Pacific ocean from cultures. – Biogeosciences **5** (1): 203-214.

Lee, J. J., G. F. Leedale & P. Bradbury (eds.) (2000). An illustrated guide to the protozoa. – 1430 pp. (two volumes); Lawrence, Kansas (Society of Protozoologists).

Legrand, C., E. Granéli & P. Carlsson (1998). Induced phagotrophy in the photosynthetic dinoflagellate *Heterocapsa triquetra*. – Aquatic microbial ecology **15** (1): 65-75.

Lessard, E. J. (1991). The trophic role of heterotrophic dinoflagellates in diverse marine environments. – Marine Microbial Foodwebs **5** (1): 49-58.

Lessard, E. J. & M. C. Murrell (1996). Distribution, abundance and size composition of heterotrophic dinoflagellates and ciliates in the Sargasso Sea near Bermuda. – Deep Sea Research I **43** (7): 1045-1065.

– (1998). Microzooplankton herbivory and phytoplankton growth in the northwestern Sargasso Sea. – Aquatic Microbial Ecology **6** (2): 173-188.

Lewis, J. (1990). The cyst-theca relationship of *Oblea rotunda* (Diplosalidaceae, Dinophyceae). – British Phycological Journal **25**: 339-351.

Loeblich III, A. R. (1965). Dinoflagellate nomenclature. – Taxon **14**: 15-18.

Lund, J. W. G., C. Kipling & E. D. Le Cren (1958). The inverted microscope method of estimating algal numbers and the statistical basis of estimations by counting. – Hydrobiologia **11** (2): 143-170.

Lundholm, N., Ø. Moestrup, G. R. Hasle & K. Hoef-Emden (2003). A study of the *Pseudo-nitzschia pseudodelicatissima/cuspidata* complex (Bacillariophyceae): What is *P. delicatissima*? – Journal of Phycology **39** (4): 797-813.

MacKenzie, L. (1991). Toxic and noxious phytoplankton in Big Glory Bay, Stewart Island New Zealand. – Journal of Applied Phycology **3** (1): 19-34.

Makino, W., K. Ito, Y. Oshima & J. Urabe (2008). Effects of *Protoceratium reticulatum* yessotoxin on feeding rates of Acartia hudsonica: A bioassay using artificial particles coated with purified toxin. – Harmful Algae **7**: 639-645.

Mann, D. G. (1999). The species concept in diatoms. – Phycologia **38** (6): 437-495.

Manton, I. & H. A. Von Stosch (1966). Observations on the fine structure of the marine centric diatom *Lithodesmium undulatum*. – Journal of the Royal Microsopical Society **85**: 119-134.

Marchant, H. J., F. J. Scott & A. T. Davidson (2005). Haptophytes: Order Prymnesiales. – Pp. 255-275 in Scott, F. J. & H. J. Marchant (eds.): Antarctic marine protists. Canberra and Hobart (Australian Biological Resources Study: Australian antarctic division).

Marret, F. & K. A. F. Zonneveld (2003). Atlas of modern organic-walled dinoflagellate cyst distribution. – Review of Palaeobotany and Palynology: 1-200.

McDermott, G. & R. Raine (2006). The dinoflagellate genus *Ceratium* in Irish shelf seas. – viii + 1-86 pp.; Galway (Martin Ryan Institute).

McQuoid, M. R. & A. Godhe (2004). Recruitment of coastal planktonic diatoms from benthic versus pelagic cells: Variations in bloom development and species composition. – Limnology and Oceanography **49** (4): 1123-1133.

McQuoid, M. R. & L. A. Hobson (1995). Importance of resting stages in diatom seasonal succession. – Journal of Phycology **31** (1): 44-50.

– (1996). Diatom resting stages. – Journal of Phycology **32** (6): 889-902.

Medlin, L. K. & A. Zingone (2007). A taxonomic review of the genus *Phaeocystis*. – Biogeochemistry **83** (1-3): 3-18.

Menden-Deuer, S., E. J. Lessard & J. Satterberg (2001). Effect of preservation on dinoflagellate and diatom cell volume and consequences for carbon biomass predictions. – Marine Ecology Progress Series **222**: 41-50.

Miller, W. I., III & Collier, A. (1978). Ultrastructure of the frustule of *Triceratium favus* (Bacillariophyceae). – Journal of Phycology **14** (1): 56-62.

Mills, K. E. & I. Kaczmarska (2006). Autogamic behavior and sex cell structure in *Thalassiosira angulata* (Bacillariophyta). – Botanica Marina **49** (5-6): 417-430.

Mitrovic, S. M., M. F. Amandi, L. McKenzie, A. Furey & K. J. James (2004). Effects of selenium, iron and cobalt addition to growth and yessotoxin production of the toxic marine dinoflagellate *Protoceratium reticulatum* in culture. – Journal of Experimental Marine Biology and Ecology **313** (2): 337-351.

Moestrup, Ø. & P. J. Hansen (1988). On the occurrence of the potentially toxic dinoflagellates *Alexandrium tamarense* (= *Gonyaulax excavata*) and *A. ostenfeldii* in Danish and Faroese waters. – Ophelia **28** (3): 195-213.

Moestrup, Ø. & H. Thomsen (1990). *Dictyocha speculum* (Silicoflagellata, Dictyophyceae). – 57 pp.; Copenhagen (Royal Danish Academy of Sciences and Letters).

Moestrup, Ø. & H. A. Thomsen (2003). Taxonomy of toxic haptophytes (prymnesiophytes).– Pp. 433-463 in Hallegraeff, G. M., D. M. Anderson & A. D. Cembella (eds.): Manual on Harmful Microalgae. Paris (UNESCO Publishing).

Montresor, M., S. Sgrosso, G. Procaccini & W. H. C. F. Kooistra (2003). Intraspecific diversity in *Scrippsiella trochoidea* (Dinophyceae): Evidence for cryptic species. – Phycologia **42** (1): 56-70.

Morill, L. C. & A. R. Loeblich III (1981). A survey of body scales in dinoflagellates and a revision of *Cachonina* and *Heterocapsa* (Pyrrhophyta). – Journal of Plankton Research **3** (1): 53-65.

Murray, S., M. F. Jørgensen, N. Daugbjerg & L. Rhodes (2004). *Amphidinium* revisited. II Resolving species boundaries in the *Amphidinium operculatum* species complex (Dinophyceae), including the descriptions of *Amphidinium trulla* sp. nov. and *Amphidinium gibbosum*. comb. nov. – Journal of Phycology **40**(2): 366-382.
Nagai, S., Y. Matsuyama, H. Takayama & Y. Kotani (2002). Morphology of *Polykrikos kofoidii* and *P. schwartzii* (Dinophyceae, Polykrikaceae) cysts obtained in culture. – Phycologia **41**(4): 319-327.
Nakata, K. (1987). The fine structure of two marine diatom species of the family Cymatosiraceae. – Bulletin of the Tokai Regional Fisheries Research Laboratory **121**: 41-45.
Nehring, S. (1997). Dinoflagellate resting cysts from recent German coastal sediments. – Botanica Marina **40**(4): 307-324.
– (1998). Establishment of thermophilic phytoplankton species in the North Sea: biological indicators of climate change? – ICES Journal of Marine Science **55**: 818-823.
– (2003). Alien species in the North Sea: invasion success and climate warming. – Ocean Challenge **13**(3): 12-16.
Occhipinti-Ambrogi, A. (2007). Global change amd marine communities: Alien species and climate change. – Marine Pollution Bulletin **55**: 342-352.
Okolodkov, Y. B. (2005). *Protoperidinium* Bergh (Dinoflagellata) in the southeastern Mexican Pacific Ocean: part 1. – Botanica Marina **48**(4): 284-296.
Okolodkov, Y. B. & J. D. Dodge (1995). Redescription of planktonic dinoflagellate *Peridiniella danica* (Paulsen) comb. nov. and its distribution in the N.E. Atlantic. – European Journal of Phycology **30**(4): 299-306.
Olli, K. (1998). Temporary cyst formation of *Heterocapsa triquetra* (Dinophyceae) in natural populations. – Marine Biology **145**: 1-8.
Olseng, C., L.-J. Naustvoll & E. Paasche (2002). Grazing by heterotrophic dinoflagellate *Protoperidinium steinii* on a *Ceratium* bloom. – Marine Ecology Progress Series **225**: 161-167.
Orlova, T. Y., I. V. Stonik, N. A. Aizdaicher, S. S. Bates, C. Leger & J. Fehling (2008). Toxicity, morphology and distribution of *Pseudo-nitzschia calliantha*, *P. multistriata* and *P. multiseries* (Bacillariophyta) from northwestern Sea of Japan. – Botanica Marina **51**(4): 297-306.
Paulsen, O. (1908). XVIII Peridiniales. – Pp. 1-124 in Brandt K. & C. Apstein (eds.): Nordisches Plankton, Band 8 (Botanischer Teil); Kiel und Leipzig (Lipsius und Tischer).
Pennick, N. C. & K. J. Clarke (1977). The ocurrence of scales in the peridinian dinoflagellate *Heterocapsa triquetra* (Ehrenb.) Stein. – British Phycological Journal **12**(1): 63-66.
Pertola, S., M. A. Faust, H. Kuosa & G. Hällfors (2003). Morphology of *Prorocentrum minimum* (Dinophyceae) in the Baltic Sea in Chesapeake Bay: Comparison of cell shapes and thecal ornamentation. – Botanica Marina **46**(5): 477-486.
Pfiester, L. A. & D. M. Anderson (1987). Dinoflagellate reproduction. – Pp. 611-648 in Taylor, F. J. R. (ed.): The Biology of Dinoflagellates. Oxford (Blackwell Scientific Publications).
Pomeroy, L. R. (1974). Oceans food web, a changing paradigm. – Bioscience **24**: 499-504.
Redfield, A. C., B. H. Ketchum & F. A. Richards (1963). The influence of organisms on the composition of seawater. – Pp. 26-77 in Hill, M. N. (ed.): The Sea; vol. 2. New York (Wiley Interscience).
Reguera, B., I. Bravo & S. Fraga (1995). Autoecology and some life history stages of *Dinophysis acuta* Ehrenberg. – Journal of Plankton Research **17**(5): 999-1015.
Rhodes, L., P. McNabb, M. F. de Salas, L. Briggs, V. Beuzenberg & M. Gladstone (2006). Yessotoxin production by *Gonyaulax spinifera*. – Harmful Algae **5**(2): 148-155.
Ribeiro, S., N. Lundholm, A. Amorim & M. Ellegaard (2010). *Protoperidinium minutum* (Dinophyceae) from Portugal: cyst-theca relationship and phylogenetic position on the basis of single cell SSU and LSU rDNA sequencing. – Phycologia. **49**(1): 48-63.
Rincé, Y. & G. Paulmier (1986). Données nouvelles sur la distribution de la diatomée marine *Coscinodiscus wailesii* Gran & Angst (Bacillariophyceae). – Phycologia **25**: 73-79.
Rines, J. E. B. (1999). Morphology and taxonomy of *Chaetoceros contortus* Schütt 1895, with preliminary observations on *Chaetoceros compressus* Lauder, 1864. – Botanica marina **42**: 539-551.
Rines, J. E. B. & P. E. Hargraves (1988). The *Chaetoceros* Ehrenberg (Bacillariophyceae) flora of Narragansett Bay, Rhode Island, USA. – Bibliotheca Phycologica **79**: 1-196.
Richardson, T. L. & G. A. Jackson (2007). Small phytoplankton and carbon export from the surface ocean. – Science **315**: 838-840.
Robinson, G. A., T. D. Budd, A. W. G. John & P. C. Reid (1980). *Coscinodiscus nobilis* (Grunow) in continuous plankton records 1977-1978. – Journal of the Marine Biological Association of the United Kingdom **60**(3): 675-680.
Ross, R. & P. A. Sims (1973). Observations on family and generic limits in the Centrales. – Beiheft zur Nova Hedwigia **45**: 97-130.
Roth, L. E. & A. de Francisco (1977). The marine diatom *Striatella unipunctata*. 2. Siliceous structures and the formation of intercalary bands. – Cytobiologie **14**(2): 207-221.
Round, F. E. (1973). On the diatom genera *Stephanopyxis* Ehr. and *Skeletonema* Grey and their classification in a revised system of the Centrales. – Botanica Marina **16**(3): 148-154.
Round, F. E., R. M. Crawford & D. G. Mann (1990). The diatoms: biology and morphology of the genera. – 747 pp.; London, UK (Cambridge University Press).
Roy, S., R. P. Harris & S. A. Poulet (1989). Inefficient feeding by *Calanus helgolandicus* and *Temora longicornis* on *Coscinodiscus wailesii*: quantitative estimation using chlorophyll-type pigment and effects on dissolved free amino acids. – Marine Ecology Progress Series **52**: 145-153.
Sar, E. A., I. Sunesen, R. Jahn, W. Kusber & A. S. Lavigne (2007). Revision of *Odontella atlantica* (Frenguelli) Sar comb. et stat. nov. with comparison to two related species, *O. rhombus* (Ehrenb.) Kutz. and *O. rhomboides* R. Jahn et Kusber. – Diatom Research **22**(2): 341-353.
Sarno, D., W. H. C. F. Kooistra, L. K. Medlin, I. Percopo & A. Zingone (2005). Diversity in the genus *Skeletonema* (Bacillariophyceae). II An assessment of the taxonomy of *S. costatum*-like species with the description of four new species. – Journal of Phycology **41**(1): 151-176.
Sawai, Y., T. Nagumo & K. Toyoda (2005). Three extant species of *Paralia* (Bacillariophyceae) along the coast of Japan. – Phycologia **44**(5): 517-529.
Schiller, J. (1918). Über neue *Prorocentrum*- und *Exuviaella*-Arten aus der Adria. – Archiv für Protistenkunde **38**: 250-262.
Schmid, A. M. (2007). The 'paradox' diatom *Bacillaria paxillifer* (Bacillariophyta) revisited. – Journal of Phycology **43**(1): 139-155.
Schnepf, E., R. Meier & G. Drebes (1988). Stability and deformation of diatom chloroplasts during food uptake of the parasitic dinoflagellate, *Paulsenella* (Dinophyta). – Phycologia **27**(2): 283-290.
Shevchenko, O. G., T. Y. Orlova & D. U. Hernandez-Becerril (2006). The genus *Chaetoceros* (Bacillariophyta) from Peter the Great bay Sea of Japan. – Botanica Marina **49**(3): 236-258.
Sicko-Goad, L., E. F. Stoermer & J. P. Kociolek (1989). Diatom resting cell rejuvenation and formation: Time course, species records and distribution. – Journal of Plankton Research **11**(2): 385-389.
Skovgaard, A. (1996). Mixotrophy in *Fragilidium subglobosum* (Dinophyceae): Growth and grazing responses as functions of light intensity. – Marine Ecology Progress Series **143**. 247-253.

Skovgaard, A., P. J. Hansen & D. K. Stoecker (2000). Physiology of the mixotrophic dinoflagellate *Fragilidium subglobosum*. I Effects of phagotrophy and irradiance on photosynthesis and carbon content. – Marine Ecology Progress Series **201**: 129-136.

Smayda, T. J. (2007). Reflections on the ballast water dispersal - harmful algal bloom paradigm. – Harmful Algae **6**(4): 601-622.

Sournia, A. (1967). Contribution à la connaissance des péridiniens microplanctoniques du Canal de Mozambique. – Bulletin du Muséum d'histoire naturelle **39**(2): 417-438.

Sournia, A., M. J. Chrétiennot-Dinet & M. Ricard (1991). Marine plankton: how many species in the world oceans? – Journal of Plankton Research, **13**(5): 1093-1099

Stabell, O. B., R. T. Aanesen & H. C. Eilertsen (1999). Toxic peculiarities of the marine alga *Phaeocystis pouchetii* detected by in vivo and in vitro bioassay methods. – Aquatic Toxicology **44**(4): 279-288.

Steidinger, K. A. & K. Tangen (1997). Dinoflagellates. – Pp. 387-584 in Tomas, C. R. (ed.): Identifying marine phytoplankton. San Diego (Academic Press).

Stein, F. (1883). Die Naturgeschichte der arthrodelen Flagellaten. Der Organismus der Infusionstiere. III Pt 2.: 1-30.

Stoecker, D. K. (2007). Mixotrophy among dinoflagellates. – Journal of Eukaryotic Microbiology **46**: 397-401.

Stoecker, D. K., A. Li, D. W. Coats, D. E. Gustavson & M. K. Nannen (1997). Mixotrophy in the dinoflagellate *Prorocentrum minimum*. – Marine Ecology Progress Series **152**: 1-12.

Stoecker, D. K., M. W. Silver, A. E. Michaels & L. H. Davis (2005). Obligate mixotrophy In *Laboea strobila*, a ciliate which retains chloroplasts. – Marine Biology **99**(3): 415-423.

Strom, S. L. & E. J. Buskey (1993). Feeding, growth and behaviour of the thecate, heterotrophic dinoflagellate *Oblea rotunda*. – Limnology and Oceanography **38**(5): 965-977.

Sundström, B. G. (1986). The marine diatom genus *Rhizosolenia*: A new approach to the taxonomy. – 117 pp. + plates, PhD; Lund, (Lund University).

Sunesen, I. & E. A. Sar (2007). Marine diatoms from Buenos Aires coastal waters (Argentina). IV. *Rhizosolenia* s. str., *Neocalyptrella*, *Pseudosolenia*, *Proboscia*. – Phycologia **46**(6): 628-643.

Syvertsen, E. E. (1977). *Thalassiosira rotula* and *T. gravida*: ecology and morphology. – Beiheft zur Nova Hedwigia **54**: 99-112.

– (1979). Resting spore formation in clonal cultures of *Thalassiosira antarctica* Comber; *T. nordenskioeldii* Cleve and *Detonula confervacea* (Cleve) Gran. – Beiheft zur Nova Hedwigia **64**: 41-63.

Syvertsen, E. E. & G. R. Hasle (1982). The marine planktonic diatom *Lauderia annulata* Cleve, with particular reference to the processes. – Bacillaria **5**: 243-256.

– (1983). The diatom genus *Eucampia*: morphology and taxonomy. – Bacillaria **6**: 169-210.

Takano, H. (1984). Scanning electron microscopy of diatoms. 7. *Odontella aurita* and *O. longicruris*. – Bulletin of the Tokai Regional Fisheries Research Laboratory **113**: 79-85.

Takano, Y. & T. Horiguchi (2006). Acquiring scanning electron microscopical, light microscopical and multiple gene sequence data from a single dinoflagellate cell. – Journal of Phycology **42**(1): 251-256.

Taylor, D. L. (1971). Taxonomy of some common *Amphidinum* species. – British Phycological Journal **6**(2): 129-133.

Taylor, F. J. R. (1972). Unpublished observations on the thecate stage of the dinoflagellate genus *Pyrocystis* by the late C. A. Kofoid and Josephine Michener. – Phycologia **11**(1): 47-55.

Taylor, F. J. R., M. Hoppenrath & J. F. Saldarriaga (2008). Dinoflagellate diversity and distribution. – Biodiversity and conservation **17**(2): 407-418.

Tett, P. & E. D. Barton (1995). Why are there about 5000 species of phytoplankton in the sea? – Journal of Plankton Research, **17**(8): 1693-1704

Throndsen, J. (1997). The planktonic marine flagellates. – Pp. 591-730 in C. R. Tomas (ed.): Identifying Marine Phytoplankton. San Diego (Academic Press).

Toriumi, S. & J. Dodge (1993). Thecal apex structure in the Peridiniaceae (Dinophyceae). – European Journal of Phycology **28**(1): 39-45.

Trigueros, J. M., A. Ansotegui & E. Orive (2000). Remarks on morphology and ecology of recurrent dinoflagellate species in the estuary of Urdaibai (Northern Spain). – Botanica Marina **43**(1): 93-103.

Tuchmann, N. C., M. A. Schollett, S. T. Rier & P. Geddes (2006). Differential heterotrophic utilization of organic compounds by diatoms and bacteria under light and dark conditions. – Hydrobiologia **561**(1): 167-177.

Uchida, T. (2001). The role of cell contact in the life cycle of some dinoflagellate species. – Journal of Plankton Research **23**(8): 889-891.

Underdahl, B., O. M. Skulberg, E. Dahl & T. Aune (1989). Disastrous bloom of *Chrysochromulina polylepis* (Prymnesiophyceae) in Norwegian coastal waters 1988 – Mortality of marine biota. – Ambio **18**: 265-270.

Utermöhl, H. (1931). Neue Wege in der quantitativen Erfassung des Planktons (mit besonderer Berücksichtigung des Ultraplanktons). – Verhandlungen der internationalen Vereinigung theoretischer und angewandter Limnologie **5**: 567-596.

Van De Meene, A. M. & J. D. Pickett-Heaps (2002). Valve morphogenesis in the centric diatom *Proboscia alata* Sundstrom. – Journal of Phycology **38**(2): 351-363.

– (2004). Valve morphogenesis in the centric diatom *Rhizosolenia setigera* (Bacillariophyceae, Centrales) and its taxonomic implications. – European Journal of Phycology **39**(1): 93-104.

Veldhuis, M. & H. J. W. de Baar (2005). Iron resources and oceanic nutrients: advancement of global environment simulations. – Journal of Sea Research **53**(1): 1-6.

Von Stosch, H. A. (1969). Dinoflagellaten aus der Nordsee I. Über *Cachonina niei* Loeblich (1968), *Gonyaulax grindleyi* Reinecke (1967) und eine Methode zur Darstellung von Peridineenpanzern. – Helgoländer Wissenschaftliche Meeresuntersuchungen **19**(4): 558-568.

– (1969). Dinoflagellaten aus der Nordsee. II. *Helgolandinium subglobosum* gen. et. spec. nov. – Helgoländer Wissenschaftliche Meeresuntersuchungen **19**(4): 569-577.

– (1977). Observations on *Bellerochea* and *Streptotheca*, including descriptions of three new planktonic diatom species. – Beiheft zur Nova Hedwigia **54**: 113-66.

– (1980). The two *Lithodesmium* species (Centrales) of European waters. – Bacillaria **3**: 7-20.

Von Stosch, H. A., G. Theil & K. V. Kowallik (1973). Entwicklungsgeschichtliche Untersuchungen an zentrischen Diatomeen. V. Bau und Lebenszyklus von *Chaetoceros didymum*, mit Beobachtungen über einige andere Arten der Gattung. – Helgoländer Wissenschaftliche Meeresuntersuchungen **25**(2-3): 384-445.

Villareal, T. A. & G. A. Fryxell (1983a). Temperature effects on the valve structure of the bipolar diatoms *Thalassiosira antarctica* and *Porosira glacialis*. – Polar Biology **2**(3): 163-169.

– (1983b). The genus *Actinocyclus* (Bacillariophyceae): Frustule morphology of *A. sagittulus* sp. nov. and two related species. – Journal of Phycology **19**(4): 452-466.

– (1990). The diatom *Porosira* Joerg.: Cingulum patterns and resting spore morphology. – Botanica Marina **33**(5): 415-422.

Wall, D. & B. Dale (1968). Modern dinoflagellate cysts and evolution of the Peridiniaceae. – Micropaleontology **14**(3): 265-304.

– (1971). A reconsideration of living and fossil *Pyrophacus* Stein, 1883 (Dinophyceae). – Journal of Phycology **7**(3): 221-235.

Watanabe, T., J. Tanaka & T. Nagumo (2006). Morphology of marine Diatom *Cerataulus turgidus* (Ehrenberg). – Bulletin of the Nippon Dental University **35**: 61-64.

Werner, D. (1971). Der Entwicklungscyclus mit Sexualphase bei der marinen Diatomee *Coscinodiscus asteromphalus*. – Archiv für Mikrobiologie **80**: 43-49.

White, A. W. (1974). Uptake of organic, compounds by two facultatively heterotrophic marine centric diatoms. – Journal of Phycology **10** (4): 433-438.

Williams, D. M. (2007). Classification and diatom systematics: the past, the present and the future. – Pp. 57-91 in Brodie, J. & J. Lewis (eds.): Unravelling the algae the past, present and future of algal systematics. Boca Raton, Fla. (CRC Press, Taylor and Francis group).

Wiltshire, K. H. & B. F. J. Manly (2004). The warming trend at Helgoland Roads, North Sea: phytoplankton response. – Helgoland Marine Research **58**(4): 269-273.

Wolff, W. J. & K. Reise (2002). Oyster imports as a vector for the introduction of alien species into northern and western European coastal waters. – Pp. 193-205 in Leppäkoski, E., S. Gollasch & S. Olenin (eds.): Invasive aquatic species of Europe – Distribution, impacts and management. Dordrecht (Kluwer Academic Publishers).

Wyatt, T. & I. R. Jenkinson (1997). Notes on *Alexandrium* population dynamics. – Journal of Plankton Research **19**(5): 551-575.

Zarauz, L. & X. Irigoien (2008). Effects of Lugol's fixation on the size structure of natural nano-microplankton samples, analyzed by means of an automatic counting method. – Journal of Plankton Research **30**(11): 1297-1303.

Zingone, A., M. Montresor & D. Marino (1998). Morphological variability of the potentially toxic dinoflagellate *Dinophysis sacculus* (Dinophyceae) and its taxonomic relationships with *D. pavillardii* and *D. acuminata*. – European Journal of Phycology **33**(3): 259-273.

Zingone, A., I. Percopo, P. A. Sims & D. Sarno (2005). Diversity in the genus *Skeletonema* (Bacillariophyceae). I. A re-examination of the type material of *S. costatum* with the description of *S. grevillei* sp. nov. – Journal of Phycology **41**(1): 140-150.

Zonneveld, K. A. F. & B. Dale (1994). The cyst-motile relationships of *Protoperidinium monospinum* (Paulsen) Zonneveld et Dale comb. nov. and *Gonyaulax verior* (Dinophyta, Dinophyceae) from the Oslo Fjord (Norway). – Phycologia **33**(5): 359-368.

Web resources

PLANKTON*NET
http://planktonnet.awi.de
A web resource with over 6000 images of phyto- and zooplankton and with additional resources such as descriptions, taxonomic and ecological literature as well as illustrated glossaries dealing with terminology specific to dinoflagellates, diatom and coccolithophorids

ALGAEBASE
http://algaebase.org
One of the most comprehensive taxonomic resources on algae (both micro- and macroalgae) in the world. It also hosts a literature database and many images

ASLO Image database
http://www.aslo.org/photopost/
A general image database related to marine and freshwater habitats. It includes everything from landscape shots to images of sampling equipment and images of organisms

Identifying harmful marine dinoflagellates
http://botany.si.edu/references/dinoflag
A comprehensive resource on harmful dinoflagellates with detailed taxon description references and many images including light micrographs and scanning electron micrographs. It is also available on CD (http://www.eti.uva.nl/products/catalogue/cd_detail.php?id=127&referrer=catalogue) from ETI Biodiversity informatics and as book.

CEDIT – Centre of Excellence for dinophyte taxonomy
http://www.dinophyta.org
A comprehensive online resource about the taxonomy of dinoflagellates, including for instance literature lists, links to pdfs with original descriptions and a list of external resources.

IOW – Image gallery of Baltic phytoplankton
http://www.io-warnemuende.de/gallery-of-baltic-microalgae.html
An online image gallery of common phytoplankton species in the Baltic. It covers many different taxon groups such as diatoms, dinoflagellates, cryptophytes and prasinophytes.

The Harmful Algal Bloom Programme
http://www.ioc-unesco.org/hab/
A resource that covers all groups of harmful algae. It contains a wide variety of resources from the short species information sheets via maps illustrating toxic events to lists of HAB experts.

The planktonic ciliate project
http://www.liv.ac.uk/ciliate/
This site contains species pages for a large number of planktonic ciliates. The image material provided covers live material but also images of cells fixed with Lugol or the Protargol method. It also contains a comprehensive illustrated glossary and a methodology section (see also the sister site dedicated to harmful phytoplankton http://www.liv.ac.uk/hab/).

Checklist of Skagerrak phytoplankton
http://www.smhi.se/oceanografi/oce_info_data/plankton_checklist/ssshome.htm
A regional website with species pages containing images, taxonomic information and references for the phytoplankton species (diatoms, dinoflagellates and other flagellates) of the Skagerrak.

Encyclopedia of Life
http://www.eol.org
An emerging online taxonomic website with the ambitious goal of producing one page with comprehensive taxonomic information for every species on the planet, including the phytoplankton. It includes links to a variety of external resources such as the Biodiversity Heritage Library.

Index of Taxa

Page numbers in **boldface** refer to detailed descriptions with figures; page numbers in *italics* refer to additional figures.

D